Advanced Mathematics 2

A unified course in pure mathematics,
mechanics and statistics

Also by L. K. Turner and D. Knighton
Advanced Mathematics 1

A list of the contents of Book 1 may be found on p. 456.

Advanced Mathematics 2

A unified course in pure mathematics,
mechanics and statistics

L. K. Turner D. Knighton F. J. Budden

510
TUR

LONGMAN GROUP LIMITED
*Longman House, Burnt Mill, Harlow, Essex CM20 2JE, England
and Associated Companies throughout the World*

*Distributed in the United States by Longman Inc,
The Longman Building, 95 Church Street, White Plains, NY 10601*

© Longman Group Limited 1975, 1986
*All rights reserved; no part of this publication
may be reproduced, stored in a retrieval system,
or transmitted in any form or by any means, electronic,
mechanical, photocopying, recording or otherwise,
without the prior written permission of the Publishers.*

First published 1975
Second edition 1986
ISBN 0 582 35514 1

Set in 10/12 pt Times New Roman

*Printed in Great Britain
by The Bath Press, Avon*

Acknowledgements

We are grateful to the following Examinations Boards for permission to reproduce questions from past examination papers in mathematics:

The Associated Examining Board; University of Cambridge Local Examinations Syndicate; Joint Matriculation Board; University of London Schools Examinations Board; University of Oxford Delegacy of Local Examinations; Oxford and Cambridge Schools Examinations Board; Welsh Joint Education Committee.

We would also like to take this opportunity to state that the above-mentioned Boards accept no responsibility whatsoever for the accuracy or method of working in the answers given.

Contents

Sections marked * may be delayed until a second reading.

Preface   **vii**

Notation   **viii**

11 Further vectors   **1**
11.1   Points of subdivision   1
11.2   Centres of mass   6
* 11.3   Vector products   16
* 11.4   Plane kinematics   22

12 Complex numbers   **29**
12.1   Introduction   29
12.2   Geometric representation   35
12.3   Complex conjugates: solution of equations   52
12.4   De Moivre's theorem   57
12.5   Loci in the Argand diagram   69
* 12.6 $e^{i\theta}$ 76
* 12.7   Interlude: hyperbolic functions   80
* 12.8   Functions of a complex variable   88

13 Differential equations   **105**
13.1   Introduction   105
13.2   Step-by-step solutions   108
13.3   First order equations with separable variables   117
* 13.4   Linear equations of first order   128
* 13.5   Linear equations with constant coefficients   134

14 Further mechanics   **146**
14.1   Elasticity: Hooke's law and elastic energy   146
14.2   Oscillations: simple harmonic motion   150
14.3   Elasticity: Newton's law of impact   163

14.4   Further statics: moments, couples and equilibrium   166
*14.5 Further dynamics: moments of inertia and rotation   182
14.6   Dimensions   195

15 Probability distributions and further statistics   **201**
15.1   Probability distributions and generators   201
15.2   Continuous probability distributions   213
15.3   The Normal distribution   223
15.4   The Poisson distribution   235
15.5   Samples   247
*15.6 Significance and confidence   263
*15.7 Correlation and regression   275

16 Matrices and transformations, determinants and linear equations **305**
16.1   Matrices   305
16.2   Linear transformations in two dimensions   313
16.3   Linear transformations in three dimensions   327
*16.4 Inverse matrices   339
*16.5 Determinants   351
*16.6 Systems of linear equations   365
*16.7 Systematic reduction: inverse matrices   375

Epilogue   **387**

Revision exercises   **393**

Answers to exercises   **418**

Contents of Book 1   **456**

Index   **459**

Preface to the second edition

The aim of this course remains that of its first edition: to provide within two books an Advanced level programme in pure mathematics, mechanics and statistics which combines directness and economy with the maximum possible clarity.

We have taken this opportunity to reflect the increasing importance of differential equations and of numerical work, to extend many of the exercises and the miscellaneous problems at the end of each chapter, and also to introduce sets of revision exercises. By these means we hope to serve the needs both of those who need more practice and of those who look for a demanding challenge.

Finally, we thank Laurice Suess and Andrew Ransom of Longman for all their assistance, and once again our wives and families for their great forbearance.

L.K.T.
D.K.
F.J.B.

Notation

Vectors

AB, CD, ...	displacement vectors
a, b, ...	free vectors, or position vectors, with moduli $a, b, c,$ ...
â, b̂, ...	unit vectors in the directions **a, b,** ...
$\mathbf{a} + \mathbf{b}, \mathbf{a} - \mathbf{b}$	vector sum and difference
i, j, k	unit vectors, with right-hand set of mutually perpendicular axes
$\mathbf{r} = x\mathbf{i} + y\mathbf{j} + z\mathbf{k}$	vector **r** with components x, y, z
$\mathbf{v} = \dfrac{d\mathbf{r}}{dt} = \dot{\mathbf{r}}$	velocity
$\mathbf{a} = \dfrac{d\mathbf{v}}{dt} = \ddot{\mathbf{r}}$	acceleration
$\mathbf{v}_P(Q) = \dfrac{d}{dt}(\mathbf{PQ})$	velocity of Q relative to P
$\mathbf{a}_P(Q) = \dfrac{d}{dt}\mathbf{v}_P(Q)$	acceleration of Q relative to P
$\omega = \dot{\theta}$	angular velocity
$\dot{\omega} = \ddot{\theta}$	angular acceleration
$\mathbf{r} = \dfrac{\alpha\mathbf{a} + \beta\mathbf{b}}{\alpha + \beta}$	**r** divides **a, b** in the ratio $\beta:\alpha$
$\mathbf{g} = \dfrac{1}{n}\sum_1^n \mathbf{r}_i$	centroid of $\mathbf{r}_i (i = 1, \ldots, n)$
$\bar{\mathbf{r}} = \dfrac{\sum_1^n m_i \mathbf{r}_i}{\sum_1^n m_i}$	centre of mass of m_i at $\mathbf{r}_i (i = 1, \ldots, n)$
$\begin{aligned}\mathbf{a}.\mathbf{b} &= ab\cos\theta \\ &= a_1 b_1 + a_2 b_2 + a_3 b_3\end{aligned}$	scalar product of **a, b**

$$\mathbf{a} \times \mathbf{b} = (ab \sin \theta)\mathbf{\hat{n}}$$
$$= (a_2b_3 - a_3b_2)\mathbf{i}$$
$$+ (a_3b_1 - a_1b_3)\mathbf{j}$$
$$+ (a_1b_2 - a_2b_1)\mathbf{k}$$

vector product of $\mathbf{a}$, $\mathbf{b}$

Mechanics

$\mathbf{F}$

force $\mathbf{F}$

$\mathbf{R} = \mathbf{P} + \mathbf{Q}$

resultant force $\mathbf{R} = \mathbf{P} + \mathbf{Q}$

$\mathbf{g}$

acceleration due to gravity
($g \approx 9.8 \, \text{m s}^{-2}$ on Earth)

$\mathbf{F}, \mathbf{N}, \mathbf{R}$

frictional component ($\mathbf{F}$) and
normal component ($\mathbf{N}$) of
resultant reaction ($\mathbf{R}$)

μ

coefficient of friction $\left(\dfrac{F}{N} \leqslant \mu \right)$

λ

angle of friction ($= \tan^{-1} \mu$)

$\mathbf{I} = \mathbf{F}t$

impulse of constant force $\mathbf{F}$ in time t

$\mathbf{I} = \displaystyle\int_{t_1}^{t_2} \mathbf{F} \, \mathrm{d}t$

impulse of force $\mathbf{F}$ in interval (t_1, t_2)

$m\mathbf{v}$

momentum of mass m moving with
velocity $\mathbf{v}$

e

coefficient of restitution

$W = \mathbf{F}.\mathbf{s}$

work done by constant force $\mathbf{F}$ in
displacement $\mathbf{s}$

$W = \displaystyle\int \mathbf{F}.\mathrm{d}\mathbf{s}$

work done by variable force $\mathbf{F}$

$\frac{1}{2}m\mathbf{v}^2$

kinetic energy of mass m moving
with velocity $\mathbf{v}$

$P = \dfrac{\mathrm{d}W}{\mathrm{d}t} = \mathbf{F}.\mathbf{v}$

power of force $\mathbf{F}$
($=$ rate of doing work)

$k \quad (T = kx)$

stiffness of a spring

$\lambda \quad \left(T = \dfrac{\lambda x}{l} \right)$

elastic modulus of spring or string
of natural length l

$$\ddot{x} = -\omega^2 x$$
$$\dot{x}^2 = \omega^2(a^2 - x^2)$$
$$x = a\sin(\omega t + \varepsilon)$$

standard equations of simple harmonic motion with amplitude a and period $2\pi/\omega$

$$\mathbf{G} = \mathbf{r} \times \mathbf{F}$$

moment about O of a force $\mathbf{F}$ acting at point $\mathbf{r}$

$$I = \sum mr^2$$
$$G(= 6.67 \times 10^{-11} \text{ in SI})$$

moment of inertia

constant of gravitation

Logarithmic, exponential, and hyperbolic functions

$$\ln x = \log_e x = \int_1^x \frac{dt}{t} \quad (x > 0)$$

natural (or Naperian) logarithmic function

$$e\,(\approx 2.718)$$
$$e^x \text{ (or } \exp x)$$
$$\sinh x = \tfrac{1}{2}(e^x - e^{-x})$$
$$\cosh x = \tfrac{1}{2}(e^x + e^{-x})$$
$$\sinh^{-1} x = \ln[x + \sqrt{(x^2 + 1)}]$$
$$\cosh^{-1} x = \ln[x + \sqrt{(x^2 - 1)}]$$

base of natural logarithms ($\ln e = 1$)
exponential function
hyperbolic sine
hyperbolic cosine
inverse hyperbolic functions

Coefficients in binomial series

$$\binom{n}{r} = \frac{n(n-1)(n-2)\ldots(n-r+1)}{r!} \quad (n \in \mathbb{R}, r \in \mathbb{Z}^+)$$
$$= {}^nC_r \quad (n, r \in \mathbb{Z}^+)$$

Complex numbers

$$z = x + iy$$
$$= r(\cos\theta + i\sin\theta) = r\operatorname{cis}\theta$$
$$\operatorname{Re} z = x$$
$$\operatorname{Im} z = y$$
$$|z| = r = \sqrt{(x^2 + y^2)}$$
$$\arg z = \theta \quad (-\pi < \theta \leqslant \pi)$$
$$z^* = x - iy = r\operatorname{cis}(-\theta)$$

complex number z

real part of z
imaginary part of z
modulus of z
argument of z
complex conjugate of z

Probability and statistics

$$\mu = \mathrm{E}(X) = \sum p_r x_r$$

mean or expected value of random variable X, distributed with probability p

$$\sigma^2 = \text{Var}(X) = E[(X - \mu)^2]$$
$$= E(X^2) - \mu^2$$
$$= \sum p_r x_r^2 - \mu^2$$

variance of random variable x

σ

standard deviation

$G(t) = \sum p_r t^r$

probability generator

$\mu = G'(1)$

$\sigma^2 = G''(1) + G'(1) - [G'(1)]^2$

$\mu = np$

mean and variance of binomial distribution

$\sigma^2 = npq$

$F(x)$

cumulative probability, or distribution, function

$f(x) = F'(x)$

probability density function

$$\Rightarrow F(X) = \int_{-\infty}^{x} f(x)\,dx$$

$$\mu = \int xf(x)\,dx$$

mean and variance of continuous distribution with p.d.f. $f(x)$

$$\sigma^2 = \int (x - \mu)^2 f(x)\,dx$$

$$= \int x^2 f(x)\,dx - \mu^2$$

$$\phi(x) = \frac{1}{\sqrt{2\pi}\,\sigma} e^{-(x-\mu)^2/2\sigma^2}$$

probability density function of Normal distribution with mean μ and standard deviation σ

$$\phi(x) = \frac{1}{\sqrt{2\pi}} e^{-x^2/2}$$

Normal distribution with mean 0 and standard deviation 1

$\phi(x)$

cumulative probability function of Normal distribution

$$p_r = \frac{\mu^r}{r!} e^{-\mu}$$

Poisson distribution with mean μ

$E(m) = \mu$

expected value, variance and standard deviation of distribution of mean m of sample size n (drawn from population with mean μ and standard deviation σ)

$$\text{Var}(m) = \frac{\sigma^2}{n}$$

$$\frac{\sigma}{\sqrt{n}}$$

$$E(s^2) = \frac{n-1}{n}\sigma^2$$

expected value of variance s^2 of sample size n

$$\Rightarrow \hat{\sigma} = \sqrt{\left(\frac{n}{n-1}\right)}s$$

unbiassed estimate for standard deviation of a population when sample size n has standard deviation s

For a bivariate distribution (x_i, y_i):

$$\bar{x} = \frac{\sum x_i}{n}, \quad \bar{y} = \frac{\sum y_i}{n}$$

means of x, y

$$s_x^2 = \frac{\sum (x_i - \bar{x})^2}{n} = \frac{\sum x_i^2}{n} - \bar{x}^2$$

variance of x

$$s_y^2 = \frac{\sum (y_i - \bar{y})^2}{n} = \frac{\sum y_i^2}{n} - \bar{y}^2$$

variance of y

$$s_{xy} = \frac{\sum (x_i - \bar{x})(y_i - \bar{y})}{n}$$

covariance of x, y

$$= \frac{\sum x_i y_i}{n} - \bar{x}\bar{y}$$

$$\frac{s_{xy}}{s_x^2}$$

regression coefficient of y on x

$$\frac{s_{xy}}{s_y^2}$$

regression coefficient of x on y

$$r = \frac{s_{xy}}{s_x s_y}$$

correlation coefficient

$$\rho = 1 - \frac{6\sum d_i^2}{n(n^2 - 1)}$$

Spearman's rank correlation coefficient

$$\tau = \frac{\delta}{\frac{1}{2}n(n-1)}$$

Kendall's rank correlation coefficient

Matrices and determinants

$\mathbf{A}, \mathbf{B}, \mathbf{C}, \ldots$	matrices
$\det \mathbf{A}$ or Δ	determinant of square matrix $\mathbf{A}$
$\mathbf{A}^T, \mathbf{B}^T, \mathbf{C}^T, \ldots$	transposed matrices
$\mathbf{A}^*, \mathbf{B}^*, \mathbf{C}^*, \ldots$	adjoint matrices

$\mathbf{A}^{-1}, \mathbf{B}^{-1}, \mathbf{C}^{-1}, \ldots$ inverse matrices

A_{ij} cofactors of a_{ij} in det $\mathbf{A}$

$\mathbf{AB}$ product matrix

Sets

$\mathbb{Z}$	integers
$\mathbb{Q}$	rational numbers
$\mathbb{R}$	real numbers
$\mathbb{C}$	complex numbers
$\mathbb{Z}^{+}, \mathbb{Q}^{+}, \mathbb{R}^{+}$	positive integers etc.
$\mathbb{Z}^{*}, \mathbb{Q}^{*}, \mathbb{R}^{*}$	$\mathbb{Z}, \mathbb{Q}, \mathbb{R}$, but omitting zero

11 Further vectors

11.1 Points of subdivision

Points of balance

Suppose that two masses are attached to the ends of a light rod AB, 2 kg at A and 3 kg at B.

It is found from elementary experiments that such a rod will balance at the point P, where

$$2AP = 3PB$$

so we call this the *point of balance*.

If we now denote the position vectors (from an origin O) of A, B, P by $\mathbf{a}, \mathbf{b}, \mathbf{p}$,

then $\quad 2(\mathbf{p} - \mathbf{a}) = 3(\mathbf{b} - \mathbf{p})$

$\Rightarrow \qquad\qquad 5\mathbf{p} = 2\mathbf{a} + 3\mathbf{b}$

$\Rightarrow \qquad\qquad \mathbf{p} = \dfrac{2\mathbf{a} + 3\mathbf{b}}{5}$

More generally, let us consider two masses, α at A and β at B

Then their 'point of balance' would be at P, where

$$\alpha AP = \beta PB$$

So $\quad \alpha(\mathbf{p} - \mathbf{a}) = \beta(\mathbf{b} - \mathbf{p})$

$\Rightarrow \quad (\alpha + \beta)\mathbf{p} = \alpha\mathbf{a} + \beta\mathbf{b}$

$\Rightarrow \qquad\qquad \mathbf{p} = \dfrac{\alpha\mathbf{a} + \beta\mathbf{b}}{\alpha + \beta}$

It is clear that only the ratio $\alpha:\beta$ is important. For if α and β are both multiplied by the same number κ,

$$\kappa\alpha\mathbf{AP} = \kappa\beta\mathbf{PB} \quad \Rightarrow \quad \alpha\mathbf{AP} = \beta\mathbf{PB}$$

and $\quad \mathbf{p} = \dfrac{\kappa\alpha\mathbf{a} + \kappa\beta\mathbf{b}}{\kappa\alpha + \kappa\beta} = \dfrac{\alpha\mathbf{a} + \beta\mathbf{b}}{\alpha + \beta}$

so that the position of P is unchanged.

When, however, the ratio changes, P will move along the line AB; and if either of the numbers α, β were allowed to be negative, P would fall outside the segment AB and be either beyond A or beyond B.

Example 1

a) $\alpha = -2, \beta = +3$
So $\quad -2\mathbf{AP} = 3\mathbf{PB} \quad$ and P is beyond B.

Furthermore, $\quad \mathbf{p} = \dfrac{-2\mathbf{a} + 3\mathbf{b}}{-2 + 3} = -2\mathbf{a} + 3\mathbf{b}$

b) $\alpha = +2, \beta = -1$
So $\quad 2\mathbf{AP} = -\mathbf{PB} \quad$ and P is beyond A.

Furthermore, $\quad \mathbf{p} = \dfrac{2\mathbf{a} - \mathbf{b}}{2 - 1} = 2\mathbf{a} - \mathbf{b}$

The only exceptional case is when $\alpha = -\beta$.

Then $\quad \alpha\mathbf{AP} = -\alpha\mathbf{PB} \quad \Rightarrow \quad \mathbf{AP} = -\mathbf{PB}$

which is impossible when A, B are distinct

(and $\quad \mathbf{p} = \dfrac{\alpha\mathbf{a} - \alpha\mathbf{b}}{\alpha - \alpha} = \dfrac{\alpha\mathbf{a} - \alpha\mathbf{b}}{0}, \quad$ which is meaningless).

Exercise 11.1a

1 If A, B and P have position vectors $\mathbf{a}$, $\mathbf{b}$ and $\mathbf{p}$, find $\mathbf{p}$ in terms of $\mathbf{a}$ and $\mathbf{b}$ and illustrate by a sketch when:

    a) $\mathbf{AP} = \mathbf{PB}$ **b)** $3\mathbf{AP} = \mathbf{PB}$ **c)** $\mathbf{AP} = 3\mathbf{PB}$

    d) $3\mathbf{AP} = -\mathbf{PB}$ **e)** $\mathbf{AP} = -3\mathbf{PB}$

2   If A, B have position vectors **a**, **b**, illustrate the points whose position
vectors are:

a) $\frac{4}{5}\mathbf{a} + \frac{1}{5}\mathbf{b}$ **b)** $\frac{3}{5}\mathbf{a} + \frac{2}{5}\mathbf{b}$ **c)** $\frac{2}{5}\mathbf{a} + \frac{3}{5}\mathbf{b}$

d) $\frac{1}{5}\mathbf{a} + \frac{4}{5}\mathbf{b}$ **e)** $2\mathbf{a} - \mathbf{b}$ **f)** $3\mathbf{a} - 2\mathbf{b}$

g) $4\mathbf{a} - 3\mathbf{b}$ **h)** $2\mathbf{b} - \mathbf{a}$ **i)** $3\mathbf{b} - 2\mathbf{a}$

j) $4\mathbf{b} - 3\mathbf{a}$

3   The points A and B have position vectors $6\mathbf{i} - 5\mathbf{j}$ and $4\mathbf{i} - 3\mathbf{j}$ respectively.
Show that the mid-point M of AB is collinear with the two points X, Y
whose position vectors are $2\mathbf{i} - 6\mathbf{j}$ and $11\mathbf{i}$ respectively.   (OC)

Points of subdivision

If $\alpha\mathbf{AP} = \beta\mathbf{PB}$

then $\alpha AP = \beta PB$

and $\dfrac{AP}{PB} = \dfrac{\beta}{\alpha}$

so P is said to *divide* AB in the ratio $\beta : \alpha$.

Example 2

a) $\alpha = 2, \beta = 3$

So $2\mathbf{AP} = 3\mathbf{PB}$

$\Rightarrow$ $\dfrac{AP}{PB} = \dfrac{3}{2}$

and P divides AB in the ratio $3:2$.

b) $\alpha = -2, \beta = 3$

So $2\mathbf{AP} = -3\mathbf{PB}$

$\Rightarrow$ $\dfrac{AP}{PB} = -\dfrac{3}{2}$

and P divides AB in the ratio $3:-2$.

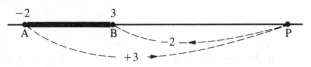

c) $\alpha = 2, \beta = -1$

So $-2\mathbf{AP} = \mathbf{PB}$

$\Rightarrow \qquad \dfrac{AP}{PB} = -\dfrac{1}{2}$

and P divides AB in the ratio $-1:2$.

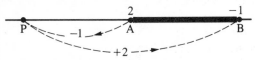

Summarising, we see that the following statements are all equivalent:

P is the point of balance of α at A and β at B

$\alpha\mathbf{AP} = \beta\mathbf{PB}$

$$\mathbf{p} = \dfrac{\alpha\mathbf{a} + \beta\mathbf{b}}{\alpha + \beta}$$

P divides AB in the ratio $\beta:\alpha$

Example 3

Find the position vectors of the points P, Q which divide AB in the ratio $1:2$ and $-5:2$ respectively.

a) P divides AB in the ratio $1:2$

$\Rightarrow$ P is the 'point of balance' of 2 at A and 1 at B

$\Rightarrow \quad \mathbf{p} = \dfrac{2\mathbf{a} + 1\mathbf{b}}{2 + 1} = \tfrac{2}{3}\mathbf{a} + \tfrac{1}{3}\mathbf{b}$

b) Q divides AB in the ratio $5:-2$

$\Rightarrow$ Q is the 'point of balance' of -2 at A and 5 at B

$\Rightarrow \quad \mathbf{q} = \dfrac{-2\mathbf{a} + 5\mathbf{b}}{-2 + 5} = -\tfrac{2}{3}\mathbf{a} + \tfrac{5}{3}\mathbf{b}$

Exercise 11.1b

1   The vertices of △ABC have position vectors **a**, **b**, **c**. Find, in terms of **a**, **b**, **c**, the position vectors of
   a) D, E, F, the mid-points of BC, CA, AB respectively;
   b) the point X which divides the line AD in the ratio 2:1, and the corresponding points Y, Z dividing BE, CF in the same ratio.
   Hence show that the medians AD, BE, CF are concurrent (at a point called the *centroid* of △ABC).

2   The vertices A, B, C of △ABC have position vectors **a**, **b**, **c**. Find, in terms of **a**, **b** and **c**, the position vectors of
   a) P, which divides BC in the ratio 3:2;
   b) Q, which divides CA in the ratio 1:3;
   c) R, which divides AB in the ratio 2:1;
   d) X, which divides AP in the ratio 5:1;
   e) Y, which divides BQ in the ratio 2:1;
   f) Z, which divides CR in the ratio 1:1.
   Hence show that AP, BQ, CR are concurrent.

3   In no. **2** find also the position vectors of
   a) U, which divides BC in the ratio $-3:2$;
   b) V, which divides CA in the ratio $-1:3$;
   c) W, which divides AB in the ratio $-2:1$.
   Hence show that U also divides QR in the ratio $-3:4$; V also divides RP in the ratio $-5:3$; and W also divides PQ in the ratio $-4:5$.

4   If the points A, B, C have position vectors **a**, **b**, **c** from an origin O, show that the equation

   $$\mathbf{r} = t\mathbf{a} + (1 - t)\mathbf{b}$$

   where t is a parameter, represents the straight line AB, and find the equation of BC.
   Find the equation of the straight line joining L the mid-point of OA to M the mid-point of BC. Find also the position vector of the point in which the line LM meets the straight line joining the mid-point of OB to the mid-point of AC.   (L)

5   OABC is a square of side 2a. **i**, **j** are unit vectors along OA, OC. The mid-point of AB is L; the mid-point of BC is M; OL, AM meet at P; BP meets OA at N. Show that the segment OP can be measured by the vector $\lambda(2a\mathbf{i} + a\mathbf{j})$ and also by the vector $2a\mathbf{i} + \mu(2a\mathbf{j} - a\mathbf{i})$. Hence determine λ and μ. Prove that ON $= \frac{2}{3}$OA.   (OC)

11.2 Centres of mass

In the last section, when masses α and β were placed at two points whose position vectors are **a** and **b**, we referred to

$$\mathbf{p} = \frac{\alpha\mathbf{a} + \beta\mathbf{b}}{\alpha + \beta}$$

as their 'point of balance'.

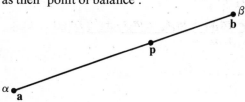

Let us now suppose that we have a number of masses:

$$m_1, m_2, m_3, m_4 \ldots$$

at $\mathbf{r}_1, \mathbf{r}_2, \mathbf{r}_3, \mathbf{r}_4 \ldots$ respectively.

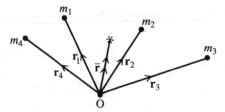

Let G be the point which has position vector $\bar{\mathbf{r}}$, where

$$\bar{\mathbf{r}} = \frac{m_1\mathbf{r}_1 + m_2\mathbf{r}_2 + m_3\mathbf{r}_3 + m_4\mathbf{r}_4 + \cdots}{m_1 + m_2 + m_3 + m_4 + \cdots}$$

Now G is a definite point, but there is a serious risk that its position will depend on our choice of origin, O.

To discover whether or not this is so, we must take another origin O′ and see whether such a calculation based on O′ leads to a different point G′. Let us suppose that from the new origin O′, O has position vector **c**, and $m_1, m_2, \ldots$ have position vectors $\mathbf{r}'_1, \mathbf{r}'_2, \ldots$.

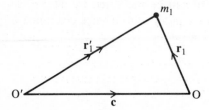

Then $\mathbf{r}'_1 = \mathbf{c} + \mathbf{r}_1,$ etc.

So the position vector of G', calculated from O', is

$$\bar{\mathbf{r}}' = \frac{m_1\mathbf{r}'_1 + m_2\mathbf{r}'_2 + m_3\mathbf{r}'_3 + \cdots}{m_1 + m_2 + m_3 + \cdots}$$

$$= \frac{m_1(\mathbf{c} + \mathbf{r}_1) + m_2(\mathbf{c} + \mathbf{r}_2) + m_3(\mathbf{c} + \mathbf{r}_3) + \cdots}{m_1 + m_2 + m_3 + \cdots}$$

$$= \frac{(m_1 + m_2 + m_3 + \cdots)\mathbf{c} + (m_1\mathbf{r}_1 + m_2\mathbf{r}_2 + m_3\mathbf{r}_3 + \cdots)}{m_1 + m_2 + m_3 + \cdots}$$

$$= \mathbf{c} + \frac{m_1\mathbf{r}_1 + m_2\mathbf{r}_2 + m_3\mathbf{r}_3 + \cdots}{m_1 + m_2 + m_3 + \cdots}$$

So $\bar{\mathbf{r}}' = \mathbf{c} + \bar{\mathbf{r}}$

Hence $\mathbf{O'G'} = \mathbf{O'O} + \mathbf{OG} = \mathbf{O'G}$

$\Rightarrow$ G and G' are identical

So if masses $m_1, m_2, m_3 \ldots$ are at points $\mathbf{r}_1, \mathbf{r}_2, \mathbf{r}_3 \ldots$, the point

$$\bar{\mathbf{r}} = \frac{m_1\mathbf{r}_1 + m_2\mathbf{r}_2 + m_3\mathbf{r}_3 + \cdots}{m_1 + m_2 + m_3 + \cdots}$$

is *not* dependent on our choice of origin, but only on the given system of masses, and is therefore called their *centre of mass*. In the particular case when the masses are all equal this becomes

$$\bar{\mathbf{r}} = \frac{1}{n}(\mathbf{r}_1 + \mathbf{r}_2 + \cdots + \mathbf{r}_n)$$

which is also known as the *centroid* of the given points. So the centroid of $\mathbf{r}_1$ and $\mathbf{r}_2$ is at $\frac{1}{2}(\mathbf{r}_1 + \mathbf{r}_2)$, and of $\mathbf{r}_1, \mathbf{r}_2, \mathbf{r}_3$ is at $\frac{1}{3}(\mathbf{r}_1 + \mathbf{r}_2 + \mathbf{r}_3)$.

Finally, in terms of components, the centre of mass is

$$\bar{x}\mathbf{i} + \bar{y}\mathbf{j} + \bar{z}\mathbf{k} = \frac{m_1(x_1\mathbf{i} + y_1\mathbf{j} + z_1\mathbf{k}) + m_2(x_2\mathbf{i} + y_2\mathbf{j} + z_2\mathbf{k}) + \cdots}{m_1 + m_2 + \cdots}$$

$$= \frac{m_1x_1 + m_2x_2 + \cdots}{m_1 + m_2 + \cdots}\mathbf{i} + \frac{m_1y_1 + m_2y_2 + \cdots}{m_1 + m_2 + \cdots}\mathbf{j}$$

$$+ \frac{m_1z_1 + m_2z_2 + \cdots}{m_1 + m_2 + \cdots}\mathbf{k}$$

So the centre of mass has coordinates

$$\bar{x} = \frac{m_1 x_1 + m_2 x_2 + \cdots}{m_1 + m_2 + \cdots} = \frac{\sum mx}{\sum m}$$

$$\bar{y} = \frac{m_1 y_1 + m_2 y_2 + \cdots}{m_1 + m_2 + \cdots} = \frac{\sum my}{\sum m}$$

$$\bar{z} = \frac{m_1 z_1 + m_2 z_2 + \cdots}{m_1 + m_2 + \cdots} = \frac{\sum mz}{\sum m}$$

Example 1

A light rectangular frame of length 4 m, breadth 3 m and height 2 m has masses 1 kg, 2 kg, 3 kg attached at corners as shown in the figure. Where is their centre of mass?

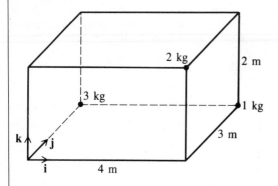

Taking unit vectors **i**, **j**, **k** as shown, the three masses are:

  1 kg at the point $4\mathbf{i} + 3\mathbf{j}$
  2 kg at the point $4\mathbf{i} + 2\mathbf{k}$
  3 kg at the point $3\mathbf{j}$

So their centre of mass is at

$$\bar{\mathbf{r}} = \frac{1 \times (4\mathbf{i} + 3\mathbf{j}) + 2 \times (4\mathbf{i} + 2\mathbf{k}) + 3 \times 3\mathbf{j}}{1 + 2 + 3}$$

$$= \frac{12\mathbf{i} + 12\mathbf{j} + 4\mathbf{k}}{6}$$

$$= 2\mathbf{i} + 2\mathbf{j} + \tfrac{2}{3}\mathbf{k}$$

Alternatively, taking axes Ox, Oy, Oz in the same directions as $\mathbf{i}$, $\mathbf{j}$, $\mathbf{k}$, the calculation could be set out:

m	x	y	z	mx	my	mz
1	4	3	0	4	3	0
2	4	0	2	8	0	4
3	0	3	0	0	9	0
$\sum m = 6$				$\sum mx = 12$	$\sum my = 12$	$\sum mz = 4$

So the centre of mass has coordinates

$$\bar{x} = \frac{\sum mx}{\sum m} = 2 \qquad \bar{y} = \frac{\sum my}{\sum m} = 2 \qquad \bar{z} = \frac{\sum mz}{\sum m} = \tfrac{2}{3}$$

and is therefore at the point $(2, 2, \tfrac{2}{3})$.

Example 2

If a square of side a is removed from the corner of a uniform square lamina of side $2a$, where is the centre of mass of the remaining piece?

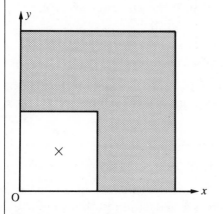

We can regard the remaining piece as being composed of a large square of mass 4 together with a smaller square of negative mass -1. So, using axes Ox, Oy as shown, we have:

	m	x	y	mx	my
large square	4	a	a	$4a$	$4a$
small square	-1	$\tfrac{1}{2}a$	$\tfrac{1}{2}a$	$-\tfrac{1}{2}a$	$-\tfrac{1}{2}a$
final lamina	$\sum m = 3$			$\sum mx = \tfrac{7}{2}a$	$\sum my = \tfrac{7}{2}a$

So $\bar{x} = \dfrac{\sum mx}{\sum m} = \dfrac{7a}{6}$, $\bar{y} = \dfrac{\sum my}{\sum m} = \dfrac{7a}{6}$

and the centre of mass lies at $(\tfrac{7}{6}a, \tfrac{7}{6}a)$.

Exercise 11.2a

1   If the vertices of a tetrahedron ABCD have position vectors **a, b, c, d**, find the position vectors of
   a) the centroid P of $\triangle$BCD;
   b) the point dividing AP in the ratio 3:1. What do you deduce about this and the three other similar lines?
   c) the mid-point of the line joining the mid-points of AB and CD.
   Hence find seven lines associated with the tetrahedron which are all concurrent. At what point do they meet?

2   If G, G′ are the centroids of $\triangle$ABC, $\triangle$A′B′C′, prove that

   $$\mathbf{AA'} + \mathbf{BB'} + \mathbf{CC'} = 3\mathbf{GG'}$$

3   Find the centres of mass of
   a) 1 kg at **a**, 2 kg at **b**, 3 kg at **c**;
   b) 1 g at (0, 0, 0), 4 g at (1, 4, 2), 3 g at (3, −1, −2), 2 g at (−4, 1, 6);
   c) 1 kg at **i** + 2**j**, 4 kg at **j** − 2**k**, 5 kg at −**i** + 3**j** + **k**.

4   If a cube of side 1 m is removed from the corner of a uniform solid cube of side 2 m, how far has its centre of mass been displaced?

5   A mass of 2 kg is placed at the point (4, 1). Find the masses which should be placed at (−1, 3) and (−2, −1) in order that the centroid of the three masses should be at the origin.   (SMP)

Continuous distributions of mass

So far, we have limited our investigations to systems consisting of a finite number of discrete masses. When there is a *continuous* distribution of mass, the method is very similar, except that we shall usually begin by dividing the given body into a series of elements.

Example 3

Uniform triangular lamina
We first divide the lamina into a set of elemental strips parallel to one of its sides. Now the centre of mass of each strip is at its mid-point, and these are

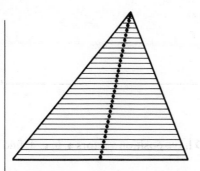

all collinear on a median. Hence the centre of mass of the lamina lies on this median, and similarly on the other two medians. It is therefore situated at the *centroid* of the given triangle.

Example 4

Find the centre of mass of:
a) a uniform lamina (or plate) in the shape of the area between $y = x^2$, $y = 0$ and $x = 1$;
b) a uniform solid formed by rotating this area about the x-axis.

a) Uniform lamina

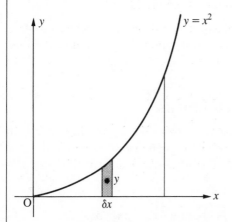

Suppose that the lamina has surface density (i.e. mass per unit area) σ, and that it is split into narrow strips of width δx and height y.

Then the strip shown has mass $m = \sigma y\,\delta x$ which can be regarded as concentrated at the point $(x, \tfrac{1}{2}y)$.

So $\bar{x} = \dfrac{\sum mx}{\sum m} = \dfrac{\sum(\sigma y\,\delta x)x}{\sum(\sigma y\,\delta x)} = \dfrac{\sum xy\,\delta x}{\sum y\,\delta x}$

$\Rightarrow$ $\bar{x} = \dfrac{\displaystyle\int_0^1 xy\,dx}{\displaystyle\int_0^1 y\,dx} = \dfrac{\displaystyle\int_0^1 x^3\,dx}{\displaystyle\int_0^1 x^2\,dx} = \dfrac{\frac{1}{4}}{\frac{1}{3}} = \dfrac{3}{4}$

and $\bar{y} = \dfrac{\sum my}{\sum m} = \dfrac{\sum(\sigma y\,\delta x)\frac{1}{2}y}{\sum(\sigma y\,\delta x)} = \dfrac{\frac{1}{2}\sum y^2\,\delta x}{\sum y\,\delta x}$

$\Rightarrow$ $\bar{y} = \dfrac{\frac{1}{2}\displaystyle\int_0^1 y^2\,dx}{\displaystyle\int_0^1 y\,dx} = \dfrac{\frac{1}{2}\displaystyle\int_0^1 x^4\,dx}{\displaystyle\int_0^1 x^2\,dx} = \dfrac{\frac{1}{10}}{\frac{1}{3}} = \dfrac{3}{10}$

So the centre of mass is at $(\frac{3}{4}, \frac{3}{10})$.

b) Uniform solid

Suppose that the solid has volume density ρ and that it is sliced into thin discs of thickness δx and radius y.

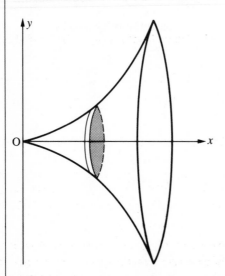

Then the mass of the disc shown $= m = \rho(\pi y^2\,\delta x) = \rho\pi y^2\,\delta x$.

Now if the centre of mass is at $(\bar{x}, \bar{y})$ we see by symmetry that $\bar{y} = 0$.

Also $\bar{x} = \dfrac{\sum mx}{\sum m} = \dfrac{\sum(\rho\pi y^2\,\delta x)x}{\sum \rho\pi y^2\,\delta x} = \dfrac{\sum xy^2\,\delta x}{\sum y^2\,\delta x}$

$$\Rightarrow \quad \bar{x} = \frac{\displaystyle\int_0^1 xy^2\,\mathrm{d}x}{\displaystyle\int_0^1 y^2\,\mathrm{d}x} = \frac{\displaystyle\int_0^1 x^5\,\mathrm{d}x}{\displaystyle\int_0^1 x^4\,\mathrm{d}x} = \frac{\frac{1}{6}}{\frac{1}{5}} = \frac{5}{6}$$

So the centre of mass is at $(\frac{5}{6}, 0)$.

Example 5

Find the centre of mass of:
a) a uniform semi-circular wire of radius a;
b) a uniform semi-circular plate of radius a.

a) Semi-circular wire
Take an axis Ox along the line of symmetry and consider a small section of wire which subtends angle $\delta\theta$ at the centre. The length of this section is $a\,\delta\theta$, so if its line density is ρ it will have mass $m = \rho a\,\delta\theta$.
 Also the length of the wire is πa, so its total mass will be $\rho\pi a$.

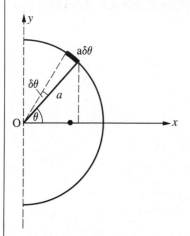

By symmetry, $\bar{y} = 0$

and $\bar{x} = \dfrac{\sum mx}{\sum m} = \dfrac{\sum (\rho a\,\delta\theta)a\cos\theta}{\rho\pi a} = \dfrac{a}{\pi}\sum \cos\theta\,\delta\theta$

$$\Rightarrow \quad \bar{x} = \frac{a}{\pi}\int_{-\pi/2}^{+\pi/2} \cos\theta\,\mathrm{d}\theta = \frac{a}{\pi}\Big[\sin\theta\Big]_{-\pi/2}^{+\pi/2} = \frac{2a}{\pi}$$

So the centre of mass of the wire is at $(2a/\pi, 0)$, i.e. approximately at $(0.64a, 0)$.

b) Semi-circular plate

Method 1

Suppose that the plate is divided into strips of length $2y$ and width δx and that its surface density is σ.

Then the mass of each strip is $\sigma(2y\,\delta x)$ and the total mass of the plate is $\frac{1}{2}\pi a^2 \sigma$.

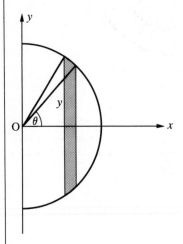

So $\bar{x} = \dfrac{\sum mx}{\sum m} = \dfrac{\sum (2\sigma y\,\delta x)x}{\frac{1}{2}\pi a^2 \sigma} = \dfrac{4}{\pi a^2}\sum xy\,\delta x$

$\Rightarrow \quad \bar{x} = \dfrac{4}{\pi a^2}\int_0^a xy\,\mathrm{d}x = \dfrac{4}{\pi a^2}\int_0^a x\sqrt{(a^2 - x^2)}\,\mathrm{d}x$

$$= \dfrac{4}{\pi a^2}\left[-\tfrac{1}{3}(a^2 - x^2)^{3/2}\right]_0^a$$

$$= \dfrac{4}{\pi a^2}\left[0 - (-\tfrac{1}{3}a^3)\right] = \dfrac{4a}{3\pi}$$

Also, by symmetry, $\bar{y} = 0$.

So the centre of mass is at $(4a/3\pi, 0)$; or, approximately, $(0.42a, 0)$.

Method 2

This last result can be obtained more easily if we make use of the previous result for a semi-circular wire. For the semi-circular plate can be dissected into a large number of sectors and each of these sectors is approximately a triangle with its centre of mass at a distance $2a/3$ from O.

In this way we see that the uniform plate of radius a has the same centre of mass as a uniform wire of radius $2a/3$.

Hence $\bar{x} = \dfrac{2(2a/3)}{\pi} = \dfrac{4a}{3\pi}$

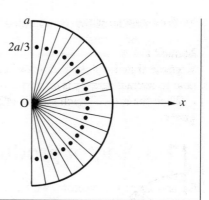

Exercise 11.2b

1   Find the coordinates of the centres of mass of uniform laminae bounded by:
   a) $y = x^3, \quad y = 0, \quad x = 2$
   b) $y = x^2, \quad y = 0, \quad x = 1, \quad x = 2$
   c) $y^2 = 4x, \quad x = 9$
   d) $y = \dfrac{ax}{h}, \quad y = 0, \quad x = h$

2   Find the coordinates of the centres of mass of the uniform solids formed by rotating the areas of no. **1** through $360°$ about the x-axis.

3   By rotating a quadrant of the curve $y = \sqrt{(a^2 - x^2)}$ about the x-axis, find the position of the centre of mass of a uniform solid hemisphere.

4   By slicing a uniform tetrahedron parallel to one of its faces, show that its centre of mass lies on the line joining the centroid of this face to the opposite vertex. Hence show that the centre of mass
   a) of a uniform tetrahedron lies at the centroid of its four vertices;
   b) of a uniform cone (or pyramid) is at one-quarter of its height above the base (see no. **2 (d)**);
   c) of a uniform hemisphere shell is half-way from its centre to the middle-point of the shell (use the result of no. **3** and part **(b)**).

5   A sphere of radius a is cut into two portions by a plane which is distant $\frac{1}{2}a$ from the centre of the sphere. Show by integration that the volume of the smaller of the two portions is $5\pi a^3/24$.
   Calculate the distance of the centre of gravity of this smaller portion from the centre of the sphere.   (OC)

6   Use Simpson's rule to find the approximate value of the x-coordinate of

the centroid of the area bounded by the *x*-axis, the line $x = 4$, and a curve joining the points

x	0	1	2	3	4
y	0	1.05	2.21	3.50	4.92

(OC)

*11.3 Vector products

If **a** and **b** are two vectors, we have seen that we can associate with them a scalar quantity, known as their scalar product. We now ask ourselves whether there is also a convenient *vector product* which we could usefully consider.

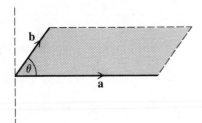

It is clear that the two vectors **a** and **b** fix a particular parallelogram, whose area is easily seen to be $ab \sin \theta$. Furthermore, the most natural direction to associate with this parallelogram would be the one which is perpendicular, or normal, to its plane. As for the particular *sense* (up or down this line), we can easily resolve our dilemma by following the movement of a right-handed screw which is turning from **a** to **b**.

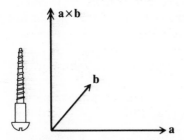

We therefore define the *vector product*, which we shall write **a × b**, as the vector

whose magnitude is $ab \sin \theta$;
whose direction is perpendicular to **a** and **b**; and
whose sense is determined by a right-hand screw turning from **a** to **b**.

The most common use of vector products occurs in mechanics, when a force **F** acts through a point whose position vector is **r** and whose moment about O is defined as **r × F**:

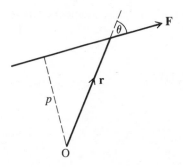

In this figure it is seen that the magnitude of this moment is $rF \sin \theta = Fp$, and so is equal to the product of the force and the length of the perpendicular arm from O. The moment itself is a vector of this magnitude which acts perpendicular to the plane of **r** and **F**. In our example, therefore, by the right-hand screw rule, this moment acts *into* the paper and represents a turning effect which is in the plane of the paper and clockwise.

As with the scalar product, the properties of the vector product follow from its definition. Some are expected, like

$$\mathbf{a} \times (\lambda\mathbf{b}) = (\lambda\mathbf{a}) \times \mathbf{b} = \lambda(\mathbf{a} \times \mathbf{b})$$

and can easily be demonstrated in a figure. Others are more surprising,

like $\mathbf{a} \times \mathbf{a} = \mathbf{0}$ (since $aa \sin 0 = 0$)
and $\mathbf{a} \times \mathbf{b} = -\mathbf{b} \times \mathbf{a}$

(because a right-hand screw turning from **a** to **b** moves *against* a right-hand screw turning from **b** to **a**). So vector products are *not* commutative.

Lastly, however, vector products (like scalar products) *are* distributive with respect to addition:

$$\mathbf{a} \times (\mathbf{b} + \mathbf{c}) = \mathbf{a} \times \mathbf{b} + \mathbf{a} \times \mathbf{c}$$

We shall not establish this until the end of this section, though in passing we remark on its importance if **P** and **Q** are two forces.

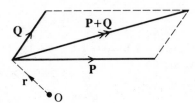

For the result means that

$$\mathbf{r} \times (\mathbf{P} + \mathbf{Q}) = \mathbf{r} \times \mathbf{P} + \mathbf{r} \times \mathbf{Q}$$

i.e. that the sum of the moments of two forces is equal to the moment of their resultant (see section 14.4).

Vector product in component form

If $\mathbf{a} = a_1\mathbf{i} + a_2\mathbf{j} + a_3\mathbf{k}$
and $\mathbf{b} = b_1\mathbf{i} + b_2\mathbf{j} + b_3\mathbf{k}$

we can easily find the vector product $\mathbf{a} \times \mathbf{b}$.
   For, firstly, we notice that

$$\mathbf{i} \times \mathbf{i} = \mathbf{j} \times \mathbf{j} = \mathbf{k} \times \mathbf{k} = 0$$
and that $\mathbf{i} \times \mathbf{j} = \mathbf{k} = -\mathbf{j} \times \mathbf{i}$
$$\mathbf{j} \times \mathbf{k} = \mathbf{i} = -\mathbf{k} \times \mathbf{j}$$
$$\mathbf{k} \times \mathbf{i} = \mathbf{j} = -\mathbf{i} \times \mathbf{k}$$

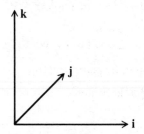

So $\mathbf{a} \times \mathbf{b} = (a_1\mathbf{i} + a_2\mathbf{j} + a_3\mathbf{k}) \times (b_1\mathbf{i} + b_2\mathbf{j} + b_3\mathbf{k})$
$$= (a_2b_3 - a_3b_2)\mathbf{i} + (a_3b_1 - a_1b_3)\mathbf{j} + (a_1b_2 - a_2b_1)\mathbf{k}$$

Example

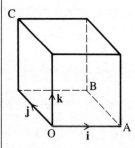

The position vectors of three points A, B, C are $\mathbf{a} = \mathbf{i}, \mathbf{b} = \mathbf{i} + \mathbf{j}, \mathbf{c} = \mathbf{j} + \mathbf{k}$.
Use vector products to find the angles between:
a) the planes OBC and OAC;
b) the line OB and the plane ABC.

a) Two vectors in plane OBC are **b** and **c**, so the normal to the plane is along

$$\begin{aligned}
\mathbf{b} \times \mathbf{c} &= (\mathbf{i} + \mathbf{j}) \times (\mathbf{j} + \mathbf{k}) \\
&= \mathbf{i} \times \mathbf{j} + \mathbf{i} \times \mathbf{k} + \mathbf{j} \times \mathbf{j} + \mathbf{j} \times \mathbf{k} \\
&= \mathbf{k} - \mathbf{j} + 0 + \mathbf{i} = \mathbf{i} - \mathbf{j} + \mathbf{k}
\end{aligned}$$

and similarly the normal to OAC is along

$$\mathbf{a} \times \mathbf{c} = \mathbf{i} \times (\mathbf{j} + \mathbf{k}) = \mathbf{k} - \mathbf{j} = -\mathbf{j} + \mathbf{k}$$

Now the angle θ between two planes is the angle between their normals, so

$$\cos \theta = \frac{(\mathbf{i} - \mathbf{j} + \mathbf{k}) \cdot (-\mathbf{j} + \mathbf{k})}{\sqrt{3} \times \sqrt{2}} = \frac{2}{\sqrt{6}} = \sqrt{\tfrac{2}{3}}$$

$$\Rightarrow \quad \theta = 35.3°$$

b) Similarly, two vectors in plane ABC

are $\mathbf{AB} = \mathbf{b} - \mathbf{a} = \mathbf{j}$
and $\mathbf{AC} = \mathbf{c} - \mathbf{a} = -\mathbf{i} + \mathbf{j} + \mathbf{k}$

So the normal to the plane is along

$$\mathbf{j} \times (-\mathbf{i} + \mathbf{j} + \mathbf{k}) = \mathbf{i} + \mathbf{k}$$

Now $\mathbf{OB} = \mathbf{b} = \mathbf{i} + \mathbf{j}$

Hence the angle between **OB** and this normal is ϕ, where

$$\cos \phi = \frac{(\mathbf{i} + \mathbf{j}) \cdot (\mathbf{i} + \mathbf{k})}{\sqrt{2} \times \sqrt{2}} = \tfrac{1}{2}$$

$$\Rightarrow \quad \phi = 60°$$

So the angle between **OB** and the plane itself is $90° - 60° = 30°$.

Scalar and vector triple products

We now proceed to ask if there is any way in which we can speak of the product of *three* vectors, **a**, **b**, **c**.

Firstly, it is clear that the expression $\mathbf{a} \cdot (\mathbf{b} \cdot \mathbf{c})$ is meaningless, since $\mathbf{b} \cdot \mathbf{c}$ is not a vector but a scalar and so cannot form a scalar product with **a**. By contrast, however, $\mathbf{a} \cdot (\mathbf{b} \times \mathbf{c})$ does exist and, as it is a scalar quantity, is called a *scalar triple product*. Furthermore, $\mathbf{a} \times (\mathbf{b} \times \mathbf{c})$ also exists and as it is a vector quantity is called a *vector triple product*.

Though these quantities are extremely important in applied mathematics, we cannot investigate them further and simply note the connection between scalar triple products and volumes.

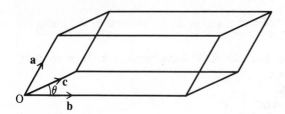

Suppose that a *parallelepiped* is formed from the three vectors **a**, **b**, **c**, as shown in the figure.

Then **b** × **c** is a vector of magnitude $bc \sin \theta$, which is equal to the area of the base parallelogram, in a direction normal to the base.

So the magnitude of **a** . (**b** × **c**)
= (projection of **a** ⊥ **b** and **c**) × $bc \sin \theta$
= perpendicular height × area of base
= volume of parallelepiped

Hence   volume of parallelepiped = **a** . (**b** × **c**)

and it clearly follows that:

volume of parallelepiped is zero   ⟺   **a**, **b**, **c** are coplanar
⟺   **a** . (**b** × **c**) = 0

Furthermore, in components:

$$\mathbf{a} = a_1\mathbf{i} + a_2\mathbf{j} + a_3\mathbf{k}$$
$$\mathbf{b} \times \mathbf{c} = (b_2c_3 - b_3c_2)\mathbf{i} + (b_3c_1 - b_1c_3)\mathbf{j} + (b_1c_2 - b_2c_1)\mathbf{k}$$
$$\Rightarrow \quad \mathbf{a} . (\mathbf{b} \times \mathbf{c}) = a_1(b_2c_3 - b_3c_2) + a_2(b_3c_1 - b_1c_3) + a_3(b_1c_2 - b_2c_1)$$

So   **a**, **b**, **c** are coplanar
$$\Leftrightarrow \quad a_1(b_2c_3 - b_3c_2) + a_2(b_3c_1 - b_1c_3) + a_3(b_1c_2 - b_2c_1) = 0$$

which is a result to which we shall return in chapter 16.

Exercise 11.3

1   If $\mathbf{a} = \mathbf{i} - \mathbf{j}$, $\mathbf{b} = \mathbf{i} + \mathbf{j} + \mathbf{k}$, $\mathbf{c} = \mathbf{k}$, calculate **b** × **c**, **c** × **a**, **a** × **b**.

2   Denoting corresponding points in no. **1** by capital letters, find:
   a) vectors which are normal to the planes OBC, OCA, ABC;
   b) the angle between the planes OBC and OCA;
   c) the angles between these planes and the plane ABC;
   d) the angles between these planes and the line AB.

3   In no. **1** find whether:
   a) **a** . (**b** × **c**) = **b** . (**c** × **a**) = **c** . (**a** × **b**)
   b) **a** × (**b** × **c**) = (**a** × **b**) × **c**

c) $\mathbf{a} \times (\mathbf{b} \times \mathbf{c}) = (\mathbf{a}.\mathbf{c})\mathbf{b} - (\mathbf{a}.\mathbf{b})\mathbf{c}$
Test these results with other vectors $\mathbf{a}$, $\mathbf{b}$, $\mathbf{c}$.

4 Let the three sides of a triangle ABC have vectors $\mathbf{a}$, $\mathbf{b}$, $\mathbf{c}$ so that $\mathbf{a} + \mathbf{b} + \mathbf{c} = \mathbf{0}$. Hence prove
a) $\mathbf{b} \times \mathbf{c} = \mathbf{c} \times \mathbf{a} = \mathbf{a} \times \mathbf{b}$
b) the sine formula for $\triangle ABC$

The distributive law for vector products †

$\mathbf{a} \times (\mathbf{b} + \mathbf{c}) = \mathbf{a} \times \mathbf{b} + \mathbf{a} \times \mathbf{c}$

Firstly, we project $\mathbf{b}$ and $\mathbf{c}$ on to the plane perpendicular to $\mathbf{a}$, and let their projections be $\mathbf{b}_1$ and $\mathbf{c}_1$.

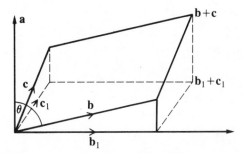

If the angle between $\mathbf{a}$ and $\mathbf{b}$ is θ, then $\mathbf{a} \times \mathbf{b}$ has magnitude $ab \sin \theta$.

But $\quad \mathbf{a} \times \mathbf{b}_1$ has magnitude ab_1
$$= ab \cos (\tfrac{1}{2}\pi - \theta)$$
$$= ab \sin \theta$$

Also, as $\mathbf{a}$, $\mathbf{b}$ and $\mathbf{b}_1$ are coplanar, $\mathbf{a} \times \mathbf{b}$ and $\mathbf{a} \times \mathbf{b}_1$ both lie along the perpendicular to this plane.

So $\qquad \mathbf{a} \times \mathbf{b} = \mathbf{a} \times \mathbf{b}_1$
similarly $\quad \mathbf{a} \times \mathbf{c} = \mathbf{a} \times \mathbf{c}_1$ $\qquad$ (1)
and $\quad \mathbf{a} \times (\mathbf{b} + \mathbf{c}) = \mathbf{a} \times (\mathbf{b}_1 + \mathbf{c}_1)$

We now concentrate on the plane perpendicular to $\mathbf{a}$, of which we can take a bird's eye view (see p. 22).

Now $\quad \mathbf{a} \times \mathbf{b}_1 \quad$ has magnitude ab_1 and is perpendicular to $\mathbf{b}_1$
$\qquad \mathbf{a} \times \mathbf{c}_1 \quad$ has magnitude ac_1 and is perpendicular to $\mathbf{c}_1$

So the vectors $\mathbf{a} \times \mathbf{b}_1$, $\mathbf{a} \times \mathbf{c}_1$, $\mathbf{a} \times (\mathbf{b}_1 + \mathbf{c}_1)$ are formed by first multiplying

† Or, more strictly, that the operation of vector multiplication distributes over that of vector addition.

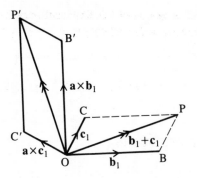

$\mathbf{b}_1, \mathbf{c}_1, \mathbf{b}_1 + \mathbf{c}_1$ by a and then rotating through a right angle.

But it is clear that $\mathbf{a} \times (\mathbf{b}_1 + \mathbf{c}_1)$ is the diagonal of the parallelogram formed by $\mathbf{a} \times \mathbf{b}_1$ and $\mathbf{a} \times \mathbf{c}_1$.

So $\mathbf{a} \times (\mathbf{b}_1 + \mathbf{c}_1) = \mathbf{a} \times \mathbf{b}_1 + \mathbf{a} \times \mathbf{c}_1$

from which it follows by (1) that

$$\mathbf{a} \times (\mathbf{b} + \mathbf{c}) = \mathbf{a} \times \mathbf{b} + \mathbf{a} \times \mathbf{c}$$

* 11.4 Plane kinematics

If a point P is moving in a plane, then we have already seen that (relative to a fixed point O) its position vector $\mathbf{r}$, velocity $\mathbf{v}$ and acceleration $\mathbf{a}$ can all be expressed in component form:

$$\mathbf{r} = x\mathbf{i} + y\mathbf{j}$$
$$\mathbf{v} = \dot{\mathbf{r}} = \dot{x}\mathbf{i} + \dot{y}\mathbf{j}$$
$$\mathbf{a} = \ddot{\mathbf{r}} = \ddot{x}\mathbf{i} + \ddot{y}\mathbf{j}$$

These, of course, are based on Cartesian axes Ox, Oy, and the question now arises whether we can possibly obtain similar results using polar coordinates.

Let us start by supposing that at time t the point P has (relative to a fixed point O and a fixed base-line), position vector $\mathbf{r}$ and polar coordinates (r, θ), where r, θ are functions of t:

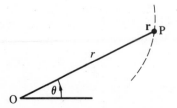

We now let **p** and **q** be unit vectors along and perpendicular to **OP**; so that as P moves, **p** and **q** change direction, rotating with angular velocity $\dot{\theta}$.

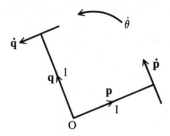

As **p** and **q** are unit vectors, we immediately see that $\dot{\mathbf{p}} = \dot{\theta}\mathbf{q}$ and $\dot{\mathbf{q}} = -\dot{\theta}\mathbf{p}$

But $\mathbf{r} = r\mathbf{p}$
So $\dot{\mathbf{r}} = \dot{r}\mathbf{p} + r\dot{\mathbf{p}} = \dot{r}\mathbf{p} + r\dot{\theta}\mathbf{q}$
Furthermore, $\ddot{\mathbf{r}} = \ddot{r}\mathbf{p} + \dot{r}\dot{\mathbf{p}} + \dot{r}\dot{\theta}\mathbf{q} + r\ddot{\theta}\mathbf{q} + r\dot{\theta}\dot{\mathbf{q}}$
             $= \ddot{r}\mathbf{p} + \dot{r}\dot{\theta}\mathbf{q} + \dot{r}\dot{\theta}\mathbf{q} + r\ddot{\theta}\mathbf{q} - r\dot{\theta}^2\mathbf{p}$
             $= (\ddot{r} - r\dot{\theta}^2)\mathbf{p} + (r\ddot{\theta} + 2\dot{r}\dot{\theta})\mathbf{q}$

Summarising,

$$\mathbf{r} = r\mathbf{p}$$
$$\dot{\mathbf{r}} = \dot{r}\mathbf{p} + r\dot{\theta}\mathbf{q}$$
$$\ddot{\mathbf{r}} = (\ddot{r} - r\dot{\theta}^2)\mathbf{p} + (r\ddot{\theta} + 2\dot{r}\dot{\theta})\mathbf{q}$$

Hence the *radial* and *transverse* components of the velocity and acceleration of P can be denoted:

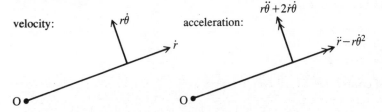

In the particular case when P is moving in a circle of radius a,

$r = a \quad \Rightarrow \quad \dot{r} = \ddot{r} = 0$

Hence the components of velocity and acceleration become

velocity:

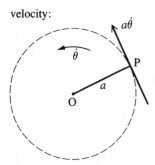

acceleration:

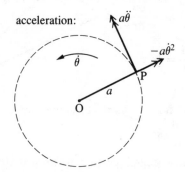

Furthermore, if P is moving with *constant* angular velocity $\dot{\theta}$ it follows that $\ddot{\theta} = 0$, so that the acceleration of P is again shown to be $a\dot{\theta}^2$ directed towards O.

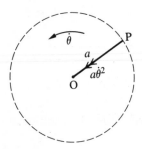

Example

In 1611 Johann Kepler announced that each planet moves so that a line from the Sun 'sweeps out' area at a constant rate (i.e. so that equal areas are swept in equal times). Deduce that the planet's acceleration must be along this line.

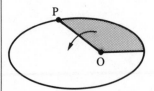

If we suppose that in time δt the planet moves from P to P' and the additional area swept out is δA, then with the usual notation,

$$\delta A \approx \tfrac{1}{2} r^2 \, \delta\theta$$

$$\Rightarrow \quad \frac{\delta A}{\delta t} \approx \tfrac{1}{2} r^2 \frac{\delta\theta}{\delta t}$$

$$\Rightarrow \quad \frac{\mathrm{d}A}{\mathrm{d}t} = \tfrac{1}{2} r^2 \dot\theta$$

So if $\mathrm{d}A/\mathrm{d}t$ is constant, it follows that

$$\frac{\mathrm{d}}{\mathrm{d}t}(r^2\dot\theta) = 0$$

$$\Rightarrow \quad r^2\ddot\theta + 2r\dot r\dot\theta = 0$$
$$\Rightarrow \quad r\ddot\theta + 2\dot r\dot\theta = 0$$

So the transverse component of the acceleration is zero, and the acceleration must be entirely radial. It was this discovery, although not expressed in this notation, that halted the search for *transverse* forces *propelling* the planets, and caused Newton to propose forces *towards* the Sun *holding* them in orbit.

Exercise 11.4

1 The polar coordinates r, θ of a particle after time t are given by

$$r = \mathrm{e}^{2t} \qquad \theta = 2t$$

a) Sketch its path, and state its polar equation.
b) Find the radial and transverse components of velocity and acceleration after time t.
c) Show that its direction of motion is always inclined at $45°$ to the radius and that its radial acceleration is zero.

2 A ship sails away from a lighthouse keeping the light on its port quarter at a constant angle α. Show that:

a) $\dfrac{r\dot\theta}{\dot r} = \tan\alpha$

b) $\dfrac{\mathrm{d}\theta}{\mathrm{d}r} = \dfrac{\tan\alpha}{r}$

c) if $r = a$ when $\theta = 0$,

then $r = a\,\mathrm{e}^{\theta\cot\alpha}$.
Sketch this path.

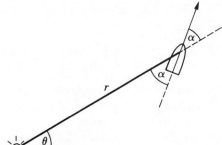

3   Describe the curve given by

$$\mathbf{r} = \begin{pmatrix} 5\cos(t^2) \\ 5\sin(t^2) \end{pmatrix}, \quad \text{where } t \text{ denotes time.}$$

Find $\mathbf{a}(= \ddot{\mathbf{r}})$ and deduce the radial and transverse components of the acceleration.

A boy slides down a banister such that his path is given by

$$\mathbf{r} = \begin{pmatrix} 5\cos(t^2) \\ 5\sin(t^2) \\ kt^2 \end{pmatrix}$$

where the units are metres and k is a constant. Indicate on a diagram the shape of the banister and the magnitudes of the boy's acceleration components in the directions **a)** towards the axis of the curve, **b)** vertically downwards, **c)** in a direction perpendicular to those in **a)** and **b)**. (*Note.* The unit vectors for $\mathbf{r}$ are horizontal south, horizontal west, and vertically downwards.)   (SMP)

4   A bead is threaded on a smooth wire in the shape of a cycloid $x = a(\theta - \sin\theta)$, $y = a(1 + \cos\theta)$, which is fixed in a vertical plane with the positive y-axis as the upward vertical. The bead is released from the position given by $\theta = 0$. Prove that, in the subsequent motion
a) $d\theta/dt$ is constant;
b) the acceleration vector has constant magnitude.   (OC)

Miscellaneous problems

1   Let O be the centre of the circumcircle of $\triangle ABC$ whose vertices have position vectors $\mathbf{a}$, $\mathbf{b}$ and $\mathbf{c}$. Show that the point H, with position vector $\mathbf{h} = \mathbf{a} + \mathbf{b} + \mathbf{c}$, is the orthocentre of $\triangle ABC$ (i.e. AH $\perp$ BC, BH $\perp$ CA, CH $\perp$ AB) and hence show that the centroid G divides OH in the ratio $1:2$.

2   The sides BC, CA, AB of $\triangle ABC$ are divided in the same ratio by P, Q, R respectively. Investigate the centroid of $\triangle PQR$.

3   O is a point in the plane of a triangle ABC and AO, BO, CO meet BC, CA, AB respectively in points P, Q, R. Using bold letters to denote position vectors from O, prove
a) that it is possible to find scalars α, β, γ such that

$$\alpha\mathbf{a} + \beta\mathbf{b} + \gamma\mathbf{c} = \mathbf{0}$$

b) that **p** can be expressed as

$$-\frac{\alpha}{\beta + \gamma}\mathbf{a} \equiv \frac{\beta\mathbf{b} + \gamma\mathbf{c}}{\beta + \gamma}$$

c) that P divides BC in the ratio γ/β;

d) *Ceva's theorem*, that

$$\frac{BP}{PC} \times \frac{CQ}{QA} \times \frac{AR}{RB} = 1$$

Finally, verify Ceva's theorem in the three cases when O is the centroid, the orthocentre, the circumcentre, of $\triangle ABC$.

4   If a line cuts the sides BC, CA, AB (extended if necessary) of a triangle ABC prove *Menelaus' theorem*, that

$$\frac{BP}{PC} \times \frac{CQ}{QA} \times \frac{AR}{RB} = -1$$

5   ABC, A'B'C' are two skew lines and AB:BC = A'B':B'C'. Prove that the mid-points of AA', BB', CC' are collinear.

6   The vertices of a tetrahedron are A, B, C, D and the points P, Q, R, S divide the segments AB, BC, CD, DA in the ratios $p:1, q:1, r:1, s:1$ where p, q, r, s are all positive. Show that P, Q, R, S are coplanar if $pqrs = 1$.
      Show that, if also $p = r$ and $q = s$, then PQRS is a parallelogram.

7   A square piece of paper ABCD of side a is folded along the diagonal BD so that the planes of the triangles ABD, CBD are perpendicular. Find the shortest distance between the edges AB and CD.

8   Prove that if the sum of the squares of two opposite edges of a tetrahedron is equal to the sum of the squares of another pair of opposite edges, then the remaining pair of opposite edges are perpendicular.

9   OA, OB, OC are three lines through the point O and the angles BOC, COA, AOB are α, β, γ respectively. Calculate $\cos^2 \theta$, where θ is the angle between the line OA and the plane OBC.

10   Kepler's first and second laws state that each planet **a)** moves in an ellipse with the sun at a focus, **b)** sweeps out area at a constant rate. In the usual notation, these can be written:

a) $r = \dfrac{l}{1 + e\cos\theta}$ (see Book 1, p. 506)

b) $r^2\dot{\theta} = h$ (see p. 25)

Prove that

$$\dot{r} = \frac{eh\sin\theta}{l} \quad \text{and} \quad \ddot{r} - r\dot{\theta}^2 = -\frac{h(1 + e\cos\theta)^2}{l^3}$$

and hence that each planet is attracted to the sun by a force which is inversely proportional to the square of its distance.

12 Complex numbers

'I met a man recently who told me that, so far from believing in the square root of minus one, he did not even believe in minus one. This is at any rate a consistent attitude. There are certainly many people who regard $\sqrt{2}$ as something perfectly obvious, but jib at $\sqrt{-1}$. This is because they think that they can visualise the former as something in physical space, but not the latter. Actually, $\sqrt{-1}$ is a much simpler concept.'

E. C. Titchmarsh, *Mathematics for the General Reader*

12.1 Introduction

Complex numbers arise initially in the solution of quadratic equations. We have seen in chapter 2 (Book 1) that the solutions of the quadratic equation $ax^2 + bx + c = 0$ are given by

$$x = \frac{-b \pm \sqrt{(b^2 - 4ac)}}{2a}$$

So the equation $x^2 - 2x - 1 = 0$ has two solutions:

$$x = \frac{2 \pm \sqrt{(4 + 4)}}{2} = 1 \pm \sqrt{2}$$

However, when we attempt to solve the equation $x^2 - 6x + 13 = 0$, we find that

$$x = \frac{6 \pm \sqrt{(36 - 52)}}{2}$$

$$= \frac{6 \pm \sqrt{-16}}{2}$$

$$= \frac{6 \pm 4\sqrt{-1}}{2}$$

$$= 3 \pm 2\sqrt{-1}$$

and since the square root of a negative number cannot be evaluated, it appears that this equation has no solutions.

Now it would be unfortunate to have to suffer the irregularity of some quadratic equations having two solutions, and others having none. How much neater if *all* quadratic equations were to have two solutions! We will therefore *invent* a 'number' $\sqrt{-1}$, which we call i (sometimes j is used instead), and suppose that it obeys all the usual laws of algebra. We can then say that the solutions of $x^2 - 6x + 13 = 0$ are $3 \pm 2i$, and we refer to numbers of this kind as *complex numbers*. With such an invention, all quadratic equations can now be said to have two solutions, either real or complex.

This extension of our number system is, as Professor Titchmarsh implied in the quotation above, no more of an innovation than the extension of the integers $\mathbb{Z}$ to the rational numbers $\mathbb{Q}$, or of the rational numbers to the real numbers $\mathbb{R}$; and just as an integer is a special case of a rational number, and a rational number is a special case of a real number, so a real number is a special case of a complex number

i.e. $\mathbb{Z} \subset \mathbb{Q} \subset \mathbb{R} \subset \mathbb{C}$

However, in referring to $\sqrt{-1}$ as a number we must not think of it in the same sense as hitherto, as a length or measure, but rather in the sense that it *behaves* like ordinary numbers.

Now the invention of i to provide solutions for all quadratic equations may possibly give the impression that further inventions may be necessary to solve cubic equations, and possibly that even more weird and wonderful numbers might have to be dreamed up to deal with equations of higher degree; and what further miracles of man's genius would be needed to assign meanings to such expressions as $(2 - i)^i$ and $\sqrt[i]{\{\sin (5 - 2i)^{2/3}\}}$? Happily, such further inventions are not necessary; the extension of the number system by the single step of adjoining the square root of -1 enables all these situations to be covered! The roots of equations of any degree of complication, and numbers of any degree of 'complexity', can be expressed in the same simple form $x + iy$, where x and y are real, just like the roots of the quadratic equation $x^2 - 6x + 13 = 0$ that we examined at the outset. It is this remarkable property of playing a unifying role in mathematics that endows i with such interest and importance.

We have therefore defined a complex number to be $z = x + iy$, where x, y are real, i is a symbol which obeys all the usual rules of algebra, and i^2 can be replaced by -1.

x is called the *real part* of z, written $x = \operatorname{Re} z$, and y the *imaginary part* of z, written $y = \operatorname{Im} z$.

Example 1

Express each of the following in the form $x + iy$:
a) $(5 + 2i) + (1 - 3i)$ b) $(5 + 2i) - (1 - 3i)$
c) $(5 + 2i) \times (1 - 3i)$ d) $(5 + 2i) \times (5 - 2i)$

a) $(5 + 2i) + (1 - 3i) = (5 + 1) + (2 - 3)i = 6 - i$

b) $(5 + 2i) - (1 - 3i) = (5 - 1) + (2 + 3)i = 4 + 5i$

c) $(5 + 2i) \times (1 - 3i) = 5 - 15i + 2i - 6i^2$
$$= 5 - 13i - (6 \times -1)$$
$$= 11 - 13i$$

d) $(5 + 2i) \times (5 - 2i) = 25 - 10i + 10i - 4i^2$
$$= 25 + 4$$
$$= 29$$

Although addition, subtraction and multiplication of complex numbers are straightforward, division presents more of a problem. However part **d)** of example 1 gives a clue as to how division may be tackled.

Example 2

Express $\dfrac{4 - i}{2 + 3i}$ in the form $x + iy$.

Multiplying both numerator and denominator by $2 - 3i$, which is referred to as the *complex conjugate* of $2 + 3i$ (i.e. the complex number with the sign of its imaginary part changed) we obtain

$$\frac{4 - i}{2 + 3i} = \frac{(4 - i)(2 + 3i)}{(2 + 3i)(2 - 3i)}$$

$$= \frac{8 - 14i + 3i^2}{4 - 9i^2}$$

$$= \frac{5 - 14i}{13} = \frac{5}{13} - \frac{14i}{13}$$

Examples 1 and 2 illustrate the fact that the repeated processes of addition, subtraction, multiplication and division by numbers other than zero always lead to another complex number (a real number being a special case of a complex number), and that the system is self-contained, or *closed*.

Moreover we note that two complex numbers are equal if and only if their real and imaginary parts are separately equal.

For if $$z_1 = z_2$$
then $$x_1 + iy_1 = x_2 + iy_2$$
$\Rightarrow$ $$x_1 - x_2 = i(y_2 - y_1)$$
$\Rightarrow$ $$(x_1 - x_2)^2 = -(y_2 - y_1)^2$$

$$\Rightarrow \quad (x_1 - x_2)^2 + (y_2 - y_1)^2 = 0$$
$$\Rightarrow \quad x_1 = x_2 \quad \text{and} \quad y_1 = y_2$$

(since each of the squares of real numbers must be positive or zero). The converse is trivial.

We also have the corollary that

$$x + iy = 0 \quad \Leftrightarrow \quad x = 0 \text{ and } y = 0$$

Example 3

Find $\sqrt{i}$ in the form $x + iy$.

If $\quad x + iy = \sqrt{i}$ then $\quad (x + iy)^2 = i$

Therefore $\quad\quad\quad x^2 + 2xyi + i^2 y^2 = i$

$\Rightarrow \quad\quad\quad\quad (x^2 - y^2) + 2xyi = i$

Comparing real and imaginary parts:

$$x^2 - y^2 = 0 \quad \text{and} \quad 2xy = 1$$
$$\Rightarrow \quad y = \pm x \quad \text{so} \quad \pm 2x^2 = 1 \quad (1)$$

The negative sign in (1) can be ignored since x^2 must be positive,

so $\quad x = y = \pm \dfrac{1}{\sqrt{2}}$

$\sqrt{i}$ therefore has two possible values, $\pm\left(\dfrac{1}{\sqrt{2}} + \dfrac{1}{\sqrt{2}}i\right)$, and we can readily check that

$$\left(\frac{1}{\sqrt{2}} + \frac{1}{\sqrt{2}}i\right)^2 = \frac{1}{2} + i + \frac{1}{2}i^2 = i$$

Example 4

Express each of the following in the form $x + iy$:

a) $(1 + \sqrt{3}i)^3$ **b)** $\dfrac{7 - 3i}{(2 - 4i)^2}$ **c)** $(1 + i)^{-3} - (1 - i)^{-3}$

a) $(1 + \sqrt{3}i)^2 = 1 + 2\sqrt{3}i + 3i^2 = -2 + 2\sqrt{3}i$

$\Rightarrow \quad (1 + \sqrt{3}i)^3 = (-2 + 2\sqrt{3}i)(1 + \sqrt{3}i)$

$\quad\quad\quad\quad\quad\quad = -2 - 2\sqrt{3}i + 2\sqrt{3}i + 6i^2$

$\quad\quad\quad\quad\quad\quad = -8$

b) $\dfrac{7 - 3i}{(2 - 4i)^2} = \dfrac{7 - 3i}{4 - 16i + 16i^2} = \dfrac{7 - 3i}{-16i - 12} = \dfrac{7 - 3i}{-4(3 + 4i)}$

$$= \frac{(7 - 3i)(3 - 4i)}{-4(3 + 4i)(3 - 4i)}$$

$$= \frac{21 - 37i + 12i^2}{-4(9 - 16i^2)} = \frac{9 - 37i}{-100} = -0.09 + 0.37i$$

c) $(1 + i)^{-3} - (1 - i)^{-3} = \dfrac{1}{1 + 3i + 3i^2 + i^3} - \dfrac{1}{1 - 3i + 3i^2 - i^3}$

$$= \frac{1}{-2 + 2i} - \frac{1}{-2 - 2i}$$

$$= \frac{1}{2}\left(\frac{1}{1 + i} - \frac{1}{1 - i}\right)$$

$$= \frac{1}{2}\left(\frac{1 - i - 1 - i}{(1 + i)(1 - i)}\right) = \frac{-2i}{4} = -\tfrac{1}{2}i$$

Exercise 12.1

1   Express, in the form $x + iy$, the sum and the difference of each of the following pairs of complex numbers:
  a) $2 + 3i$, $1 + 4i$
  b) $6 + 5i$, $4 - 3i$
  c) $5 + i$, $5 - i$
  d) $-1 - i$, $-2 - 5i$
  e) $3 + 2i$, $1 - i$
  f) $4 + 2i$, $3 + 5i$

2   Express the following in the form $x + iy$:
  a) $5i(1 - 2i)$
  b) $(4 + 3i)(2 - i)$
  c) $(7 - 2i)(-1 - 3i)$
  d) $(1 - 2i)^2$
  e) $(2 - 3i)^2 - (4 + i)^2$
  f) $(1 + 2i)^3$

3   Express in the form $x + iy$:
  a) $\dfrac{1}{2 + i}$

  b) $\dfrac{2}{3 - i}$

  c) $\dfrac{1}{(i - 2)(1 - 3i)}$

  d) $\dfrac{7 + 4i}{3 - 2i}$

  e) $\dfrac{i + 1}{i - 1}$

  f) $(1 + 2i)^{-3}$

  g) $\dfrac{(2 - i)^2 (1 + i)}{1 - 3i}$

  h) $\dfrac{1}{\cos\theta - i\sin\theta}$

4 Find the real numbers x and y given that

$$\frac{1}{x + iy} = 2 - 3i \tag{L}$$

5 Find integers p, q such that $(3 + 7i)(p + iq)$ is purely imaginary.

6 If $z = x + iy$ $(x, y \in \mathbb{R})$, express in the form $X + iY$:
   a) z^3 **b)** z^{-2} **c)** $z + z^{-1}$

7 Show that $(4 - 5i)(2 + 3i) = 23 + 2i$, and deduce factors of $23 - 2i$. Hence show that

$$(4^2 + 5^2)(2^2 + 3^2) = 23^2 + 2^2$$

Use a similar method to express $(4^2 + 7^2)(9^2 + 5^2)$ as the sum of two squares.

8 The complex number z satisfies $\dfrac{z}{z + 2} = 2 - i$. Find the real and imaginary parts of z.   (O)

9 Solve the following equation for z:

$$\frac{z}{3 + 4i} + \frac{z - 1}{5i} = \frac{5}{3 - 4i}$$

10 Find real values of x and y such that $(1 + 3i)x + (2 - 5i)y + 2i = 0$.

11 Find the square roots of
   a) $3 + 4i$ **b)** $5 - 12i$ **c)** $-8 + 6i$ **d)** $-i$

12 Verify that $x = 2$, $y = -1$ satisfy the simultaneous equations

$$x^3 - 3xy^2 = 2, \quad 3x^2y - y^3 = -11$$

Hence find the cube root of $2 - 11i$.

13 Find real values of x and y to satisfy

$$\frac{x}{1 - 2i} + \frac{y}{3 - 2i} = \frac{-5 + 6i}{1 + 8i}$$

14 The complex numbers

$$z_1 = 1 + ia, \quad z_2 = a + ib$$

where a and b are real, are such that $z_1 - z_2 = 3i$. Find a and b, and show that

$$\frac{1}{z_1} + \frac{1}{z_2} = \frac{7 - i}{10}$$

Hence, or otherwise, find

$$\frac{z_1{}^2 - z_2{}^2}{z_1 z_2}$$

in the form $x + iy$, where x and y are real.   (JMB)

15 Given that $z = \cos\theta + i\sin\theta$, where $z \neq -1$, show that

$$\frac{2}{1+z} = 1 - i\tan\tfrac{1}{2}\theta$$ (L)

16 Solve the equations:
   a) $z^2 + 4z + 29 = 0$ **b)** $4z^2 - 12z + 25 = 0$
   c) $z^2 + 2iz + 1 = 0$ **d)** $z^2 + iz = 2$

17 Find the quadratic equations whose roots are:
   a) $3 + \sqrt{5}$, $3 - \sqrt{5}$ **b)** $-3 + i$, $-3 - i$
   c) $2 + \sqrt{3}i$, $2 - \sqrt{3}i$ **d)** $-i$, $1 + 2i$

18 Find the sum and the product of the roots z_1, z_2 of the equations
   a) $z^2 - 5z + 7 = 0$ **b)** $z^2 + z + 7 = 0$
   Obtain in each case the value of $z_1^2 + z_2^2$. What do you notice, and how do
   you account for it?

19 Sum the series:
   a) $1 + i + i^2 + i^3 + \cdots + i^n$
   b) $1 - 2i + 4i^2 - 8i^3 + \cdots + (-2i)^{n-1}$
   c) $1 + 2i + 3i^2 + 4i^3 + \cdots + ni^{n-1}$

20 Express $\dfrac{x}{x^2 + 1}$ in partial fractions in the form $\dfrac{A}{x + i} + \dfrac{B}{x - i}$.

   Repeat for the expression $\dfrac{1}{x^2 + 1}$.

12.2 Geometric representation

The Argand diagram

Any complex number $z = x + iy$ can clearly be regarded as an ordered pair of
real numbers (x, y). It seems natural, therefore, to represent complex numbers
by the points in a plane, using coordinate axes Ox, Oy, and by assigning the
complex number $x + iy$ to the point whose coordinates are (x, y). This sets up
a one-to-one correspondence between the complex numbers and the points of
the Cartesian plane (see over):

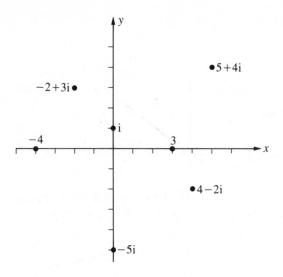

This method of representing the complex numbers is called the *Argand diagram* after the French mathematician J. R. Argand (1768–1822). Ox and Oy are called the 'real' and 'imaginary' axes, for those complex numbers of the form $x + 0$i (i.e. the real numbers) are represented by points of the x-axis, whose role is thus seen to be simply that of the real number line; while those complex numbers whose 'real parts' are zero appear in the Argand diagram on the y-axis, and are sometimes loosely referred to as 'purely imaginary', among them being the number i itself.

Addition

We can now see that complex numbers are added together in just the same way as 2-dimensional vectors:

$$(5 + 2i) + (1 - 3i) = 6 - i$$

$$\begin{pmatrix} 5 \\ 2 \end{pmatrix} + \begin{pmatrix} 1 \\ -3 \end{pmatrix} = \begin{pmatrix} 6 \\ -1 \end{pmatrix}$$

Complex numbers, therefore, can be represented by position vectors as well as by points in the plane. It must be emphasised however that a complex number z corresponds not only to the particular vector from the origin to the point, but to the whole class of equivalent vectors which are equal and parallel to this position vector.

So addition of complex numbers on the Argand diagram looks just like addition of vectors, using either the parallelogram or equivalent triangle method.

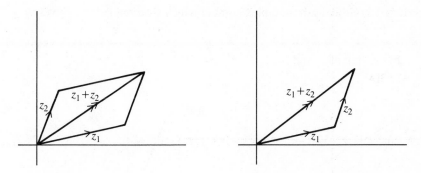

Furthermore, extension of the triangle method allows the addition of as many complex numbers as we like, by forming their associated vector polygon.

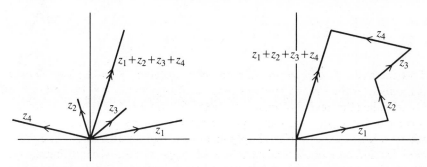

Negation

$-z$ is clearly the complex number whose corresponding vector is equal and opposite to that of z.

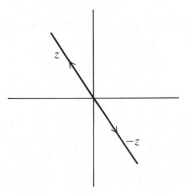

Subtraction

Subtraction can now be interpreted in two ways, and $z_1 - z_2$ can be considered as either **a)** the complex number which when added to z_2 gives z_1:

$$(z_1 - z_2) + z_2 = z_1$$

or   **b)** the sum of z_1 and $-z_2$:

$$z_1 - z_2 = z_1 + (-z_2)$$

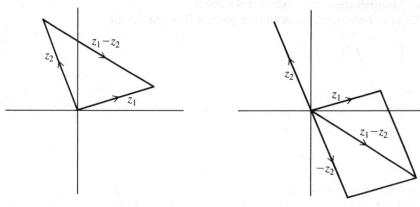

The first of these two geometrical interpretations tends to be the more useful.

Example 1

Three of the vertices of a parallelogram ABCD in the Argand diagram have associated complex numbers $a = 1 + 3i$, $b = 2 + i$, $c = 5 + 4i$. Find the complex number d, representing the vertex D.

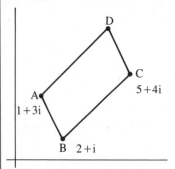

The complex number representing the vector **CD** is the same as that representing **BA**, which is $(1 + 3i) - (2 + i) = -1 + 2i$.

Hence
$$\begin{aligned} d &= c + (-1 + 2i) \\ &= (5 + 4i) + (-1 + 2i) \\ &= 4 + 6i \end{aligned}$$

Multiplication

The geometric effect of multiplication by complex numbers will be dealt with more fully later in this section, but we can at this stage examine a few special cases.

a) Multiplication by a positive real number
This can be interpreted as an enlargement from the origin.

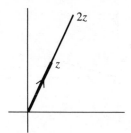

b) Multiplication by −1
This produces a reversal in direction, or a rotation of 180°.

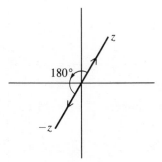

c) Multiplication by i
$$i \times (x + iy) = -y + ix$$

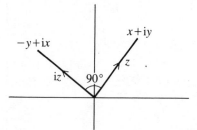

The associated vectors $\begin{pmatrix} x \\ y \end{pmatrix}$ and $\begin{pmatrix} -y \\ x \end{pmatrix}$, which are of equal length, have a scalar product of zero, and are therefore perpendicular.

So multiplication by i can be interpreted as a rotation of any complex number through 90° about the origin. Hence two successive multiplications by i will give a rotation of 180°, and since $i^2 = -1$, this geometric interpretation is consistent with that for multiplication by -1.

Example 2

A square has its centre at $2 + 3i$, and one vertex at the origin. Find the other vertices.

Let M be the centre of the square, and O, A, B, C the vertices.

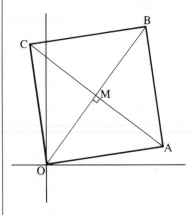

Vector **MO** is then associated with the complex number $-2 - 3i$. **MA** is the vector **MO** rotated through 90°.

So $\quad$ **MA** $= i(-2 - 3i)$
$$= -2i - 3i^2$$
$$= 3 - 2i$$

Hence $\quad$ **OA** $=$ **OM** $+$ **MA** $= (2 + 3i) + (3 - 2i) \quad = 5 + i$
and $\quad$ **OB** $= 2$**OM** $\qquad = 2(2 + 3i) \qquad\qquad = 4 + 6i$
Now $\quad$ **MC** $= -$**MA** $\qquad = -3 + 2i$
so $\qquad$ **OC** $=$ **OM** $+$ **MC** $= (2 + 3i) + (-3 + 2i) = -1 + 5i$

Therefore the other vertices of the square are at $5 + i$, $4 + 6i$ and $-1 + 5i$.

Triangle inequalities

The *modulus* of a complex number, written $|z|$, is the length of its associated vector.

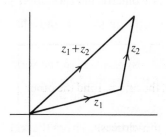

Since z_1, z_2 and $z_1 + z_2$ form the three sides of a triangle, it is immediately obvious that

$$|z_1 + z_2| \leqslant |z_1| + |z_2|$$

(and the equality will occur only if the triangle collapses into a straight line).

More generally, this inequality can readily be extended to give

$$|z_1 + z_2 + \cdots + z_n| \leqslant |z_1| + |z_2| + \cdots + |z_n|$$

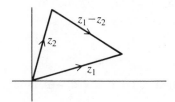

Similarly $|z_1| \leqslant |z_1 - z_2| + |z_2|$

$\Rightarrow$ $|z_1 - z_2| \geqslant |z_1| - |z_2|$

and $|z_2| \leqslant |z_1 - z_2| + |z_1|$

$\Rightarrow$ $|z_1 - z_2| \geqslant |z_2| - |z_1|$

so that $|z_1 - z_2| \geqslant ||z_1| - |z_2||$

Exercise 12.2a

1   Plot the following points on the Argand diagram:
   a) $1 - i$ b) $-2 + 3i$ c) $-2 - 3i$ d) $4 - i$
   e) $-12 + 5i$ f) $\cos \frac{1}{3}\pi + i \sin \frac{1}{3}\pi$ g) $\cos \pi + i \sin \pi$ h) $-\frac{1}{2}\sqrt{3} + \frac{1}{2}i$

2   In each of the following, mark on the Argand diagram the complex numbers $z_1, z_2, z_1 + z_2$ and $z_1 - z_2$:
   a) $z_1 = 2 - i$, $z_2 = 1 + 3i$

b) $z_1 = 4 + 2i, \quad z_2 = 3i$
c) $z_1 = 1 + 2i, \quad z_2 = 1 - 2i$
d) $z_1 = -1 + 4i, \quad z_2 = 3 - 2i$
e) $z_1 = 6 - 5i, \quad z_2 = -3 + i$

3 For each part of no. **2**, check that $|z_1 + z_2| \leqslant |z_1| + |z_2|$ and that $|z_1 - z_2| \geqslant ||z_1| - |z_2||$.

4 Using the pairs of complex numbers in no. **2**, express $z_1 z_2$ in the form $x + iy$, and in each case plot z_1, z_2 and $z_1 z_2$ on the Argand diagram.

   Measure, or calculate, the distances of z_1, z_2 and $z_1 z_2$ from the origin. What appears to be the connection between these distances?

   Measure or calculate the angle (measured anti-clockwise) from the real axis to the position vector represented by each of the complex numbers z_1, z_2 and $z_1 z_2$. What appears to be the connection between these angles?

5 If $z = 1 + i$, simplify $z^2, z^3, z^4, \ldots$, and mark these points on the Argand diagram.

6 If $z_1 = 2 - i, z_2 = 1 + 3i$, mark on the Argand diagram the points

$$z_1, \quad z_2, \quad z_1 + z_2, \quad z_1 - z_2, \quad z_2 - z_1, \quad z_1 z_2, \quad \frac{z_1}{z_2}, \quad \frac{z_2}{z_1}, \quad z_1^2$$

7 Let $z = -5 + 12i$. Plot the following points on the Argand diagram:

$$z, \quad iz, \quad z - 1, \quad 3z, \quad -3z, \quad \frac{1}{z}, \quad z^2$$

Mark also the points $z', -z'$ and zz' where $z' = -5 - 12i$.

8 ABCD is a rhombus with $\mathbf{AC} = 2\mathbf{BD}$. If $b = 3 + i, d = 1 - 3i$, find a and c.

9 The centre of a square is the point $-2 + i$, and one vertex is $1 + 3i$. Find the others.

10 ABCD is a square, with $a = 3 + i, b = 4 - 2i$. Find possible positions of C and D.

11 ABCD is a square with $a = -2 + 5i, c = 4 - 8i$. Find b and d.

12 The points P and Q are represented on the Argand diagram by the complex numbers $p = 3 + 2i$ and $q = -1 + 8i$. Find the complex numbers which represent
   a) the mid-point of PQ;
   b) the point which divides PQ in the ratio $2:1$;
   c) the point which divides PQ in the ratio $1:3$.

13 Find the centroid of the triangle ABC where $a = -2i, b = 4 + i, c = -10 - 5i$.

14 ABCD is any quadrilateral. Prove by using complex numbers that the lines joining the mid-points of AB and CD, of AC and BD, and of AD and BC all bisect each other.

15 ABCD is a parallelogram. P is the mid-point of BC, BD and AP intersect at F. Prove by using complex numbers that F trisects both BD and AP.

16 Using only the definition $|z|^2 = x^2 + y^2$, prove that $|z_1 + z_2| \leqslant |z_1| + |z_2|$.

17 If $|z_1| = |z_2|$, prove that $(z_1 + z_2)/(z_1 - z_2)$ is purely imaginary. Interpret this result geometrically.

18 Prove that $|z_1 + z_2|^2 + |z_1 - z_2|^2 = 2|z_1|^2 + 2|z_2|^2$, and give a geometrical interpretation of this result.

Modulus–argument form

We have seen that it is possible to represent a complex number by a point, or position vector, in a plane. But besides using Cartesian coordinates (x, y), it is also possible to identify a point by its polar coordinates (r, θ).

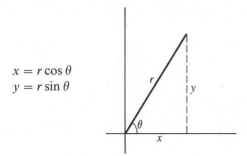

$x = r \cos \theta$
$y = r \sin \theta$

So we can also write a complex number in polar form:

$z = x + iy$
$\quad = r \cos \theta + ir \sin \theta$
$\quad = r(\cos \theta + i \sin \theta)$

which for convenience we will frequently abbreviate to $r \operatorname{cis} \theta$.
r is the modulus of z and θ is called the *argument*, written $\arg z$.

$$r = |z| = \sqrt{(x^2 + y^2)} \quad \text{and} \quad \theta = \arg z = \tan^{-1}(y/x)$$

It must be emphasised that, although in polar coordinates it is possible for r to be negative, here r is essentially positive.

Furthermore, when finding the argument of a complex number there are clearly an infinite number of possibilities; for instance $\arg i$ could be given as $\pi/2$ (the obvious value), or $5\pi/2$, $-3\pi/2$, etc. Of these we usually select the *principal value* θ where $-\pi < \theta \leqslant \pi$.

Example 3

Express in polar form   **a)** $1 + i$ **b)** $4 - 3i$ **c)** $-\sqrt{3} + i$

a)

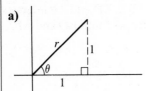

$$r = \sqrt{(1^2 + 1^2)} = \sqrt{2}$$
$$\tan \theta = 1 \quad \Rightarrow \quad \theta = \frac{\pi}{4}$$

So $1 + i = \sqrt{2}\left(\cos\frac{\pi}{4} + i\sin\frac{\pi}{4} \right)$

b)

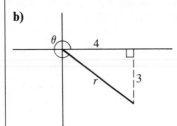

$$r = \sqrt{(4^2 + 3^2)} = 5$$
$$\tan \theta = -\tfrac{3}{4} \quad \Rightarrow \quad \theta \approx -0.64$$

So $4 - 3i = 5\{\cos(-0.64) + i\sin(-0.64)\}$
or more conveniently, $5 \operatorname{cis}(-0.64)$

c)

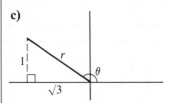

$$r = \sqrt{(1 + 3)} = 2$$
$$\tan \theta = -\frac{1}{\sqrt{3}} \quad \Rightarrow \quad \theta = \frac{5\pi}{6}$$

So $-\sqrt{3} + i = 2\left(\cos\frac{5\pi}{6} + i\sin\frac{5\pi}{6} \right)$

Example 4

Express the complex number with $|z| = 4$ and $\arg z = \pi/3$ in the form $z = x + iy$.

$$x = 4 \cos \frac{\pi}{3} = 2$$

$$y = 4 \sin \frac{\pi}{3} = 2\sqrt{3}$$

so $z = 2 + 2\sqrt{3}i$

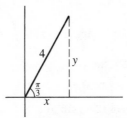

Example 5

If $z = r(\cos \theta + i \sin \theta)$, express $\dfrac{1}{z}$ in modulus–argument form.

$$\frac{1}{z} = \frac{1}{r(\cos \theta + i \sin \theta)}$$

$$= \frac{\cos \theta - i \sin \theta}{r(\cos \theta + i \sin \theta)(\cos \theta - i \sin \theta)}$$

$$= \frac{\cos \theta - i \sin \theta}{r(\cos^2 \theta + \sin^2 \theta)}$$

$$= \frac{1}{r}(\cos \theta - i \sin \theta)$$

So $\left| \dfrac{1}{z} \right| = \dfrac{1}{r}$ and $\arg\!\left(\dfrac{1}{z} \right) = -\theta$

Example 6

If $z = \cos \theta + i \sin \theta$, express in modulus–argument form

a) $1 - z$ b) $\dfrac{2z}{1 - z^2}$

a) $1 - z = 1 - \cos \theta - i \sin \theta$

$$= 2 \sin^2 \frac{\theta}{2} - 2i \sin \frac{\theta}{2} \cos \frac{\theta}{2}$$

$$= 2\sin\frac{\theta}{2}\left[\sin\frac{\theta}{2} - i\cos\frac{\theta}{2}\right]$$

$$= 2\sin\frac{\theta}{2}\left[\cos\left(\frac{\theta}{2} - \frac{\pi}{2}\right) + i\sin\left(\frac{\theta}{2} - \frac{\pi}{2}\right)\right]$$

So $|1 - z| = 2\sin\dfrac{\theta}{2}$ and $\arg(1 - z) = \dfrac{\theta}{2} - \dfrac{\pi}{2}$

b) $\dfrac{2z}{1 - z^2} = \dfrac{2(\cos\theta + i\sin\theta)}{1 - (\cos\theta + i\sin\theta)^2}$

$$= \frac{2(\cos\theta + i\sin\theta)}{1 - \cos^2\theta - 2i\sin\theta\cos\theta + \sin^2\theta}$$

$$= \frac{2(\cos\theta + i\sin\theta)}{2\sin\theta(\sin\theta - i\cos\theta)}$$

$$= \frac{(\cos\theta + i\sin\theta)(\sin\theta + i\cos\theta)}{\sin\theta(\sin\theta - i\cos\theta)(\sin\theta + i\cos\theta)}$$

$$= \frac{i(\sin^2\theta + \cos^2\theta)}{\sin\theta(\sin^2\theta + \cos^2\theta)}$$

$$= i\,\mathrm{cosec}\,\theta$$

which has modulus $\mathrm{cosec}\,\theta$, and argument $\pi/2$.

Multiplication and division

Multiplication of complex numbers in Cartesian form, $x + iy$, can be cumbersome, and does not lend itself to simple geometrical interpretation, whilst division is even more awkward. When modulus–argument form is used, however, much simpler results emerge.

For $\mathrm{cis}\,\theta_1 \times \mathrm{cis}\,\theta_2 = (\cos\theta_1 + i\sin\theta_1)(\cos\theta_2 + i\sin\theta_2)$

$$= (\cos\theta_1\cos\theta_2 - \sin\theta_1\sin\theta_2)$$
$$+ i(\sin\theta_1\cos\theta_2 + \cos\theta_1\sin\theta_2)$$
$$= \cos(\theta_1 + \theta_2) + i\sin(\theta_1 + \theta_2)$$
$$= \mathrm{cis}(\theta_1 + \theta_2)$$

So $\boxed{\mathrm{cis}\,\theta_1\,\mathrm{cis}\,\theta_2 = \mathrm{cis}(\theta_1 + \theta_2)}$

Hence if $z_1 = r_1\,\mathrm{cis}\,\theta_1$ and $z_2 = r_2\,\mathrm{cis}\,\theta_2$, it follows that

$$z_1 z_2 = r_1\,\mathrm{cis}\,\theta_1 \times r_2\,\mathrm{cis}\,\theta_2 = r_1 r_2\,\mathrm{cis}(\theta_1 + \theta_2)$$

So, in multiplying complex numbers, we simply multiply their moduli and add their arguments:

$$|z_1 z_2| = |z_1||z_2| \quad \text{and} \quad \arg(z_1 z_2) = \arg z_1 + \arg z_2$$

Furthermore,

$$\frac{z_1}{z_2} = \frac{r_1 \operatorname{cis} \theta_1}{r_2 \operatorname{cis} \theta_2}$$

$$= \frac{r_1 \operatorname{cis} \theta_1 \times \operatorname{cis}(-\theta_2)}{r_2 \operatorname{cis} \theta_2 \times \operatorname{cis}(-\theta_2)}$$

$$= \frac{r_1 \operatorname{cis}(\theta_1 - \theta_2)}{r_2 \operatorname{cis} 0} = \frac{r_1}{r_2} \operatorname{cis}(\theta_1 - \theta_2)$$

So, in dividing complex numbers, we divide their moduli and subtract their arguments:

$$\left|\frac{z_1}{z_2}\right| = \frac{|z_1|}{|z_2|} \quad \text{and} \quad \arg\frac{z_1}{z_2} = \arg z_1 - \arg z_2$$

Example 7

Express the following in the form $x + iy$:

a) $(1 + i)^6$ **b)** $\dfrac{i}{\sqrt{3} + i}$

a) $(1 + i) = \sqrt{2}\left(\cos\dfrac{\pi}{4} + i\sin\dfrac{\pi}{4}\right) = \sqrt{2}\operatorname{cis}\dfrac{\pi}{4}$

so by an obvious extension of the multiplication rule,

$$(1 + i)^6 = \sqrt{2}^6 \operatorname{cis}\left(6 \times \frac{\pi}{4}\right)$$

$$= 8 \operatorname{cis}\frac{3\pi}{2}$$

$$= -8i$$

b) $\dfrac{i}{\sqrt{3} + i} = \dfrac{\operatorname{cis}\dfrac{\pi}{2}}{2\operatorname{cis}\dfrac{\pi}{6}}$

$$= \tfrac{1}{2} \operatorname{cis} \left(\frac{\pi}{2} - \frac{\pi}{6} \right)$$

$$= \tfrac{1}{2} \operatorname{cis} \frac{\pi}{3}$$

$$= \frac{1}{4} + \frac{\sqrt{3}}{4} i$$

Using modulus–argument form, the geometrical interpretation of multiplication of complex numbers is now quite straightforward.

If a complex number is multiplied by $r(\cos \theta + i \sin \theta)$ then its modulus is multiplied by r, and its argument is increased by an angle θ. The multiplication of the modulus is equivalent to an enlargement, scale factor r, centre the origin; and the increase in argument is equivalent to a rotation of angle θ about the origin. This composite transformation, of enlargement and rotation, is known as a *spiral similarity*.

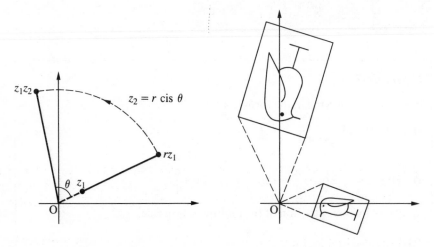

Therefore multiplication by a complex number with unit modulus effects just a rotation, as in the case of multiplication by i, which we have already seen is equivalent to a quarter-turn.

Example 8

a) Find the geometric effect of multiplication by $1 + i$.
b) Find the complex number, multiplication by which is equivalent to a rotation of $-70°$.

a) $1 + i$ has modulus equal to $\sqrt{2}$, and an argument of $\pi/4$. Therefore multiplication by $1 + i$ is equivalent to an enlargement scale factor $\sqrt{2}$, and a rotation of $\dfrac{\pi}{4}$ about the origin.

b) The required complex number must have modulus $= 1$, and argument $= -70°$. It can therefore be written in the form $\cos 70° - i \sin 70°$ or, approximately, $0.34 - 0.94i$.

Example 9

Show that if triangle ABC is equilateral with the sense of rotation from A to B to C being anti-clockwise, then $a + \omega b + \omega^2 c = 0$, where $\omega = \operatorname{cis} \frac{2}{3}\pi$, and a, b, c are the complex numbers associated with points A, B, C.

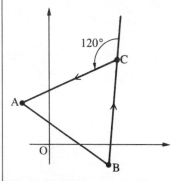

Since triangle ABC is equilateral, **CA** is obtained from **BC** by a rotation of $120° = \frac{2}{3}\pi$ about C.

So $\qquad a - c = (c - b) \operatorname{cis} \frac{2}{3}\pi$

$\Rightarrow \qquad a - c = \omega(c - b)$

$\Rightarrow \quad a + \omega b - (1 + \omega)c = 0 \qquad\qquad\qquad\qquad (1)$

Now $\omega^3 = 1 \;\Rightarrow\; \omega^3 - 1 = 0 \;\Rightarrow\; (\omega - 1)(\omega^2 + \omega + 1) = 0$

But $\omega \neq 1$, so $\omega^2 + \omega + 1 = 0 \;\Rightarrow\; 1 + \omega = -\omega^2$

So equation (1) can be written $\quad a + \omega b + \omega^2 c = 0$

Exercise 12.2b

1 Find the modulus and argument of each of the following complex numbers:

- **a)** $1 - i$
- **b)** $-2 + 2i$
- **c)** $-1 - i$
- **d)** -3
- **e)** $4i$
- **f)** $3 + i$
- **g)** $1 - 3i$
- **h)** $3 + 4i$
- **i)** $-5 + 12i$
- **j)** $2 + 5i$

2 Write the complex numbers with the following moduli and arguments, in the form $x + iy$:

a) $|z| = 4$, $\arg z = -\dfrac{\pi}{2}$ 　　　　　 b) $|z| = 5$, $\arg z = \pi$

c) $|z| = 10$, $\arg z = -\dfrac{\pi}{4}$ 　　　　 d) $|z| = 4$, $\arg z = \dfrac{\pi}{6}$

e) $|z| = 4$, $\arg z = -\dfrac{\pi}{6}$ 　　　　 f) $|z| = 1$, $\arg z = -\dfrac{3\pi}{4}$

3 If $a = 2 + i$, $b = 1 + 3i$, mark in the Argand diagram the points representing:

a) ab　 b) $\dfrac{a}{b}$　 c) $\dfrac{b}{a}$　 d) a^2　 e) b^2　 f) $a^2 + b^2$

and find the modulus and argument in each case.

4 If $a = \cos \frac{2}{3}\pi + i \sin \frac{2}{3}\pi$ and $b = 2(\cos \frac{1}{4}\pi - i \sin \frac{1}{4}\pi)$, evaluate
a) $a^2, a^3, \ldots$　 b) $b^2, b^3, \ldots$　 c) $ab, a/b$
Give the principal value of the arguments and mark each point in the Argand diagram.

5 Simplify:
a) $(\cos \frac{1}{6}\pi + i \sin \frac{1}{6}\pi) \times (\cos \frac{4}{3}\pi + i \sin \frac{4}{3}\pi)$
b) $(\cos \frac{3}{4}\pi + i \sin \frac{3}{4}\pi)^2$
c) $(\cos 80° + i \sin 80°)^3$
d) $(\cos \frac{7}{10}\pi + i \sin \frac{7}{10}\pi)^5$

6 A complex number z has modulus 5 and argument $\frac{1}{3}\pi$. Write down the modulus and argument of the following:
a) iz　 b) z^3 　　　　　　　　　　　　　　　　 (SMP)

7 If $a = \cos 3\theta + i \sin 3\theta$ and $b = \cos 7\theta + i \sin 7\theta$, find the modulus and argument of the complex number $c = a + b$. (You may assume that $0 < \theta < \frac{1}{10}\pi$.) 　　　　　　　 (SMP)

8 Find, in a simple form not involving i, the quadratic equation whose roots are $r(\cos \theta \pm i \sin \theta)$, where $r > 0$. If this equation is

$$x^2 + 2px + q = 0$$

deduce that $r = \sqrt{q}$, and express $\cos \theta$ in terms of p and q. 　　　 (SMP)

9 Given that $z = \sqrt{3} + i$, find the modulus and argument of

a) z^2　 b) $\dfrac{1}{z}$

Show in an Argand diagram the points representing the complex numbers z, z^2 and $\dfrac{1}{z}$. 　　　　　　　　　　　　　　　　 (L)

10 Find the modulus and argument of the complex numbers z_1, z_2 and z_3, where

$$z_1 = (1 - i), \quad z_2 = z_1{}^3, \quad z_3 = \frac{\sqrt{3} - i}{\sqrt{3} + i}$$

Mark on an Argand diagram the points representing z_1, z_2 and z_3.   (L)

11 Show accurately on an Argand diagram the roots z_1, z_2 of the quadratic equation $z^2 - 4z + 8 = 0$, where Im $z_1 > 0$. Show also the points representing $z_1 + z_2$ and $z_1 z_2$. Express z_1 and z_2 in modulus–argument form, and hence obtain Re $(z_1{}^8)$ and Im $(z_2{}^6)$.   (L)

12 If $z = \cos \theta + i \sin \theta$, express in modulus–argument form:

a) $1 + z$ **b)** $\dfrac{z - 1}{z + 1}$ **c)** $\dfrac{z}{z^2 + 1}$ **d)** $\dfrac{z + i}{z - i}$

13 If $a = 2 - i$, $b = 4 - 3i$, find two values of c so that triangle ABC is equilateral.

14 Given that $z = \cos \theta + i \sin \theta$, prove that

$$1 - z^2 = 2 \sin \theta (\sin \theta - i \cos \theta)$$

Given also that $0 < \theta < \pi$, find the modulus and argument of

$$\frac{2}{1 - z^2} \tag{JMB}$$

15 Find the modulus and argument of each of the complex numbers z_1 and z_2, where

$$z_1 = \frac{1 + i}{1 - i}, \quad z_2 = \frac{\sqrt{2}}{1 - i}$$

Plot the points representing z_1, z_2 and $z_1 + z_2$ on an Argand diagram. Deduce from your diagram that

$$\tan \tfrac{3}{8}\pi = 1 + \sqrt{2} \tag{L}$$

16 The complex numbers z_1 and z_2 are given by

$$z_1 = \tfrac{1}{2}(1 + i\sqrt{3}), \quad z_2 = i$$

State the modulus and the argument of z_1 and of z_2. Represent the complex numbers z_1, z_2 and $z_1 + z_2$ on an Argand diagram. By using your diagram, or otherwise, show that

$$\tan \tfrac{5}{12}\pi = 2 + \sqrt{3} \tag{JMB}$$

12.3 Complex conjugates: solution of equations

We have already noted that the roots of the quadratic equation $z^2 - 6z + 13 = 0$ are $3 + 2i$ and $3 - 2i$. Pairs of complex numbers like this, which differ only in the sign of their imaginary part, are called *conjugate*.

More generally, if $z = x + iy$ we define the complex conjugate of z. denoted by z^*, as

$$z^* = x - iy$$

and on the Argand diagram z and z^* will be reflections in the real axis.

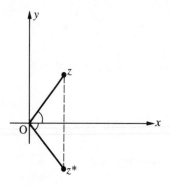

Note that:

a) When $y = 0$ (i.e. the complex number is real) then z is its own conjugate: real numbers are *self-conjugate*.

b) When $|z| = 1$, $z = \text{cis } \theta$, $z^* = \text{cis } (-\theta)$, and in this case conjugates are reciprocals.

c) Both the sum and the product of a pair of conjugates are real numbers:

$$z + z^* = (x + iy) + (x - iy) = 2x \qquad = 2 \,\text{Re}\, z$$
$$zz^* = (x + iy)(x - iy) \quad = \quad x^2 + y^2 = |z|^2$$

however, by contrast,

$$z - z^* = 2iy = 2i\text{Im}\, z$$

The reader can quickly verify that for any complex number z,

$$(z^*)^* = z \quad \text{and} \quad (-z)^* = -z^*$$
$$\text{also} \quad (z_1 \pm z_2)^* = z_1^* \pm z_2^* \quad \text{and} \quad (z_1 z_2)^* = z_1^* z_2^*$$

The result $zz^* = |z|^2$ has already proved useful in the division of complex numbers. For example,

$$\frac{1+3i}{2+i} = \frac{(1+3i)(2-i)}{(2+i)(2-i)} = \frac{5+5i}{5} = 1+i$$

As we observed at the beginning of this section, the roots of the quadratic equation $z^2 - 6z + 13 = 0$ are a pair of conjugate complex numbers. Clearly this is bound to happen when solving any quadratic equation $az^2 + bz + c = 0$, if a, b, c are real and $b^2 < 4ac$.

For $\quad z = \dfrac{-b \pm \sqrt{(b^2 - 4ac)}}{2a} = -\dfrac{b}{2a} \pm \dfrac{\sqrt{(4ac - b^2)}}{2a}i$

which is a conjugate pair provided that the coefficients a, b, c are real.

If, however, we solve a quadratic equation with *complex* coefficients, such as

$$z^2 + (-1 + i)z + (2 + i) = 0$$
then $\qquad (z - i)(z - 1 + 2i) = 0$
$$z = i \text{ or } 1 - 2i$$

and clearly the roots are *not* conjugate.

We can now prove that, for any polynomial equation with real coefficients, the complex roots occur in conjugate pairs.

Let the polynomial be

$$f(z) = a_n z^n + a_{n-1} z^{n-1} + \cdots + a_1 z + a_0$$

and suppose that z is a root, so that $f(z) = 0$. We therefore have to prove that z^* is also a root.

Now $\quad f(z^*) = a_n(z^*)^n + a_{n-1}(z^*)^{n-1} + \cdots + a_1 z^* + a_0$
$$= a_n^*(z^*)^n + a_{n-1}^*(z^*)^{n-1} + \cdots + a_1^* z^* + a_0^*$$

(since all the a_n are real and therefore self conjugate).

But $\quad a_n^*(z^*)^n = a_n^*(z^n)^* = (a_n z^n)^*$, etc.

Hence $\quad f(z^*) = (a_n z^n)^* + (a_{n-1} z^{n-1})^* + \cdots + (a_1 z)^* + a_0^*$
$$= (a_n z^n + a_{n-1} z^{n-1} + \cdots + a_1 z + a_0)^*$$
$$= \{f(z)\}^*$$
$$= 0^*$$
$$= 0$$

So whenever z is a root, so is z^*; and we see that the roots of such equations with real coefficients always occur in conjugate pairs (or as real numbers, which are self-conjugate).

Example 1

Solve the equation $z^3 - 3z^2 + 9z + 13 = 0$, given that one root is $2 + 3i$.

As the coefficients of the equation are real, the complex conjugate $2 - 3i$ must be a second root.

So if the third root is α, the sum of the roots of the equation (from the coefficient of z^2) gives:

$$\alpha + (2 + 3i) + (2 - 3i) = -(-3) = 3$$
$$\Rightarrow \qquad\qquad \alpha = -1$$

Alternatively,

$$(z - 2 - 3i)(z - 2 + 3i) = z^2 - 4z - 13$$

So, factorising,

$$z^3 - 3z^2 + 9z + 13 = (z^2 - 4z + 13)(z + 1) = 0$$

giving $\alpha = -1$ as the third root.

The fundamental theorem of algebra

The invention of complex numbers to ensure that every quadratic equation has exactly two roots may well lead us to ask whether every cubic will have three roots, every quartic four roots, and in general whether a polynomial equation of degree n will have exactly n roots. In fact this is true, and it follows from the *fundamental theorem of algebra*, which states that:

For every polynomial $f(z)$, there exists a value of z ($\in \mathbb{C}$) which satisfies the equation $f(z) = 0$.

In other words, every polynomial equation has at least one (complex) root, and this is still true even when its coefficients are themselves complex.

The proof of this theorem is well beyond the scope of this course, but it is quite easy to use the theorem to establish the number of roots of a polynomial equation.

Let $f(z)$ be a polynomial of degree n. Then by the fundamental theorem, the equation $f(z) = 0$ must have a root α, and hence $f(z)$ must have a factor $(z - \alpha)$.

So $f(z) = (z - \alpha)g(z)$,   where $g(z)$ is a polynomial of degree $n - 1$

Similarly,

$$g(z) = (z - \beta)h(z), \quad \text{where } h(z) \text{ is a polynomial of degree } n - 2$$

So that

$$f(z) = (z - \alpha)(z - \beta)h(z)$$

Clearly this process can be repeated until $f(z)$ is written as a product of n linear factors. Hence $f(z) = 0$ must have n roots (some of which may be repeated).

Now if a polynomial has real coefficients, any complex zeros must occur in conjugate pairs, and the linear factors associated with these conjugates, when multiplied together, will give real quadratic factors:

$$(z - a + bi)(z - a - bi) = z^2 - 2az + a^2 + b^2$$

It therefore follows that any polynomial with real coefficients can be written as the product of real linear and quadratic factors.

Example 2

Show that 2i is a zero of the polynomial $f(z) = z^5 - z^4 + 4z^3 - 2z^2 + 8$, and hence express $f(z)$ as a product of real factors.

$$f(2i) = 32i^5 - 16i^4 + 32i^3 - 8i^2 + 8$$
$$= 32i - 16 - 32i + 8 + 8$$
$$= 0$$

so 2i is a zero of $f(z)$.

Since 2i is a zero of $f(z)$, then $-2i$ must also be a zero, and $(z - 2i)(z + 2i) = z^2 + 4$, which must be a factor of $f(z)$.

Now $z^5 - z^4 + 4z^3 - 2z^2 + 8 = (z^2 + 4)(z^3 - z^2 + 2)$

and it follows from the factor theorem that $z + 1$ is a factor of $z^3 - z^2 + 2$:

$$z^3 - z^2 + 2 = (z + 1)(z^2 - 2z + 2)$$

Hence $f(z) = (z + 1)(z^2 + 4)(z^2 - 2z + 2)$

Exercise 12.3

1 Prove that:

a) $(z_1 + z_2)^* = z_1^* + z_2^*$ **b)** $(z_1 - z_2)^* = z_1^* - z_2^*$

c) $(z_1 z_2)^* = z_1^* z_2^*$ **d)** $\left(\dfrac{z_1}{z_2}\right)^* = \dfrac{z_1^*}{z_2^*}$

2 Prove that $z_1^* z_2 = (z_1 z_2^*)^*$, and verify by putting:

$$z_1 = r_1(\cos\theta_1 + i\sin\theta_1), \quad z_2 = r_2(\cos\theta_2 + i\sin\theta_2)$$

3 If $z = r(\cos\theta + i\sin\theta)$ show that $r = \sqrt{(zz^*)}$, $\theta = \frac{1}{2}(\arg z - \arg z^*)$ and also that $z^2 + (z^*)^2 = 2|z|^2 \cos 2\theta$.

Proves that

$$\arg z = \theta \quad \Rightarrow \quad z = z^*(\cos 2\theta + i \sin 2\theta)$$

Is the converse true?

4 If z_1 and z_2 are complex numbers such that $z_1 + z_2$ and $z_1 z_2$ are both real, prove that $z_1^* = z_2$.

5 Given that $1 - i$ is a root of $2z^3 - 7z^2 + 10z - 6 = 0$, find the other roots.

6 If $2 + i$ is a root of $2z^3 - 9z^2 + 14z - 5 = 0$, find the other roots.

7 Given that $(1 - i)$ is a root of the equation $z^3 - 4z^2 + 6z - 4 = 0$, find the other roots. (L)

8 One root of the equation $z^3 + z^2 + 4z + \lambda = 0$, where λ is a real number, is $1 - 3i$. Find the other roots and the value of λ. (L)

9 Given that $z + n (n \in \mathbb{Z}^+)$ is a factor of $z^3 + 6z^2 + 16z + 16$, find n and then factorise completely.

10 Show by substitution that $1 + 2i$ is a solution of the equation $x^3 + x + 10 = 0$. *Write down* the other complex solution. Hence, or otherwise, express the left-hand side of the equation as the product of a quadratic factor and a linear factor, both with real coefficients. (SMP)

11 Given that $2 + 3i$ is a root of the polynomial equation $P(z) = 0$, where

$$P(z) \equiv z^4 - 3z^3 + 7z^2 + 21z - 26$$

factorise $P(z)$ into linear and quadratic factors with real coefficients. Find the other 3 roots of the equation $P(z) = 0$. (L)

12 Solve the equation

$$z^4 - 3z^3 + 4z^2 - 3z + 1 = 0$$

given that one root is $\frac{1}{2}(1 - \sqrt{3}i)$. Solve also by making the substitution $w = z + 1/z$.

13 Solve the equation $z^4 - 6z^2 + 25 = 0$.

14 Solve $z^4 - 6z^3 + 23z^2 - 34z + 26 = 0$, given that one root is $1 + i$.

15 Given that $4 + 3i$ is a root of the equation

$$z^5 - 8z^4 + 25z^3 - 125z^2 + 1000z - 3125 = 0$$

write down a second root. Find the modulus and argument of each of the other roots, and show that in the Argand diagram the five points which represent the roots of the equation all lie on a circle. (JMB)

16 The equation

$$z^2 + (1 - 2i)z - (7 + i) = 0$$

has roots α, β. Find the equation with numerical coefficients whose roots are $\alpha - i$, $\beta - i$. Hence, or otherwise, find the values of α and β.   (JMB)

17 Prove that the area of the triangle OAB is $\dfrac{1}{4i}(a^*b - ab^*)$

18 Show that the equation of the circle

$$(x - \alpha)^2 + (y - \beta)^2 = \rho^2$$

can be written as

$$zz^* - p^*z - pz^* + pp^* - \rho^2 = 0$$

where $p = \alpha + i\beta$.

12.4 De Moivre's theorem

Having established methods of obtaining the sum, difference, product and quotient of complex numbers, and their interpretation on the Argand diagram, we now investigate their powers.

Since it has already been shown that

$$(\cos \theta_1 + i \sin \theta_1)(\cos \theta_2 + i \sin \theta_2) = \cos (\theta_1 + \theta_2) + i \sin (\theta_1 + \theta_2)$$

then $(\cos \theta + i \sin \theta)^2 = \cos 2\theta + i \sin 2\theta$

and $(\cos \theta + i \sin \theta)^3 = \cos 3\theta + i \sin 3\theta$

and it clearly follows (and can be proved formally by induction) that, if n is a positive integer

$$(\cos \theta + i \sin \theta)^n = \cos n\theta + i \sin n\theta$$

Furthermore, if n is a negative integer, we can let $m = -n$.

Then $(\cos \theta + i \sin \theta)^n = (\cos \theta + i \sin \theta)^{-m}$

$$= \frac{1}{(\cos \theta + i \sin \theta)^m}$$

$$= \frac{1}{\cos m\theta + i \sin m\theta}$$

$$= \cos m\theta - i \sin m\theta$$

$$= \cos(-m\theta) + i \sin (-m\theta)$$

$$= \cos n\theta + i \sin n\theta$$

Also, when $n = 0$, $(\cos \theta + i \sin \theta)^0 = 1 = \cos 0 + i \sin 0$

We therefore have de Moivre's theorem, for integer indices, that:

$$(\cos \theta + i \sin \theta)^n = \cos n\theta + i \sin n\theta$$

or $(\text{cis } \theta)^n = \text{cis } n\theta$

Example 1

Simplify $(1 + \sqrt{3}i)^{-9}$.

$$
\begin{aligned}
(1 + \sqrt{3}i)^{-9} &= [2(\cos \tfrac{1}{3}\pi + i \sin \tfrac{1}{3}\pi)]^{-9} \\
&= 2^{-9}(\cos \tfrac{1}{3}\pi + i \sin \tfrac{1}{3}\pi)^{-9} \\
&= 2^{-9}[\cos(-3\pi) + i \sin(-3\pi)] \\
&= -\tfrac{1}{512}
\end{aligned}
$$

De Moivre's theorem can also be used to obtain certain trigonometric results, as in the following examples.

Example 2

Use de Moivre's theorem to find formulae for $\cos 4\theta$ and $\sin 4\theta$ in terms of $\cos \theta$ and $\sin \theta$.

We shall use the abbreviation $c = \cos \theta$, $s = \sin \theta$.

$$
\begin{aligned}
\cos 4\theta + i \sin 4\theta &= (c + is)^4 \\
&= c^4 + 4c^3is + 6c^2i^2s^2 + 4ci^3s^3 + i^4s^4 \\
&= (c^4 - 6c^2s^2 + s^4) + i(4c^3s - 4cs^3)
\end{aligned}
$$

Equating real and imaginary parts,

$$
\begin{aligned}
\cos 4\theta &= c^4 - 6c^2s^2 + s^4 = c^4 - 6c^2(1 - c^2) + (1 - c^2)^2 \\
&= 8c^4 - 8c^2 + 1 \\
\sin 4\theta &= 4c^3s - 4cs^3 = 4cs(c^2 - s^2)
\end{aligned}
$$

Example 3

Express $\cos^4 \theta$ and $\sin^3 \theta$ in terms of sines and cosines of multiple angles.

We know that $\quad z = \cos \theta + i \sin \theta \quad \Rightarrow \quad z^{-1} = \cos \theta - i \sin \theta$

so $\qquad z + z^{-1} = 2 \cos \theta \quad$ and $\qquad z - z^{-1} = 2i \sin \theta$

Also that $\qquad z^n = \cos n\theta + i \sin n\theta \quad$ and $\quad z^{-n} = \cos n\theta - i \sin n\theta$

so that $\quad z^n + z^{-n} = 2 \cos n\theta \quad$ and $\qquad z^n - z^{-n} = 2i \sin n\theta$

Hence $\quad (z + z^{-1})^4 = z^4 + 4z^2 + 6 + 4z^{-2} + z^{-4}$

$\Rightarrow \qquad (2 \cos \theta)^4 = (z^4 + z^{-4}) + 4(z^2 + z^{-2}) + 6$

$\Rightarrow \qquad 16 \cos^4 \theta = 2 \cos 4\theta + 8 \cos 2\theta + 6$

$\Rightarrow \qquad \cos^4 \theta = \tfrac{1}{8} \cos 4\theta + \tfrac{1}{2} \cos 2\theta + \tfrac{3}{8}$

Also $\qquad (z - z^{-1})^3 = z^3 - 3z + 3z^{-1} - z^{-3}$

$$\Rightarrow \qquad (2i \sin \theta)^3 = (z^3 - z^{-3}) - 3(z - z^{-1})$$
$$\Rightarrow \qquad -8i \sin^3 \theta = 2i \sin 3\theta - 6i \sin \theta$$
$$\Rightarrow \qquad \sin^3 \theta = -\tfrac{1}{4} \sin 3\theta + \tfrac{3}{4} \sin \theta$$

Rational powers

If $n = p/q$, where p, q are integers, then

$$\left(\cos \frac{p}{q} \theta + i \sin \frac{p}{q} \theta \right)^q = \cos \left(\frac{p}{q} \theta \times q \right) + i \sin \left(\frac{p}{q} \theta \times q \right)$$
$$= \cos p\theta + i \sin p\theta$$
$$= (\cos \theta + i \sin \theta)^p$$

So $\cos \dfrac{p}{q} \theta + i \sin \dfrac{p}{q} \theta$ is a qth root of $(\cos \theta + i \sin \theta)^p$, i.e. a value of $(\cos \theta + i \sin \theta)^{p/q}$: as we shall shortly see there are a number of possible values.

Therefore if n is a fraction, de Moivre's theorem now states that:

> $\cos n\theta + i \sin n\theta$ is *one of the values* of $(\cos \theta + i \sin \theta)^n$
>
> or $\text{cis } n\theta$ is *one of the values of* $(\text{cis } \theta)^n$

Example 4

Use de Moivre's theorem to find:
a) the cube roots of i;
b) the cube roots of $2 + 2i$.

a) $\sqrt[3]{i} = i^{1/3} = \left(\cos \dfrac{\pi}{2} + i \sin \dfrac{\pi}{2} \right)^{1/3}$ so one value of $\sqrt[3]{i}$ is

$$\cos \frac{\pi}{6} + i \sin \frac{\pi}{6} = \tfrac{1}{2}\sqrt{3} + \tfrac{1}{2}i$$

However, we can also regard i as having argument $\dfrac{\pi}{2} + 2\pi = \dfrac{5\pi}{2}$,

$\dfrac{\pi}{2} + 4\pi = \dfrac{9\pi}{2}$, etc., so that $i^{1/3}$ could also be

$$\left(\cos \frac{5\pi}{2} + i \sin \frac{5\pi}{2} \right)^{1/3} = \cos \frac{5\pi}{6} + i \sin \frac{5\pi}{6} = -\tfrac{1}{2}\sqrt{3} + \tfrac{1}{2}i$$

or $\left(\cos \dfrac{9\pi}{2} + i \sin \dfrac{9\pi}{2} \right)^{1/3} = \cos \dfrac{3\pi}{2} + i \sin \dfrac{3\pi}{2} = -i$

The addition of further multiples of 2π to the argument of i cannot produce any different values of $i^{1/3}$ since

$$\left\{\cos\left(\frac{\pi}{2} + 6\pi\right) + i\sin\left(\frac{\pi}{2} + 6\pi\right)\right\}^{1/3} = \cos\left(\frac{\pi}{6} + 2\pi\right) + i\sin\left(\frac{\pi}{6} + 2\pi\right)$$

$$= \cos\frac{\pi}{6} + i\sin\frac{\pi}{6}$$

which is a repeat of the first root.

So the three values of $i^{1/3}$ are $\operatorname{cis}\frac{\pi}{6}$, $\operatorname{cis}\frac{5\pi}{6}$, $\operatorname{cis}\frac{3\pi}{2}$, which represent points equally spaced around the unit circle.

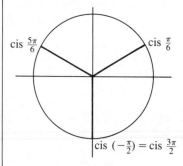

In view of the way in which the three cube roots are derived, it is clear that the cube roots of any number must have arguments which are spaced by $2\pi/3$. We can use this fact in **b)**.

b) $(2 + 2i)^{1/3} = \left(2\sqrt{2}\operatorname{cis}\frac{\pi}{4}\right)^{1/3} = \sqrt{2}\operatorname{cis}\frac{\pi}{12}$ as one value

The other two cube roots must be equally spaced around the circle of radius $\sqrt{2}$. They are therefore $\sqrt{2}\operatorname{cis}\frac{3\pi}{4}$ and $\sqrt{2}\operatorname{cis}\frac{17\pi}{12}$ (or using the principal

value of the argument $\sqrt{2}\operatorname{cis}\left(-\frac{7\pi}{12}\right)$).

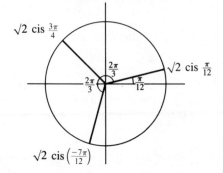

Example 5

Find the values of $(-1 + i)^{3/5}$.

Now $\quad -1 + i = \sqrt{2}\operatorname{cis}\dfrac{3\pi}{4} \quad$ or $\quad \sqrt{2}\operatorname{cis}\left(\dfrac{3\pi}{4} + 2n\pi\right)$

So $\quad (-1 + i)^{3/5} = 2^{0.3}\operatorname{cis}\left(\dfrac{9\pi}{20} + \dfrac{6n\pi}{5}\right)$

and putting $n = 0,\ 1,\ 2,\ 3,\ 4$, we obtain the five distinct values of $(-1 + i)^{3/5}$. These are shown in the diagram, where the arrows show the route followed by adding $6\pi/5\ (= 216°)$ repeatedly to the argument. Clearly the five solutions form a regular pentagon.

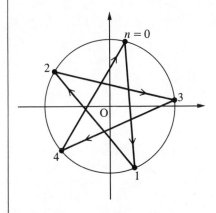

Exercise 12.4a

1 Express in the form $r(\cos\theta + i\sin\theta)$, where the argument has its principal value:

a) $(\cos\pi/5 + i\sin\pi/5)^{14}$ **b)** $(\cos 100° + i\sin 100°)^{-5}$

c) $(\cos\pi/6 + i\sin\pi/6)^{16}$ **d)** $(\cos 80° - i\sin 80°)^{7/2}$

e) $(28 + 96i)^4$ **f)** $(1 + i)^{10}$

g) $(\sqrt{3} - i)^{7/4}$ **h)** $(\sqrt{3} - i)^{20}$

i) $\sqrt[3]{(1 + i)}$ **j)** $\sqrt{(1 - \sqrt{3}i)}$

2 Express in the form $r(\cos\theta + i\sin\theta)$:

a) $(1 + \cos\theta + i\sin\theta)^6$

b) $(\sin\theta - i\cos\theta)^5$

c) $(1 + i\tan\theta)^n$

d) $\dfrac{(\cos 5\alpha + i\sin 5\alpha)^6(\cos 3\alpha - i\sin 3\alpha)^7}{(\cos\alpha + i\sin\alpha)^4(\cos 2\alpha - i\sin 2\alpha)^{-2}}$

3 Given that $z = 1 + i\sqrt{3}$, prove that $z^{14} = 2^{13}(-1 + i\sqrt{3})$ (L)

4 Use the method of mathematical induction to prove that, if θ is any angle and if n is a natural number, then

$$(\cos\theta + i\sin\theta)^n = \cos n\theta + i\sin n\theta$$

Use this result to find the value of $(1 + i)^{12}$.   (SMP)

5 Find the values of **a)** $\sqrt[4]{i}$ and   **b)** $\sqrt[5]{-i}$.

6 Find as many as possible values of $(1 + i)^{1/6}$.

7 **a)** Solve $z^3 - 1 = 0$.
 b) By writing -1 as $\cos\pi + i\sin\pi$, obtain the roots of $z^3 + 1 = 0$.

8 Obtain $\sin 3\theta$ in terms of $\sin\theta$, and $\cos 3\theta$ in terms of $\cos\theta$ by using de Moivre's theorem.

9 Find formulae for:
 a) $\cos 6\theta$ in terms of $\cos\theta$;
 b) $\sin 7\theta/\sin\theta$ in terms of $\cos\theta$;
 c) $\tan 5\theta$ in terms of $\tan\theta$;
 d) $\sin 6\theta/\sin\theta$ in terms of $\cos\theta$.

10 Express   **a)** $\cos^3\theta$ **b)** $\sin^4\theta$ **c)** $\sin^5\theta$ in terms of the sines and cosines of multiple angles.

11 Given that $z = \cos\theta + i\sin\theta$, show that
 a) $z + \dfrac{1}{z} = 2\cos\theta$
 b) $z^3 + z^{-3} = 2\cos 3\theta$
 Solve $z^3 + z^{-3} = -\sqrt{2}$ and show the six roots on an Argand diagram.
 Expand $(z + z^{-1})^6$ and hence express $\cos^6\theta$ in the form

$$A\cos 6\theta + B\cos 4\theta + C\cos 2\theta + D$$ (AEB, 1981)

12 If $\cos\alpha + \cos\beta + \cos\gamma = 0$ and $\sin\alpha + \sin\beta + \sin\gamma = 0$, prove that α, β and γ are separated by $\frac{2}{3}\pi$. Generalise in the case of the equations

$$\sum_1^n \cos\alpha_r = 0, \quad \sum_1^n \sin\alpha_r = 0$$

13 Prove that $(1 + \cos\theta - i\sin\theta)^n = 2^n \cos^n\frac{1}{2}\theta\,(\cos\frac{1}{2}n\theta - i\sin\frac{1}{2}n\theta)$.
 Hence *write down* the value of $(1 + \cos\theta + i\sin\theta)^n$ and simplify the sum of these two expressions.

14 Express $\left(\dfrac{1 + \cos\theta + i\sin\theta}{1 + \cos\theta - i\sin\theta}\right)^n$ in the form $X + iY$.   (MEI)

The *n*th roots of a complex number

We can now see in general how the *n*th roots of a complex number may be found.

If we write the number in the form $z = r(\cos\theta + i\sin\theta)$ then

$$\{r(\cos\theta + i\sin\theta)\}^{1/n} = r^{1/n}\left(\cos\frac{\theta}{n} + i\sin\frac{\theta}{n}\right)$$

is one of the roots.

To obtain the other roots we write $z = r\{\cos(\theta + 2k\pi) + i\sin(\theta + 2k\pi)\}$ where $k = 1, 2, 3, \ldots, n - 1$. Then these roots are

$$z^{1/n} = r^{1/n}\cos\left(\frac{\theta + 2k\pi}{n}\right) + i\sin\left(\frac{\theta + 2k\pi}{n}\right)$$

Alternatively, using the Argand diagram, if we start with the root $r^{1/n}\left\{\cos\dfrac{\theta}{n} + i\sin\dfrac{\theta}{n}\right\}$, the other roots can be obtained by stepping off angles of $2\pi/n$ around the circle of radius $r^{1/n}$.

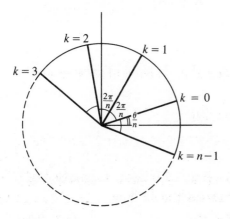

In particular, the *n*th roots of unity are given by

$$(\cos 2k\pi + i\sin 2k\pi)^{1/n} = \cos\frac{2k\pi}{n} + i\sin\frac{2k\pi}{n} \quad (k = 0, 1, 2, \ldots n - 1)$$

and these form a regular *n*-gon, in the Argand diagram, with one vertex at the point 1 on the real axis.

If we let $\omega = \cos\dfrac{2\pi}{n} + i\sin\dfrac{2\pi}{n}$, then the n roots may be written as $1, \omega, \omega^2,$ $\ldots \omega^{n-1}$ with $\omega^n = 1$.

Clearly, from the symmetry of the diagram (overleaf), the complex roots of 1 occur in conjugate pairs: ω^r and ω^{n-r}.

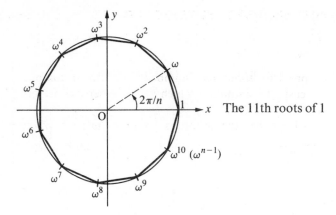

The 11th roots of 1

Also, since $\omega^n = 1$, we have $\omega^n - 1 = 0$

$\Rightarrow$ $(\omega - 1)(\omega^{n-1} + \omega^{n-2} + \ldots + \omega + 1) = 0$

and since $\omega \neq 1$, then $1 + \omega + \omega^2 + \ldots + \omega^{n-1} = 0$

Furthermore, since $1, \omega, \omega^2, \ldots, \omega^{n-1}$ are the roots of $z^n - 1 = 0$,

$$z^n - 1 = (z - 1)(z - \omega)(z - \omega^2) \ldots (z - \omega^{n-1})$$

Example 6

Solve the equation $z^6 + 1 = 0$

$z^6 = -1 = \text{cis}\,(2k\pi + \pi)$

So, by de Moivre's theorem, $z = [\text{cis}\,(2k\pi + \pi)]^{1/6} = \text{cis}\dfrac{(2k + 1)\pi}{6}$

Taking $k = 0, 1, 2, 3, 4, 5$ we obtain $z = \text{cis}\,\pi/6,\ \text{cis}\,\pi/2,\ \text{cis}\,5\pi/6,\ \text{cis}\,7\pi/6,$
$\text{cis}\,3\pi/2,\ \text{cis}\,11\pi/6$

$\Rightarrow$ $z = \pm i,\ \ \pm \text{cis}\dfrac{\pi}{6}$ or $\pm \text{cis}\dfrac{5\pi}{6}$

$= \pm i,\ \ \tfrac{1}{2}\sqrt{3} \pm \tfrac{1}{2}i,\ \ -\tfrac{1}{2}\sqrt{3} \pm \tfrac{1}{2}i$

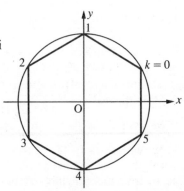

Example 7

Factorise $z^5 - 1$ over the real numbers.

Although this is a problem about real numbers, the use of complex numbers leads quite easily to a solution which would otherwise be very difficult to obtain.

The equation $z^5 - 1 = 0$ has the solutions 1, ω, ω^2, ω^3, ω^4, where $\omega = \cos \dfrac{2\pi}{5} + i \sin \dfrac{2\pi}{5}$, and $\omega^5 = 1$.

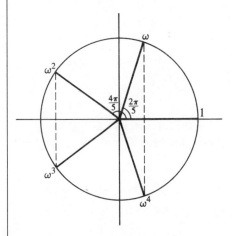

So $z^5 - 1 = (z - 1)(z - \omega)(z - \omega^2)(z - \omega^3)(z - \omega^4)$

But $(z - \omega)(z - \omega^4) = z^2 - (\omega + \omega^4)z + \omega^5$

$$= z^2 - \left(2\cos\dfrac{2\pi}{5}\right)z + 1 \quad \text{(see the diagram)}$$

and $(z - \omega^2)(z - \omega^3) = z^2 - (\omega^2 + \omega^3)z + \omega^5$

$$= z^2 - \left(2\cos\dfrac{4\pi}{5}\right)z + 1$$

So $z^5 - 1 = (z - 1)\left(z^2 - 2z\cos\dfrac{2\pi}{5} + 1\right)\left(z^2 - 2z\cos\dfrac{4\pi}{5} + 1\right)$

Exercise 12.4b

1 Find:
   a) the cube roots of $\frac{1}{8}(\cos 60° + i\sin 60°)$
   b) the fifth roots of $4\sqrt{2}(\cos 80° - i\sin 80°)$
   c) the eighth roots of $\cos 160° - i\sin 160°$

d) the ninth roots of i

e) the four values of $(1 - \sqrt{3}\,i)^{3/4}$

f) the sixth roots of $i - \sqrt{3}$

g) $(-i)^{3/4}$

h) four values of z to satisfy $z^4 + 1 = i$

2 Show that $(1 + 3i)^4 = 28 - 96i$, and find the other fourth roots of $28 - 96i$.

3 Solve the equations:

a) $z^3 + 8 = 0$

b) $z^6 + z^5 + z^4 + z^3 + z^2 + z + 1 = 0$

c) $z^4 - z^3 + z^2 - z + 1 = 0$

d) $z^{10} - 1 = 0$

4 Show that the roots of the equation $z^3 = 1$ are 1, ω and ω^2, where

$$\omega = -\frac{1}{2} + \frac{i\sqrt{3}}{2}$$

Express the complex number $5 + 7i$ in the form $A\omega + B\omega^2$, where A and B are real, and give the values of A and B in surd form.　　　(AEB, 1982)

5 Prove that

$$(\cos A + i \sin A)(\cos B + i \sin B) = \cos(A + B) + i \sin(A + B)$$

Write down the six roots of the equation

$$z^6 = \cos \alpha + i \sin \alpha$$

in the form $\cos \theta + i \sin \theta$ and show that the points representing them on the Argand diagram lie at the vertices of a regular hexagon. Show that this property is also true for the roots of the equation

$$(z - a)^6 = b$$

where a and b are two non-zero complex numbers.　　　(AEB, 1981)

6 a) z is a variable within the set of complex numbers. By factorising $z^4 - 1$, or otherwise, find the four solutions of the equation

$$z^4 = 1 \tag{1}$$

b) By writing the equation

$$(z + 1)^4 = z^4 \tag{2}$$

in the form

$$\left(1 + \frac{1}{z}\right)^4 = 1$$

or otherwise, use your answers to (1) to solve (2). Draw a diagram showing the position of the solutions in the complex number plane. Why are there only three solutions to this equation and not four? (SMP)

7 Calculate in modulus–argument form all the values of z which satisfy

$$z^5 = i$$

By equating imaginary parts of this equation, deduce that $\sin(\pi/10)$ is one solution of the equation

$$16y^5 - 20y^3 + 5y - 1 = 0$$

and find two other solutions. (SMP)

8 A complex number, z, satisfies the equation $(z - 1)^n = z^n$. Show that

$$z\left(2\sin^2\frac{k\pi}{n} - 2\,\mathrm{i}\sin\frac{k\pi}{n}\cos\frac{k\pi}{n}\right) = 1$$

where k is any integer. Hence, show that

$$z = \frac{1}{2}\left(1 + \mathrm{i}\cot\frac{k\pi}{n}\right)$$ (MEI)

In nos 9 to 13 take ω to be a complex cube root of 1.

9 Prove that:
  a) $(1 + \omega)(1 + \omega^2) = 1$
  b) $(1 + \omega^2)^3 = -1$
  c) Simplify $(1 + \omega)(1 + 2\omega)(1 + 3\omega)(1 + 5\omega)$

10 Prove by reference to the Argand diagram that $1 + 2\omega^2$ is purely imaginary.

11 Form the quadratic equations whose roots are:
  a) $3 + \omega, 3 + \omega^2$ b) $2\omega - \omega^2, 2\omega^2 - \omega$

12 Prove that $1 + \omega^n + \omega^{2n}$ takes the values 3 or 0 according as n is or is not divisible by 3.

13 From the identity
$$a^2 + b^2 + c^2 - bc - ca - ab \equiv (a + b\omega + c\omega^2)(a + b\omega^2 + c\omega)$$
$$(a, b, c \in \mathbb{C})$$
show that if $\sum a^2 = \sum bc$, then either $a - c = \omega(c - b)$ or else $a - c = \omega^2(c - b)$ and interpret in the Argand diagram.

12.5 Loci in the Argand diagram

We now consider the locus of a point P in an Argand diagram by imposing conditions on its associated complex number.

If A is a fixed point in the diagram, with complex number a, then the complex number $z - a$ corresponds to the vector **AP**.

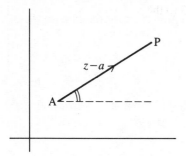

$|z - a|$ is the distance of P from A, and $\arg(z - a)$ is the angle which the vector **AP** makes with the real axis.

Therefore the locus defined by $|z - a| = k$ is a circle, centre A, radius k, and $|z - a| < k$ is the interior of the circle.

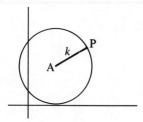

The locus defined by $\arg(z - a) = \alpha$ is the half-line originating from A at an angle α to the real axis.

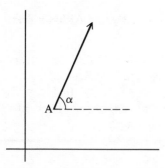

If the line is to continue through A, then we require also that $\arg(z - a) = \alpha + \pi$.

Example 1

On the same diagram show the loci given by $|z - 1 - i| = 1$ and $\arg(z - 1) = \pi/3$. Find their common points.

$|z - 1 - i| = 1$ is the circle, centre $1 + i$, radius 1. $\arg(z - 1) = \pi/3$ is the half-line, originating from 1, at an angle $\pi/3$ to the real axis.

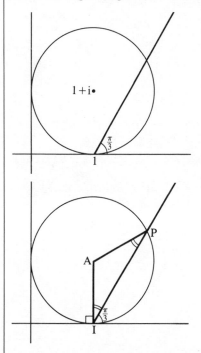

One common point of the circle and half-line is obviously on the real axis at $z = 1$. Let the other point be P.

Then $\angle\,\text{AIP} = \pi/6$, and since IAP is an isosceles triangle, $\angle\,\text{API} = \pi/6$ as well.

Therefore $\angle\,\text{IAP} = 2\pi/3$, and it follows that $\mathbf{AP}$ makes an angle $(2\pi/3) - (\pi/2) = \pi/6$ with the real axis.

Since $\qquad\qquad \mathbf{OP} = \mathbf{OA} + \mathbf{AP}$

complex number $\qquad p = (1 + i) + \left(\cos\dfrac{\pi}{6} + i\sin\dfrac{\pi}{6}\right)$

$$= 1 + \frac{\sqrt{3}}{2} + \frac{3}{2}i$$

If both A and B (with complex numbers a and b) are fixed points, and if the locus of P is defined by $|z - a| = |z - b|$ then this means that P is equidistant from A and B. So the locus is the mediator (or perpendicular bisector) of AB.

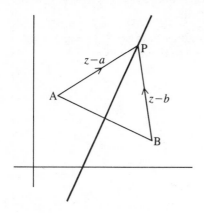

Example 2

Find the Cartesian equation of the locus defined by

$$|z + 1| = |z - 2 - i|$$

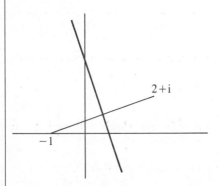

The equation can be written as $|z - (-1)| = |z - (2 + i)|$, and the locus is therefore the mediator of -1 and $2 + i$.

If $z = x + iy$, then $|z + 1| = \sqrt{\{(x + 1)^2 + y^2\}}$
and $|z - 2 - i| = \sqrt{\{(x - 2)^2 + (y - 1)^2\}}$

So $(x + 1)^2 + y^2 = (x - 2)^2 + (y - 1)^2$
$\Rightarrow$ $x^2 + 2x + 1 + y^2 = x^2 - 4x + 4 + y^2 - 2y + 1$
$\Rightarrow$ $6x + 2y = 4$
$\Rightarrow$ $3x + y = 2$

If we now, however, consider $|z - a| = \lambda|z - b|$, or $\left|\dfrac{z - a}{z - b}\right| = \lambda$ (where $\lambda \neq 1$),

then the locus is not, as might possibly be expected, another straight line, but a circle.

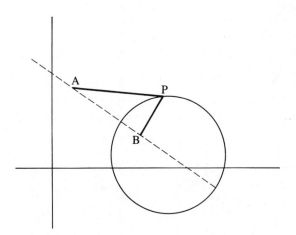

Such a circle is called a *circle of Apollonius*, and it can be located as in the following example.

Example 3

Prove that the locus defined by $|z - 1| = 2|z|$ is a circle, and find its centre and radius.

Letting $z = x + iy$, we have

$$|(x - 1) + iy| = 2|x + iy|$$
$$\Rightarrow \qquad (x - 1)^2 + y^2 = 4(x^2 + y^2)$$
$$\Rightarrow \quad 3x^2 + 3y^2 + 2x - 1 = 0$$
$$\Rightarrow \qquad x^2 + y^2 + \tfrac{2}{3}x - \tfrac{1}{3} = 0$$
$$\Rightarrow \qquad (x + \tfrac{1}{3})^2 + y^2 = (\tfrac{2}{3})^2$$

which represents a circle with centre at $(-\tfrac{1}{3}, 0)$ and radius $\tfrac{2}{3}$.

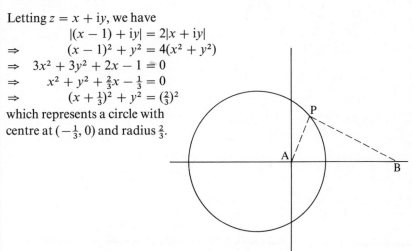

Alternatively we could have proceeded:

$$|z - 1| = 2|z|$$
$$\Rightarrow \qquad |z - 1|^2 = 4|z|^2$$
$$\Rightarrow \qquad (z - 1)(z^* - 1) = 4zz^*$$
$$\Rightarrow \qquad 3zz^* + (z + z^*) - 1 = 0$$
$$\Rightarrow \qquad zz^* + \tfrac{1}{3}(z + z^*) - \tfrac{1}{3} = 0$$
$$\Rightarrow \qquad (z + \tfrac{1}{3})(z^* + \tfrac{1}{3}) = \tfrac{4}{9}$$
$$\Rightarrow \qquad |z + \tfrac{1}{3}|^2 = \tfrac{4}{9}$$
$$\Rightarrow \qquad |z + \tfrac{1}{3}| = \tfrac{2}{3}$$

which again represents a circle, with its centre at $-\tfrac{1}{3}$ on the real axis, and radius $\tfrac{2}{3}$.

Having established that $\left| \dfrac{z - a}{z - b} \right| = \lambda$ can define a circle, it is quite surprising to discover that $\arg\left(\dfrac{z - a}{z - b} \right) = \lambda$ also defines an arc of a circle.

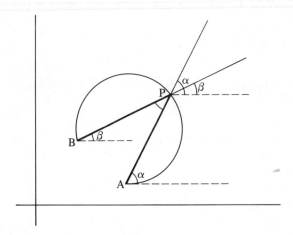

$$\arg\left(\frac{z - a}{z - b} \right) = \arg(z - a) - \arg(z - b)$$
$$= \alpha - \beta$$
$$= \angle APB$$

So if $\arg\left(\dfrac{z - a}{z - b} \right) = \lambda$, then angle APB is constant, which means that P lies on a circular arc from A to B.

Example 4

Find the centre of the circle, an arc of which is defined by $\arg\left(\dfrac{z-2}{z-i}\right) = \dfrac{\pi}{6}$.
Also find the equation which gives the remaining arc of the circle.

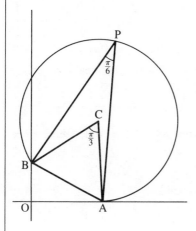

Let A, B be the points with complex numbers 2, i respectively, C the centre of the circle, and P any point on the circular arc.

Since $\angle APB = \dfrac{\pi}{6}$, it follows that $\angle ACB = \dfrac{\pi}{3}$. Therefore ABC must be an equilateral triangle, and $\angle ABC = \dfrac{\pi}{3}$.

Now $\mathbf{BA} = 2 - i$, and $\mathbf{BC}$ can be obtained by rotating $\mathbf{BA}$ through $\dfrac{\pi}{3}$.

So $\quad \mathbf{BC} = (2 - i)\left(\cos\dfrac{\pi}{3} + i\sin\dfrac{\pi}{3}\right)$

$$= (2 - i)\left(\frac{1}{2} + \frac{\sqrt{3}}{2}i\right)$$

$$= \left(1 + \frac{\sqrt{3}}{2}\right) + \left(\sqrt{3} - \frac{1}{2}\right)i$$

and $\quad \mathbf{OC} = \mathbf{OB} + \mathbf{BC}$

$$= i + \left(1 + \frac{\sqrt{3}}{2}\right) + \left(\sqrt{3} - \frac{1}{2}\right)i$$

$$= \left(1 + \frac{\sqrt{3}}{2}\right) + \left(\sqrt{3} + \frac{1}{2}\right)i$$

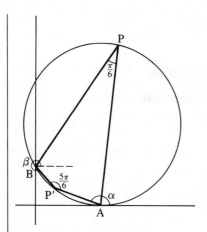

Suppose that P' is any point on the minor arc of the circle; then, since $APBP'$ is a cyclic quadrilateral:

$$\angle AP'B = \pi - \frac{\pi}{6} = \frac{5\pi}{6}$$

so $\quad \beta - \alpha = \dfrac{5\pi}{6} \quad$ or $\quad \alpha - \beta = -\dfrac{5\pi}{6}$

Therefore the minor arc is defined by $\quad \arg\left(\dfrac{z-2}{z-i}\right) = -\dfrac{5\pi}{6}$

Exercise 12.5

In nos **1–4** describe the loci defined by:

1
  a) $|z| = 2$
  b) $|z - 2i| = 3$
  c) $|z + 2 - i| = \sqrt{5}$
  d) $\arg z = 2\pi/3$
  e) $\arg(z + 2) = \pi/2$
  f) $\arg(z - 3 + 2i) = -3\pi/4$

2
  a) $|z - 1| = |z + 1|$
  b) $|z - i| = |z + 1|$

  c) $z = z^*$
  d) $\left|\dfrac{z}{z + 2 + i}\right| = 1$

  e) $|z| = 3|z - 1|$
  f) $\left|\dfrac{z + 2 - i}{z - 1}\right| = 2$

3
  a) $\arg\left(\dfrac{z - 1}{z + 1}\right) = \dfrac{\pi}{2}$
  b) $\arg\left(\dfrac{z - 2}{z - 2i}\right) = \dfrac{\pi}{4}$

  c) $\arg(z - 1) - \arg(z + i) = \pi$

4 **a)** $|z| + |z - 1 + i| = 2$
  b) $|z - 1| + |z + 1| = 4$
  c) $|z - 1| - |z + 1| = 1$

5 Find the Cartesian equations of the loci in no. **2**.

6 Shade the regions of the Argand diagram defined by

  a) $|z| < 3$ **b)** $|z + 2| > 2$ **c)** $|z + 4i| \geqslant 3$

  d) $|z - 3 + i| > 2$ **e)** $|z - 1| + |z + 1| \leqslant 4$ **f)** $\left| \dfrac{z + 2i}{z - 2i} \right| \geqslant 1$

7 By using an Argand diagram, or otherwise, find the complex number z for which $\arg z = \pi/6$ and $\arg (z - 2i) = -\pi/6$, giving your answer in the form $a + bi$. (MEI)

8 By shading in three separate Argand diagrams, show the regions in which the point representing z can lie when
  a) $|z| > 3$
  b) $|z - 2| < |z - 4|$
  c) $0 < \arg (z + 3) < \pi/6$
  Shade in another sketch the region in which z can lie when all these inequalities apply. (L)

9 Prove that, in the complex plane, the set of points

  $$\{z : |z + 4 - 2i| = 2\sqrt{5}\}$$

  is a circle through the origin.
  By drawing a careful diagram in each case, and shading the appropriate region, illustrate the following sets in the complex plane:
  a) $C = \{z : |z + 4 - 2i| \leqslant 2\sqrt{5}\}$
  b) $D = \{z : \frac{1}{3}\pi \leqslant \arg (z + 4 - 2i) \leqslant \frac{1}{2}\pi\}$
  c) $C \cap D$
  Establish that the point $z = 4i$ lies in C but not in D. (SMP)

10 The point P represents the complex number z in the Argand diagram. Describe fully the locus of P in each of the following cases:
  a) when $|z| = |z - 2|$;
  b) when $\arg (z - 2) = \arg (z + 2) + \pi/2$.
  Find the complex number z which satisfies both the above equations. (JMB)

11 From the equation $|z + 16| = 4|z + 1|$, deduce that $|z| = 4$, and interpret geometrically.

12 Show that $|z + 2i| = |2iz - 1|$ represents a circle, and find its centre and radius.

13 If $(z - \mathrm{i})/(z - 1)$ is purely imaginary, show that the locus of z is a circle centre $\frac{1}{2}(1 + \mathrm{i})$, radius $2\sqrt{2}$.

14 Sketch on the same Argand diagram the loci of z and w, where

$$|z| = |z - 4\mathrm{i}| \quad \text{and} \quad |w + 2| = 1$$

State
a) the minimum value of $|z - w|$;
b) the Cartesian form of z for which $\arg z = \pi/4$.   (AEB, 1982)

15 Sketch and label in one Argand diagram the three sets of points corresponding to values of z for which
a) $\mathrm{Re}\,(z + 1) = |z - 1|$
b) $z^4 - 1 = 0$
c) $\arg(z + \mathrm{i}) - \arg(z - \mathrm{i}) = \pi/2$
Give a precise geometrical description of each of these sets of points. (L)

16 In an Argand diagram, P represents a complex number z such that

$$2|z - 2| = |z - 6\mathrm{i}|$$

Show that P lies on a circle, and find
a) the radius of this circle;
b) the complex number represented by its centre.
If, further,

$$\arg\left(\frac{z - 2}{z - 6\mathrm{i}}\right) = \pi/2$$

find the complex number represented by P.   (L)

* 12.6 $\mathrm{e}^{\mathrm{i}\theta}$

Let us now consider the function $z = \cos\theta + \mathrm{i}\sin\theta$. Continuing to treat i like any other number, we have, by differentiation:

$$\frac{\mathrm{d}z}{\mathrm{d}\theta} = -\sin\theta + \mathrm{i}\cos\theta = \mathrm{i}(\cos\theta + \mathrm{i}\sin\theta)$$

Hence $\dfrac{\mathrm{d}z}{\mathrm{d}\theta} = \mathrm{i}z$

$\Rightarrow \qquad \mathrm{i}\dfrac{\mathrm{d}\theta}{\mathrm{d}z} = \dfrac{1}{z}$

$\Rightarrow \qquad \mathrm{i}\theta = \ln z + c$

But when $\theta = 0$, $z = 1$. So $c = 0$.

$\Rightarrow \quad i\theta = \ln z$

$\Rightarrow \quad z = e^{i\theta}$

So $\boxed{\cos \theta + i \sin \theta = e^{i\theta}}$

It may be objected that we have nowhere defined the meaning of a number such as e^z when z is complex; but the reader should not be deterred by such inhibitions! Indeed, the above paragraph may be regarded as providing, if not a definition, at least a reasonable exposition of the *meaning* of $e^{i\theta}$, making it consistent with the familiar processes of mathematics. For example, the rule for the multiplication of complex numbers:

$$\text{cis } \theta_1 \times \text{cis } \theta_2 = \text{cis } (\theta_1 + \theta_2)$$

may now be re-stated:

$$e^{i\theta_1} \times e^{i\theta_2} = e^{i(\theta_1 + \theta_2)}$$

which is just what we should expect in a well-behaved system, 'imaginary' powers of e obeying the same rules as real powers. Indeed, the rule for addition of *arguments* (see section 12.2) when multiplying complex numbers is now seen to be no more than a disguised version of the rule for addition of indices, or *logarithms*!

Again, de Moivre's theorem may now be rewritten in the perhaps more familiar form $(e^{i\theta})^n = e^{in\theta}$; and, as further support for the truth of the statement $e^{i\theta} = \text{cis } \theta$, it will be remembered that we obtained the following infinite series by using Maclaurin's expansion:

$$\cos \theta = 1 - \frac{\theta^2}{2!} + \frac{\theta^4}{4!} - \cdots \qquad \sin \theta = \theta - \frac{\theta^3}{3!} + \frac{\theta^5}{5!} - \cdots$$

Then $\cos \theta + i \sin \theta = \left(1 - \dfrac{\theta^2}{2!} + \dfrac{\theta^4}{4!} - \cdots\right) + i\left(\theta - \dfrac{\theta^3}{3!} + \dfrac{\theta^5}{5!} - \cdots\right)$

$$\text{(for all real } \theta)$$

$$= 1 + i\theta - \frac{\theta^2}{2!} - \frac{i\theta^3}{3!} + \frac{\theta^4}{4!} + \cdots$$

$$= 1 + (i\theta) + \frac{(i\theta)^2}{2!} + \frac{(i\theta)^3}{3!} + \cdots = e^{i\theta}$$

being the infinite series for e^z, with z substituted by $i\theta$. This further encourages us to accept the validity of the expansion of the exponential function as an infinite series in cases when the exponent is imaginary.

The above is neither a complete nor a conclusive argument: a great deal of analysis, involving the use of the exponential series $e^z = 1 + z + z^2/2! + \cdots$, is needed to make the definition of $e^{i\theta}$ watertight. Meanwhile, however, we shall proceed on the basis that the above broad outline makes sense.

The most remarkable result may be obtained by putting $\theta = \pi$:

$$e^{i\pi} = \cos\pi + i\sin\pi = -1 + 0i = -1$$

so we have derived the equation due to the great mathematician Euler:

$$e^{i\pi} + 1 = 0$$

which unites in one spectacular relationship the five most important numbers in mathematics: $0, 1, \pi,$ e and i. Sense or nonsense? It must be true! *Certum est quia impossibile est:* nothing so incredible could be an invention!

But the use of the $e^{i\theta}$ notation enables us to obtain other quite surprising results. For example, suppose we wish to find i^i.

We may write $i = \cos\pi/2 + i\sin\pi/2 = e^{i\pi/2}$

So $i^i = (e^{i\pi/2})^i$
$$= e^{-\pi/2} \approx 0.207, \quad \text{a real value!}$$

Also, since $\cos x + i\sin x = e^{ix}$
and $\cos x - i\sin x = e^{-ix}$

we deduce, by adding and subtracting, that

$$\cos x = \frac{e^{ix} + e^{-ix}}{2} \qquad \sin x = \frac{e^{ix} - e^{-ix}}{2i}$$

and discover the astonishing revelation that trigonometric functions, which started life as ratios of sides of a right-angled triangle, are really exponential functions in disguise.

Example 1

Express $(2 + 4i)\,e^{i\pi/3}$ in the form $x + iy$.

$$(2 + 4i)\,e^{i\pi/3} = (2 + 4i)\left(\cos\frac{\pi}{3} + i\sin\frac{\pi}{3}\right)$$

$$= (2 + 4i)\left(\frac{1}{2} + \frac{\sqrt{3}}{2}i\right)$$

$$= (1 - 2\sqrt{3}) + (2 + \sqrt{3})i$$

Example 2

Evaluate $\displaystyle\int e^{-3x}\cos 2x\,dx$ and $\displaystyle\int e^{-3x}\sin 2x\,dx$.

Integrals of this kind are usually found by two applications of integration by parts, a tedious procedure. We can alternatively use complex numbers:

Let $C = \displaystyle\int e^{-3x}\cos 2x\,dx$ and $S = \displaystyle\int e^{-3x}\sin 2x\,dx$.

Then $C + iS = \displaystyle\int e^{-3x}(\cos 2x + i\sin 2x)\,dx$

$$= \int e^{-3x}e^{2ix}\,dx$$

$$= \int e^{(-3+2i)x}\,dx$$

$$= \frac{e^{(-3+2i)x}}{-3+2i}$$

$$= \frac{-3-2i}{13}e^{-3x}e^{2ix}$$

$$= \frac{-3-2i}{13}e^{-3x}(\cos 2x + i\sin 2x)$$

Separating the real and imaginary parts, we obtain C and S:

$$C = \frac{e^{-3x}}{13}(-3\cos 2x + 2\sin 2x) \quad\text{and}\quad S = \frac{e^{-3x}}{13}(-2\cos 2x - 3\sin 2x)$$

Exercise 12.6

1   Express in the form $x + iy$:
   a) $e^{i\pi/2}$ b) $e^{-i\pi/3}$ c) $e^{3i\pi/4}$ d) $e^{-2i\pi/3}$
   e) $e^{(-1+i\pi/2)}$ f) e^{i} g) e^{-2i} h) e^{1+i}

2   Express in the form $re^{i\theta}$:
   a) $1 + i$ b) $3i$ c) -2 d) $\dfrac{1}{\sqrt{2}} - \dfrac{1}{\sqrt{2}}i$
   e) $\sqrt{3} + i$ f) $-1 + \sqrt{3}i$ g) $2 + i$ h) $3 + 4i$

3   Find the quadratic equation which has solutions $2e^{\pm i\pi/6}$.

4 Find the solutions of the following equations in the form $re^{i\theta}$:
   a) $z^3 = 1$
   b) $z^3 = -8$
   c) $z^5 = 4 + 4i$

5 Use $\cos 3x + i \sin 3x = e^{3ix}$ to integrate $e^{-2x}\cos 3x$ and $e^{-2x}\sin 3x$.

6 Using the exponential definitions of sin and cos, prove the formula

$$\sin(\theta + \phi) = \sin\theta\cos\phi + \cos\theta\sin\phi$$

7 Using the exponential definitions of sin and cos, prove the formulae

$$\cos^2 x + \sin^2 x = 1 \quad \text{and} \quad \cos^2 x - \sin^2 x = \cos 2x$$

8 Given that $z_1 = 2e^{i\pi/6}$ and $z_2 = \sqrt{2}e^{i\pi/4}$, express z_1 and z_2 in Cartesian form, and deduce the value of $\sin 75°$ in surd form.   (AEB, 1983)

9 Given that the complex numbers a, b, c are represented in the Argand diagram by the points A, B, C respectively and that

$$(b - a)/(c - a) = e^{i\pi/3}$$

show that the triangle ABC is equilateral. Write down another relation similar to the one above, and deduce that

$$a^2 + b^2 + c^2 = bc + ca + ab \qquad \text{(JMB)}$$

10 Show that $z = -i$ is one root of

$$z^3 + (2 + 3i)z^2 + 2iz + 2i = 0$$

Find the remaining roots in the form

$$z = -(1 + i) \pm re^{i\theta},$$

where you should state values for r and θ. Plot all the roots on a diagram of the complex plane.   (SMP)

*12.7 Interlude: hyperbolic functions

In the last section we discovered that $\cos x$ and $\sin x$ could be written in exponential form as

$$\cos x = \frac{e^{ix} + e^{-ix}}{2} \qquad \sin x = \frac{e^{ix} - e^{-ix}}{2i}$$

We now examine the functions which are of the same exponential form as $\cos x$ and $\sin x$, but with the 'i's removed. For reasons which will shortly become clear, they are called the hyperbolic cosine ($\cosh x$) and the hyperbolic sine ($\sinh x$).

So $\quad \cosh x = \dfrac{e^x + e^{-x}}{2} \qquad \sinh x = \dfrac{e^x - e^{-x}}{2}$

Firstly we see that

$$\sinh(-x) = \tfrac{1}{2}(e^{-x} - e^x) = -\sinh x$$
$$\text{and} \quad \cosh(-x) = \tfrac{1}{2}(e^{-x} + e^x) = \cosh x$$

So that $\cosh x$ is an even function and $\sinh x$ is an odd function. This is also clear by consideration of the graphs of e^x and e^{-x}, and of $\cosh x$ and $\sinh x$.

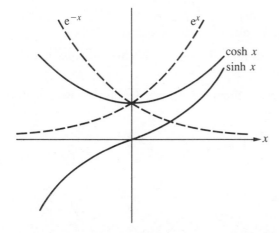

Although the graphs of the circular functions and the hyperbolic functions are quite different, there are many similarities between the two sets of functions, and the first of these becomes evident when we differentiate $\sinh x$ and $\cosh x$.

$$\frac{d}{dx}(\sinh x) = \frac{d}{dx}\left(\frac{e^x - e^{-x}}{2}\right) = \frac{e^x + e^{-x}}{2} = \cosh x$$

$$\frac{d}{dx}(\cosh x) = \frac{d}{dx}\left(\frac{e^x + e^{-x}}{2}\right) = \frac{e^x - e^{-x}}{2} = \sinh x$$

From their exponential formulae, we can obtain directly the following relations connecting the circular and hyperbolic functions:

$$\cosh ix = \cos x \qquad \sinh ix = i \sin x$$
$$\cos ix = \cosh x \qquad \sin ix = i \sinh x$$

These relationships, combined with basic trigonometric formulae, enable us to establish similar formulae for the hyperbolic functions.

For example
$$\begin{aligned}
\cosh(a + b) &= \cos i(a + b)\\
&= \cos ia \cos ib - \sin ia \sin ib\\
&= \cosh a \cosh b - (i \sinh a)(i \sinh b)
\end{aligned}$$

⇒ $\cosh(a + b) = \cosh a \cosh b + \sinh a \sinh b$

compared with $\cos(a + b) = \cos a \cos b - \sin a \sin b$

Resemblances between circular and hyperbolic functions such as this can be summarised in *Osborn's rule*, which states that the formulae correspond exactly, provided that the sign is changed whenever there is a product of two sine or sinh functions. (The change in sign of course corresponds to the implied i^2 term, as in the example above.)

We therefore have corresponding formulae such as

$\cos^2 x + \sin^2 x = 1$ and $\cosh^2 x - \sinh^2 x = 1$
$\cos 2x = 1 - 2\sin^2 x$ and $\cosh 2x = 1 + 2\sinh^2 x$
$\sin 2x = 2\sin x \cos x$ and $\sinh 2x = 2\sinh x \cosh x$

The reason for the name hyperbolic functions can now be explained by considering the equation $\cosh^2 x - \sinh^2 x = 1$.

$\cos x$ and $\sin x$ are described as circular functions because the parametric representation

$x = \cos t, \quad y = \sin t$

can be used for the circle

$x^2 + y^2 = 1$

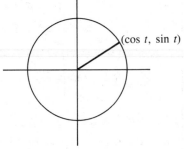

Similarly the parametric representation

$x = \cosh t, \quad y = \sinh t$

can be used for the rectangular hyperbola

$x^2 - y^2 = 1$

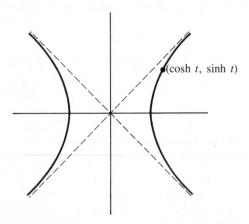

Lastly, and again by analogy with the trigonometric functions, we define

$$\tanh x = \frac{\sinh x}{\cosh x} \qquad \coth x = \frac{\cosh x}{\sinh x}$$

$$\operatorname{sech} x = \frac{1}{\cosh x} \qquad \operatorname{cosech} x = \frac{1}{\sinh x}$$

Example 1

Solve the equation $2\cosh x + \sinh x = 3$.

$$2\cosh x + \sinh x = 3$$

$$\Rightarrow \quad 2\left(\frac{e^x + e^{-x}}{2}\right) + \frac{e^x - e^{-x}}{2} = 3$$

$$\Rightarrow \qquad\qquad 3e^x + e^{-x} = 6$$

$$\Rightarrow \qquad\qquad 3e^{2x} - 6e^x + 1 = 0$$

Regarding this as a quadratic equation in e^x, we have

$$e^x = \frac{6 \pm \sqrt{(36 - 12)}}{6}$$

$$= 1 \pm \sqrt{\tfrac{2}{3}}$$

$$\Rightarrow \quad x = \ln\left(1 \pm \sqrt{\tfrac{2}{3}}\right) \approx 0.597 \text{ or } -1.696$$

Exercise 12.7a

1 Using the exponential formulae for the hyperbolic functions, prove that
   a) $\cosh 2x = 1 + 2\sinh x$
   b) $\sinh 2x = 2\sinh x \cosh x$
   c) $\cosh x - \sinh x = 1$

2 Show that the hyperbola

$$\frac{x^2}{a^2} - \frac{y^2}{b^2} = 1$$

can be represented parametrically by the hyperbolic functions

$$x = a\cosh\theta, \quad y = b\sinh\theta$$

Find $\dfrac{dy}{dx}$ in terms of θ.

3   Show that:

a) $\tanh x = \dfrac{e^{2x} - 1}{e^{2x} + 1}$

b) $\tanh x$ is an odd function

c) $\dfrac{d}{dx}(\tanh x) = \text{sech}^2\, x$

and sketch the graph of $\tanh x$.

4   Differentiate with respect to x:
   a) $\sinh 2x$ **b)** $\cosh \frac{1}{2}x$ **c)** $x \cosh x$ **d)** $\ln \sinh x$
   e) $\ln \cosh x$ **f)** $\text{sech}\, x$ **g)** $\text{cosech}\, x$ **h)** $\ln \tanh \frac{1}{2}x$

5   Find the following integrals:

   a) $\displaystyle\int \sinh x \, dx$ **b)** $\displaystyle\int \cosh 2x \, dx$ **c)** $\displaystyle\int \text{sech}^2\, x \, dx$

   d) $\displaystyle\int \tanh x \, dx$ **e)** $\displaystyle\int x \cosh x \, dx$ **f)** $\displaystyle\int x\, \text{sech}^2\, x \, dx$

6   Calculate:
   a) the area beneath the curve $y = \cosh x$ between $x = -1$ and $x = +1$;
   b) the volume formed when this area is rotated about the x-axis.

7   Use Maclaurin series to find the power series for:
   a) $\sinh x$ **b)** $\cosh x$

8   Solve the equations:
   a) $\cosh x + \sinh x = 2$
   b) $2 \cosh x - \sinh x = 2$

9   Prove that $e^{x - iy} = (\cosh x + \sinh x)(\cos y - i \sin y)$.

10   If $\sin (x + iy) = r(\cos \theta + i \sin \theta)$, prove that $r = \frac{1}{2}(\cosh 2y - \cos 2x)$ and $\tan \theta = \cot x \tanh y$.

11   Find the real and imaginary parts and the moduli of:
   a) $\sin z$ **b)** $\cos z$ **c)** $\tan z$,   where $z = x + iy$

12   If $\sinh (x + iy) = e^{\frac{1}{3}i\pi}$, find x and y.

Inverse hyperbolic functions

We see from a graph that the function $\sinh x$ sets up a 1–1 mapping of the real numbers $\mathbb{R}$ on to $\mathbb{R}$, so that its inverse function is defined throughout $\mathbb{R}$ and we can write:

$$y = \sinh x \quad \Leftrightarrow \quad x = \sinh^{-1} y$$

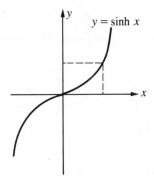

Now $y = \sinh x = \tfrac{1}{2}(e^x - e^{-x})$
$\Rightarrow$ $e^{2x} - 2y\,e^x - 1 = 0$
$\Rightarrow$ $e^x = y + \sqrt{(y^2 + 1)}$ (since $e^x > 0$)
$\Rightarrow$ $x = \ln\{y + \sqrt{(y^2 + 1)}\}$

So $\sinh^{-1} y = \ln\{y + \sqrt{(y^2 + 1)}\}$

and just as sinh is related to the exponential function, so $\sinh^{-1}$ is related to the logarithmic function.

The function $\cosh x$ is slightly more awkward, for it maps all $x \in \mathbb{R}$ on to the range $y \geqslant 1$, and each of the numbers of the range (except $y = 1$) arises from two equal and opposite members of the domain.

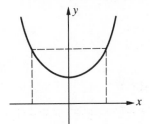

If, however, we restrict the domain to $x \geqslant 0$, we can obtain an inverse function, and write

$y = \cosh x$ $(x \geqslant 0)$ $\Rightarrow$ $x = \cosh^{-1} y$ $(y \geqslant 1)$

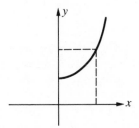

Now $y = \cosh x = \frac{1}{2}(e^x + e^{-x})$

$\Rightarrow$ $e^{2x} - 2y\,e^x + 1 = 0$

$\Rightarrow$ $(e^x)^2 - 2y\,e^x + 1 = 0$

$\Rightarrow$ $e^x = y \pm \sqrt{(y^2 - 1)}$

$\Rightarrow$ $x = \ln\{y \pm \sqrt{(y^2 - 1)}\}$

But $y - \sqrt{(y^2 - 1)} = \dfrac{1}{y + \sqrt{(y^2 - 1)}}$

$\Rightarrow$ $\ln\{y - \sqrt{(y^2 - 1)}\} = \ln\dfrac{1}{y + \sqrt{(y^2 - 1)}}$

$$= -\ln\{y + \sqrt{(y^2 - 1)}\}$$

Hence $x = \pm\ln\{y + \sqrt{(y^2 - 1)}\}$

But $x \geqslant 0$,   so $x = +\ln\{y + \sqrt{(y^2 - 1)}\}$

$\Rightarrow$ $\cosh^{-1} y = \ln\{y + \sqrt{(y^2 - 1)}\}$

We can now write these inverse hyperbolic functions as

$$\sinh^{-1} x = \ln\{x + \sqrt{(x^2 + 1)}\}$$
$$\cosh^{-1} x = \ln\{x + \sqrt{(x^2 - 1)}\}$$

Their main importance is in the calculation of two standard integrals:

a) $\displaystyle\int \frac{dx}{\sqrt{(x^2 + 1)}}$ b) $\displaystyle\int \frac{dx}{\sqrt{(x^2 - 1)}}$

a) In $\displaystyle\int \frac{dx}{\sqrt{(x^2 + 1)}}$,   put $x = \sinh u$ $\Rightarrow$ $dx = \cosh u\, du$.

Then $\displaystyle\int \frac{dx}{\sqrt{(x^2 + 1)}} = \int \frac{\cosh u\, du}{\sqrt{(\sinh^2 u + 1)}}$

$$= \int \frac{\cosh u\, du}{\cosh u}$$

$$= \int du$$

$$= u + c$$

$$= \sinh^{-1} x + c$$

b) In $\displaystyle\int \frac{dx}{\sqrt{(x^2 - 1)}}$,   put $x = \cosh u$ $\Rightarrow$ $dx = \sinh u\, du$.

Then $\displaystyle\int \frac{dx}{\sqrt{(x^2-1)}} = \int \frac{\sinh u\, du}{\sqrt{(\cosh^2 u - 1)}}$

$$= \int \frac{\sinh u\, du}{\sinh u}$$

$$= \int du$$

$$= u + c$$

$$= \cosh^{-1} x + c$$

and in general

$$\int \frac{dx}{\sqrt{(x^2 + a^2)}} = \sinh^{-1}\left(\frac{x}{a}\right) + c$$

$$\int \frac{dx}{\sqrt{(x^2 - a^2)}} = \cosh^{-1}\left(\frac{x}{a}\right) + c$$

Example 2

Evaluate $\displaystyle\int_7^{10} \frac{dx}{\sqrt{\{x^2 - 6x + 5\}}}$

$$\int_7^{10} \frac{dx}{\sqrt{\{x^2 - 6x + 5\}}} = \int_7^{10} \frac{dx}{\sqrt{\{(x-3)^2 - 4\}}}$$

$$= \left[\cosh^{-1}\left(\frac{x-3}{2}\right) \right]_7^{10}$$

$$= \cosh^{-1} 3.5 - \cosh^{-1} 2$$

$$= 1.925 - 1.317$$

$$= 0.608$$

Exercise 12.7b

1 Express as logarithms:
  a) $\sinh^{-1} \frac{3}{4}$ **b)** $\cosh^{-1} 2$ **c)** $\tanh^{-1} x$

2 Find the following integrals:

  a) $\displaystyle\int \frac{dx}{\sqrt{(x^2 + 4)}}$ **b)** $\displaystyle\int \frac{dx}{\sqrt{(x^2 - 9)}}$ **c)** $\displaystyle\int \frac{dx}{\sqrt{(4x^2 + 9)}}$

d) $\displaystyle\int \frac{\mathrm{d}x}{\sqrt{(9x^2 - 4)}}$ 　　　**e)** $\displaystyle\int \frac{\mathrm{d}x}{\sqrt{(x^2 + 2x)}}$ 　　　**f)** $\displaystyle\int \frac{\mathrm{d}x}{\sqrt{(x^2 + 2x + 2)}}$

3　Find the following integrals:

a) $\displaystyle\int \sinh^{-1} x \, \mathrm{d}x$ 　　　　　　　　　**b** $\displaystyle\int \cosh^{-1} x \, \mathrm{d}x$

c) $\displaystyle\int \sqrt{(x^2 + 1)} \, \mathrm{d}x$ 　　　　　　　　**d)** $\displaystyle\int \sqrt{(x^2 - 4)} \, \mathrm{d}x$

4　Evaluate:

a) $\displaystyle\int_0^2 \frac{\mathrm{d}x}{\sqrt{(x^2 + 1)}}$ 　　　　　　　　**b)** $\displaystyle\int_2^3 \frac{\mathrm{d}x}{\sqrt{(x^2 - 4)}}$

c) $\displaystyle\int_1^2 \sqrt{(x^2 - 1)} \, \mathrm{d}x$ 　　　　　　　**d)** $\displaystyle\int_0^1 \sqrt{(1 + 4x^2)} \, \mathrm{d}x$

5　Find $\displaystyle\int \frac{\mathrm{d}x}{a^2 - x^2}$ by two methods: the use of partial fractions, and the substitution $x = \tanh u$.

6　Prove that

$$\sinh^{-1} x = \ln\{x + \sqrt{(1 + x^2)}\}$$

Show that

$$\sinh^{-1} x - \ln x \quad\rightarrow\quad \ln 2 \text{ as } x \rightarrow \infty.$$
Hence prove that the integral

$$\int_1^\infty \left\{ \frac{1}{(1 + x^2)^{1/2}} - \frac{1}{x} \right\} \mathrm{d}x$$

exists, and find its value. 　　　　　　　　　　　　　　　　　(SMP)

*12.8 Functions of a complex variable

When x is a real variable, the function $f(x)$ is usually illustrated by a graph. Alternatively, however, it can be represented as a mapping from a set of values of x (its domain) onto a set of values of $f(x)$ (its range) by an arrow diagram.

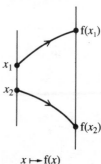

$x \longmapsto f(x)$

For instance, $f: x \mapsto 2x + 1$ can be represented by

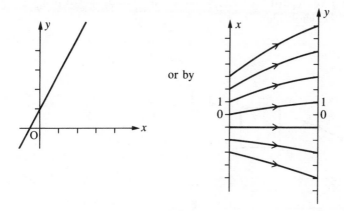

or by

If z is a complex variable, the function $f(z)$ can no longer be shown in the first of these ways, as a graph, and has to be shown as a mapping, this time from a two-dimensional set of points (its domain) in one Argand diagram onto a set of points in another Argand diagram (its range).

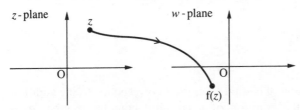

So, just as we can speak of $y = f(x)$ as a mapping $f: x \mapsto y$ of points from an x-line into a y-line, so we can also speak of $w = f(z)$ as a mapping $f: z \mapsto w$ of points from a z-plane into a w-plane.

For example, $w = 2z + 3 - i$ maps points as follows

	A	B	C	D
z	$2 + 2i$	$3 + 2i$	$3 + 3i$	$2 + 3i$
w	$7 + 3i$	$9 + 3i$	$9 + 5i$	$7 + 5i$
	A$'$	B$'$	C$'$	D$'$

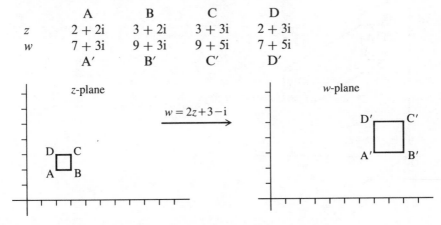

But it is rather simpler in this case to think of the z-plane and w-plane as being superimposed, and to regard $w = 2z + 3 - i$ or $z \mapsto 2z + 3 - i$ as a transformation of the whole plane.

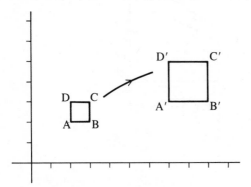

A mapping of this kind can be thought of as composed of simpler transformations:

Translation

We know that the addition of complex numbers corresponds to the addition of vectors. Therefore the addition of a constant complex number a, is equivalent to a constant displacement of all points in the complex plane. Hence $w = z + a$ represents a translation.

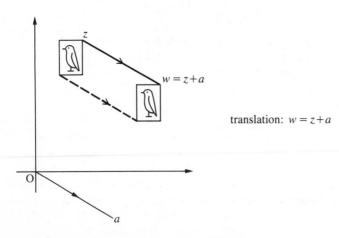

translation: $w = z + a$

Enlargement

If the centre of enlargement is the origin, this is effected by multiplying each complex number in the plane by a real number. So an enlargement is represented by $w = kz$, where k (which is real) is the enlargement scale factor.

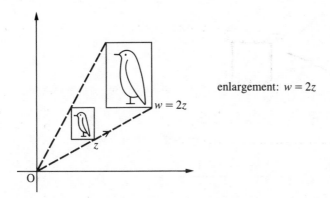

enlargement: $w = 2z$

Rotation

We have already seen in section 12.2 that the transformation $w = z(\cos \alpha + i \sin \alpha)$ has the effect of a rotation about the origin, since the modulus of z is unchanged, whilst its argument is increased by angle α.

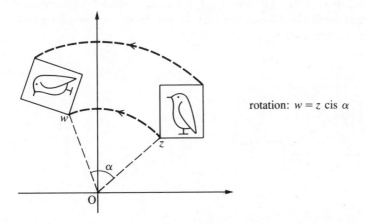

rotation: $w = z \operatorname{cis} \alpha$

Note that when $\alpha = \pi/2$, we obtain $w = iz$, and we have already seen that multiplication by i may be interpreted geometrically as a positive quarter-turn.

For a rotation of angle α about a point (A) other than the origin, we have

$$w - a = (z - a)(\cos \alpha + i \sin \alpha)$$

so that $\qquad w = (z - a)(\cos \alpha + i \sin \alpha) + a$

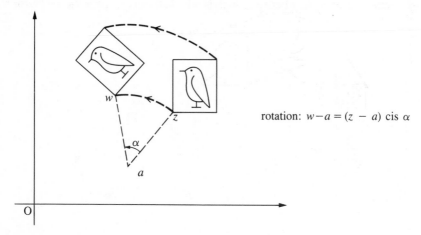

rotation: $w - a = (z - a)\,\text{cis}\,\alpha$

Reflection

Since the conjugation of any complex number is its mirror image in the real axis, reflection in $y = 0$ is represented by $w = z^*$.

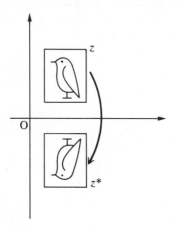

Similarly, reflection in $x = 0$ is represented by $w = -z^*$.

Reflections in other lines are given by equations of the form $w = az^* + b$, where a and b are complex numbers.

We can now return to our original mapping $w = 2z + 3 - i$, and regard it as being composed of two simpler functions: $z' = 2z$, which is an enlargement, scale factor 2, centre the origin; and $w = z' + 3 - i$, which is a translation.

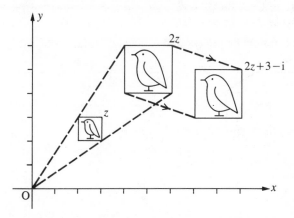

Example 1

Identify the transformation given by $w = iz + 4$.

If we break down the function into two components:

$$z' = iz \quad \text{and} \quad w = z' + 4$$

then the transformation can be regarded as a rotation of $\pi/2$ about the origin, followed by a translation of 4 units in the direction of the real axis.

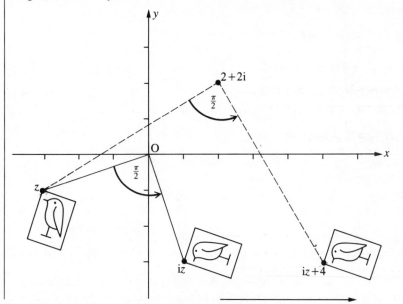

However, the function can also be written in the form

$$w = i(z - 2 - 2i) + 2 + 2i$$
$$\Rightarrow \quad w - (2 + 2i) = i\{z - (2 + 2i)\}$$

and the transformation can now be interpreted as a single rotation of $\pi/2$ about the point $2 + 2i$.

Example 2

Find the function which defines a reflection in the line $y = x$.

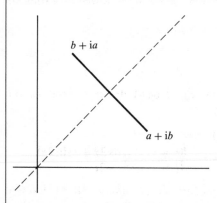

For the general complex number $z = a + ib$, we require $w = b + ia$.

Now $\quad z^* = a - ib$

and $\quad iz^* = i(a - ib)$
$$= ia - i^2 b$$
$$= b + ia$$

Therefore the required function is $w = iz^*$.

Exercise 12.8a

1 Interpret the following transformations, and find the invariant points where possible:

a) $w = z + 2 - 4i$
c) $w = 5z - 2i,$
e) $w = (3 - 4i)z$
g) $w + 2 = \frac{1}{2}(z + 2)$
i) $w = iz - 1$

b) $w = 2 - 4i - z$
d) $w = 2iz$
f) $w = 5(\cos \alpha + i \sin \alpha)z$
h) $w = z(1 - i)$

2 Write down an equation of the form $w = f(z)$ to represent the following:

a) a translation $+3$ in the direction of the imaginary axis;

b) a spiral similarity with scale factor 3 and angle of rotation $+135°$ about O;

c) a spiral similarity with scale factor 3 and angle of rotation $+135°$ about i;

d) a rotation through $-\pi/2$ about the point $1 - i$;

e) a half-turn about $4 + i$;

f) an enlargement, scale factor 4, centre the point $2 + i$;

g) an enlargement, scale factor 3, centre the origin followed by a translation -2 in the direction of the imaginary axis;

h) the same as in **g)** but in the reverse order;

i) an enlargement scale factor 2, centre the point i followed by a rotation through $-\pi/4$ about the point 2;

j) the same as in **i)** but in the reverse order.

3 Analyse the transformations:

a) $w = iz + 1 - i$ **b)** $w = iz^* + 1 - i$

4 By using complex numbers show that successive half-turns about a and b are equivalent to a translation $2\mathbf{AB}$.

5 Interpret the following transformations:

a) $w = -iz^*$ **b)** $w = z^*(\cos 2\alpha + i \sin 2\alpha)$

c) $w = iz^* - 2$ **d)** $w = i(z^* - 2)$

6 If z moves along the semi-circle connecting the points i, 1 and $-i$, describe the loci of the following:

a) $3z$ **b)** $-2z$ **c)** iz **d)** $2 + iz$ **e)** $2z^*$

The reciprocal function $w = 1/z$

Hitherto, functions have defined transformations in which shape has been preserved. We now examine briefly some aspects of the function $w = 1/z$ for which this is not the case.

Since we are dealing with a more complicated function, instead of superimposing the z and w planes, we shall present them separately, and just as the general complex number z is written $x + iy$, so w will be denoted by $u + iv$.

Now $w = \dfrac{1}{z} \Rightarrow u + iv = \dfrac{1}{x + iy}$

$$= \frac{x - iy}{x^2 + y^2}$$

so that $u = \dfrac{x}{x^2 + y^2}$ and $v = \dfrac{-y}{x^2 + y^2}$

If we now consider the line $x = c$ in the z-plane, then in the w-plane

$$u = \frac{c}{c^2 + y^2} \quad \text{and} \quad v = \frac{-y}{c^2 + y^2}$$

Hence $\quad c^2 + y^2 = \dfrac{c}{u} \quad \Rightarrow \quad y^2 = \dfrac{c}{u} - c^2$

Therefore $\qquad\qquad v^2 = \dfrac{y^2}{(c^2 + y^2)^2} = \dfrac{u^2}{c^2}\left(\dfrac{c}{u} - c^2\right) = \dfrac{u}{c} - u^2$

$\Rightarrow \qquad u^2 - \dfrac{u}{c} + v^2 = 0$

$\Rightarrow \qquad \left(u - \dfrac{1}{2c}\right)^2 + v^2 = \left(\dfrac{1}{2c}\right)^2$

This, in the w-plane, is the equation of a circle, centre at the point $1/2c$ on the real axis, and radius $1/2c$.

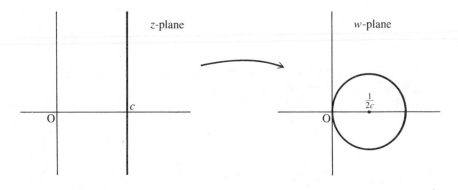

Therefore a line parallel to the imaginary axis is transformed into a circle through the origin, with its centre on the real axis. And since $1/z$ is a self-inverse function, the circle will also map on to the straight line.

However, in contrast, any straight line through the origin is transformed on to another line through the origin (in fact on to its reflection in the real axis). This follows immediately if we represent the complex numbers in modulus–argument form.

If $\quad z = r\operatorname{cis}\theta \quad$ then $\quad w = \dfrac{1}{r}\operatorname{cis}(-\theta)$

so the half-line $\theta = \alpha$ is transformed to the half-line $\theta = -\alpha$, and to complete the line (though excluding the origin), $\theta = \alpha + \pi$ is transformed to $\theta = -(\alpha + \pi)$.

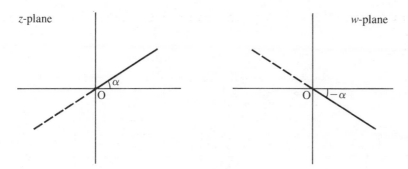

Representation in modulus–argument form also enables us to see fairly easily the effect of the transformation on circles with their centres at the origin.

The circle $r = c$ in the z-plane will be transformed into the circle $r = 1/c$ in the w-plane, and if a point travels anti-clockwise around one circle, its image will travel in a clockwise direction around the other.

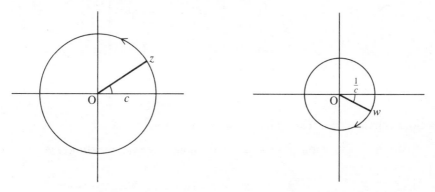

Example 3

Find the image of the circle $|z| = 1$ under the transformation

$$w = \frac{z - 1}{z + 1}$$

The function can be rewritten in the form

$$w = 1 - \frac{2}{z + 1}$$

and then broken down into simpler functions:

$$z_1 = z + 1 \qquad\qquad \text{a translation}$$

$$z_2 = \frac{1}{z_1} = \frac{1}{z + 1} \qquad\qquad \text{the reciprocal function}$$

$$z_3 = -2z_2 = \frac{-2}{z+1} \qquad \text{a half-turn and enlargement}$$

$$w = z_3 + 1 = 1 - \frac{2}{z+1} \quad \text{a translation}$$

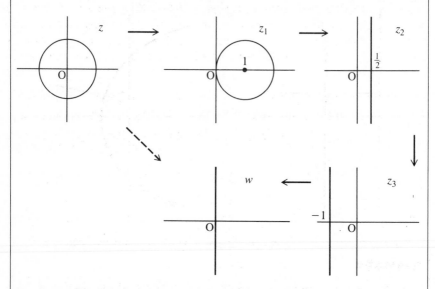

So the circle is transformed into the imaginary axis.

Alternatively we may rearrange the formula to make z the subject:

$$z = \frac{1+w}{1-w}$$

Since $|z| = 1$, it follows that $\quad |1+w| = |1-w|$

$\Rightarrow \qquad\qquad\qquad\qquad |w+1| = |w-1|$

and, as we have seen in section 12.5, the locus of w is then the mediator of the points 1 and -1; that is, the imaginary axis.

The square function $w = z^2$

Writing $w = u + iv$, we obtain

$$u + iv = (x + iy)^2 = x^2 - y^2 + 2ixy$$

so $\quad u = x^2 - y^2 \quad$ and $\quad v = 2xy$

Therefore if we consider the line $x = c$, we have

$$u = c^2 - y^2 \quad \text{and} \quad v = 2cy$$

so that $u = c^2 - \dfrac{v^2}{4c^2}$,   which is the equation of a parabola.

And it can similarly be shown that a line $y = c$ will also transform on to a parabola.

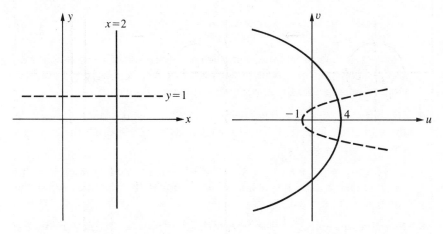

Example 4

If z traces out a complete anti-clockwise revolution of the circle $|z| = 2$, describe the path of $w = z^2$.

In modulus–argument form, if $z = 2 \operatorname{cis} \theta$, then $w = 4 \operatorname{cis} 2\theta$. So w lies on a circle of radius 4. And since its argument is twice that of z, as z makes one anti-clockwise revolution, w will complete two anti-clockwise revolutions of the circle $|w| = 4$.

Exercise 12.8b

1 If z moves along the semi-circle connecting the points i, 1 and $-$i, describe the loci of the following:
   a) z^2 **b)** $z^2 - 1$ **c)** $\sqrt{z}$

2 If z describes the semi-circle from the point 1 through the point i to the point -1, and thence the part of the real axis from -1 to $+1$, describe the loci of the following:
   a) $z + \mathrm{i}$ **b)** $2z$ **c)** $\mathrm{i}z$ **d)** z^2 **e)** $1/z$

3 If z moves along the real axis from -1 to $+1$, find the locus of $(1 - \mathrm{i}z)/(z - \mathrm{i})$.

4 The complex number z has modulus 1 and argument θ.

a) Find the modulus and argument of each of the following: $2z^2$, $\dfrac{1}{z}$

b) Describe the motion of the points representing w in each of the following cases as z moves anti-clockwise round the circle, centre O, radius 1, starting at the point $(1, 0)$:

i) $w = 2z^2$ ii) $w = \dfrac{1}{z}$ iii) $w = z - 2$ iv) $w = \dfrac{1}{z - 1}$ (MEI)

5   If $w = (z - 1)/(z + 1)$, show that when w traces out the lines $u = $ constant and $v = $ constant, the locus of z is two systems of circles. Find also the locus of w as z traces the upper half of the circle diameter -1, $+1$.

6   If $w = (z - i)/(z + i)$ and lies below the real axis, show that w lies outside the unit circle $|w| = 1$. How will w move as z travels along the real axis from $-\infty$ to $+\infty$?

7   If $w = (z - 6i)/(z + 8)$, show that the locus of z is a circle when w traces the imaginary axis, but a line when w traces the real axis. Find the equations of the z-loci.

8   Examine the effect of the transformation $w = (z + 2)/(z + i)$ on the system of concentric circles having centres at the origin in the z-plane. What is the image of the exterior of the circle $|z| = 1$?

9   If $w = (z + 1)^2$ and z traces the unit circle, show that the locus of w has polar equation $r = 2(1 + \cos\theta)$ (a *cardioid*).

10   What is the effect of $w = e^z$ on the system of lines $x = $ constant, $y = $ constant in the z-plane?

11   If $(1 - z)(1 - w) = 1$ and $\arg z = \pi/4$, find the locus of w.

12   Show that under the transformation given by

$$w = \frac{1}{z + 1} \quad (z \in \mathbb{C}, z \neq -1)$$

the circle with centre -1 and radius r is transformed to the circle with centre O and radius $1/r$. Find, and illustrate in an Argand diagram, the image of the imaginary axis under this transformation.   (JMB)

13   Describe the image of the circle $|z| = a$ under the transformation f given by $f(z) = z^2$. Find the image of the circle $|z| = 4$, excluding the point $z = 4$, under the transformation g given by

$$g(z) = \frac{4}{z - 4}$$

Find the image of the circular disc $|z| < 4$ under the transformation g.

Find the image of the circular disc $|z| < 2$ under the transformation h given by h = gf.   (JMB)

14 Two complex numbers z and w are related by

$$w = z + \frac{a^2}{z}$$

where a is a positive real number and P, Q are the points representing z, w respectively in an Argand diagram.

a) Show that if $z = 2a\,(\cos\theta + \mathrm{i}\sin\theta)$, then

$$w = \frac{5a}{2}\cos\theta + \frac{3a}{2}\mathrm{i}\sin\theta$$

Describe the loci of P and Q as θ varies.

b) Show that if P moves along the positive part of the real axis, w is real and has a minimum. Find this minimum value and the position of P when it is attained.   (MEI)

15 Show that if

$$w = \frac{2z - \mathrm{i}}{z - 2}$$

then $z = \dfrac{2w - \mathrm{i}}{w - 2}$

Deduce that the image of the circle $|z| = 1$ under the transformation $z \mapsto w$ is a circle C in the w-plane of centre $-\frac{2}{3} + \frac{2}{3}\mathrm{i}$ and radius $\frac{1}{3}\sqrt{17}$. Show that the image of the diameter of the circle $|z| = 1$ which lies on the real axis ($\arg z = 0$ or π) is a diameter of C. Find the image of the diameter of $|z| = 1$ which lies on the imaginary axis ($\arg z = \pm\pi/2$).   (SMP)

16 The complex plane is mapped on to itself by the transformation T where

$$\mathrm{T}(z) = (z + \mathrm{i})/(z + 1), \quad z \neq -1.$$

a) Find, in the form $a + b\mathrm{i}$ (a, b real), the image of $z = 2 + 4\mathrm{i}$.
b) Find the complex number which maps to $-\mathrm{i}$ under T.
c) Find, in the form $c + d\mathrm{i}$ (c, d real), those complex numbers which are mapped on to themselves by T.
d) Describe carefully, with the aid of a diagram, the image under T of the set $\{z : |z| < 1\}$.
e) Find the set of points which map into the real axis, again illustrating your answer by a sketch.   (OC)

17 The equation

$$w = \frac{z + \mathrm{i}}{z - 1}$$

defines a transformation of the complex plane. Find the sets of points in the z-plane which map into
a) the unit circle $|w| = 1$;
b) the circle $|w| = 2$;
c) the line $\arg w = \pi/2$.
Sketch each of these sets of points on diagrams of the complex z-plane.

(SMP)

Miscellaneous problems

1 Solve the equation $z^4 + 10z^2 + 169 = 0$ by first solving for z^2, and then for z in the form $a + ib$ where a and b are real. Hence or otherwise, express $z^4 + 10z^2 + 169$ as the product of two real quadratic factors.

2 Show that for real x, $\quad \dfrac{1}{1 + x^2} = \dfrac{1}{2i}\left(\dfrac{1}{x - i} - \dfrac{1}{x + i}\right).$

Deduce that

$$\frac{d^{n-1}}{dx^{n-1}}\left(\frac{1}{1 + x^2}\right) = (-1)^{n-1}(n - 1)!\frac{1}{2i}\left(\frac{1}{(x - i)^n} - \frac{1}{(x + i)^n}\right)$$

Prove that, if $x - i = r(\cos\theta - i\sin\theta)$,
then $(x - i)^{-n} = r^{-n}(\cos n\theta + i\sin n\theta)$,
and obtain a similar expression for $(x + i)^{-n}$. Hence obtain a formula for

$$\frac{d^{n-1}}{dx^{n-1}}\left(\frac{1}{1 + x^2}\right) \quad \text{in terms of } r \text{ and } \theta,$$

and deduce that for $0 < x < \frac{1}{2}\pi$,

$$\frac{d^n}{dx^n}(\tan^{-1} x) = (-1)^{n-1}\frac{(n - 1)!}{(1 + x^2)^{n/2}}\sin(n\cot^{-1} x) \qquad \text{(MEI)}$$

3 What geometrical transformation is given by

$$z' - c = (z - c)(\cos\theta + i\sin\theta), \quad \text{where } c \text{ is a complex constant?}$$

If the transformation $z \mapsto z'$ corresponds to a turn of $120°$ about the point $(1, 0)$, and the transformation $z' \mapsto z''$ corresponds to a turn of $60°$ about the point $(-3, 0)$, show that $z'' = -z + 2\sqrt{3}i$, and hence express in geometrical terms the single transformation equivalent to the two transformations in the order given.

(SMP)

4 **a)** Given that one root of the equation $z^4 - 4z^3 + 12z^2 + 4z - 13 = 0$ is $2 - 3i$, find the other three roots.
b) Given $z = 2(\cos\pi/6 + i\sin\pi/6)$ illustrate on the Argand diagram the points representing the complex numbers z, z^*, $(z^*)^2$, $(z^*)^2/|z|$.

S is the set of points in the first quadrant which lie on the circumference of the circle $|z| = 2$. Find the image of S under the mapping $z \mapsto (z*)^2/|z|$.

(MEI)

5   In each of the following cases, find all complex numbers z satisfying the equation:
a) $z^3 + z^{-3} = 1$
b) $(z + z*)^3 = 1$
c) $(z + |z|)^3 = 1$
d) $z^3 + |z|^3 = 1$
e) $(z + iz)^3 = 1$

(OC)

6   a) Assuming the result

$$\cos\theta + i\sin\theta = e^{i\theta}$$

show that

$$\cos\theta = \tfrac{1}{2}(e^{i\theta} + e^{-i\theta})$$

Hence, or otherwise, express

$$S_n = \tfrac{1}{2} + \cos\theta + \cos 2\theta + \ldots + \cos n\theta$$

as two geometric series, where $0 \leqslant \theta < 2\pi$.
    Hence, or otherwise, show that

$$S_n = \frac{\sin\left(n + \tfrac{1}{2}\right)\theta}{2\sin\left(\tfrac{1}{2}\theta\right)}, \quad \text{when} \quad \theta \neq 0$$

b) Deduce that

$$\int_{-\pi}^{\pi} \frac{\sin\left(n + \tfrac{1}{2}\right)\theta}{2\sin\left(\tfrac{1}{2}\theta\right)} \, d\theta = \pi$$

c) Investigate the behaviour of S_n/n as $n \to \infty$.   (MEI)

7   The vertices of a triangle in the Argand diagram are given by the complex numbers a, b, c. Say whether the following are true or false, with reasons:
a) The centroid is at the point $\tfrac{1}{3}(a + b + c)$.
b) The area of the triangle is $\tfrac{1}{2}\{(b - c)^2 + (c - a)^2 + (a - b)^2\}$.
c) The triangle is equilateral if and only if $(a - b)(a - c) + (b - c)^2 = 0$.

(OS)

8   Let $f(z) = \tfrac{1}{2}(z + z^{-1})$ where z is a non-zero complex number.
a) If z describes a circle $|z| = r$ where r is a positive real constant, show that if $r \neq 1$, $f(z)$ describes an ellipse in the complex number plane. What happens if $r = 1$?
b) If z describes the half-line $\arg z = \theta$, where θ is constant, prove that, in general (i.e. if certain exceptional values of θ are excluded) $f(z)$ describes one branch of a hyperbola.

c) What are the exceptional cases in **b)**? Describe what happens for these exceptional cases.   (os)

9   If $|z - 1| = 1$, prove that $\arg(z - 1) = 2\arg z = \tfrac{2}{3}\arg(z^2 - z)$, and give a geometrical interpretation of these results.   (os)

10   Show that the roots of $(z + i)^5 = (z - i)^5$ are $\cot k\pi/5$ $(k = 1, 2, 3, 4)$, and hence evaluate $\cot^2 \pi/5 + \cot^2 2\pi/5$, and $\cot \pi/5 \cot 2\pi/5$. Generalise the result and thence obtain

$$\sum_1^n \cot^2 \frac{k\pi}{2n + 1} = \tfrac{1}{3}n(2n - 1)$$

11   Prove that the roots of

$$(z + 1)^n + (z - 1)^n = 0 \quad \text{are} \quad i\cot\frac{(2k + 1)\pi}{2n} \quad (k = 0, 1, 2, \ldots, n - 1)$$

Deduce that

$$\sum_0^{n-1} \cot\frac{(2k + 1)\pi}{2n} = 0 \quad \text{and} \quad \sum_0^{n-1} \cot^2\frac{(2k + 1)\pi}{2n} = n^2 - n$$

12   Prove that $e^{2k\pi i} = 1$ for all integral values of k, and that $e^{\frac{1}{2}\pi i} = i$. Deduce that a value of i^i is $e^{-\frac{1}{2}\pi}$, and using the result $e^{2k\pi i} = 1$, find a formula for all possible values of i^i.   (cs)

13   If $\cos z = w$, prove that

$$\frac{u^2}{\cos^2 x} - \frac{v^2}{\sin^2 x} = 1 \quad \text{and} \quad \frac{u^2}{\cosh^2 y} + \frac{v^2}{\sinh^2 y} = 1$$

and discuss interpretations in terms of the Argand diagram.

14   For a rotation through α and about O, we have

$$z' = z \operatorname{cis} \alpha \quad \text{or} \quad x' + iy' = (x + iy)(\cos\alpha + i\sin\alpha)$$

Equating real and imaginary parts gives:

$$\left.\begin{array}{l} x' = x\cos\alpha - y\sin\alpha \\ y' = x\sin\alpha + y\cos\alpha \end{array}\right\}, \text{ so we get the rotation matrix } \begin{pmatrix} \cos\alpha & -\sin\alpha \\ \sin\alpha & \cos\alpha \end{pmatrix}.$$

Show that reflection in the line $y = x \tan \alpha$ can be represented

$$z' = z^* \operatorname{cis} 2\alpha$$

By equating real and imaginary parts, find the matrix for this reflection.

15   Prove Napoleon's theorem, that if equilateral triangles BCP, CAQ and ABR are drawn on the outside of any $\triangle ABC$, then their centroids form another equilateral triangle.

13 Differential equations

13.1 Introduction

We have seen that the process of integration is the reverse of differentiation and requires the discovery of a function having the given rate of change. Frequently, however, we do not know either the function or its rate of change, but merely how they are related.

When, for instance, in section 10.2 we were investigating the cooling of bathwater, we simply knew that after time t the excess temperature θ and its rate of decrease $-\dfrac{d\theta}{dt}$ were connected by the equation

$$\frac{d\theta}{dt} = -k\theta \tag{1}$$

and in the same section it was mentioned that the current x flowing in a particular electrical circuit obeyed the equation

$$\frac{d^2x}{dt^2} + 2\frac{dx}{dt} + 5x = 0 \tag{2}$$

Similarly, the rate of growth of the mass m of an organism frequently depends on the equation

$$\frac{dm}{dt} = km \tag{3}$$

and the rate of spread of a disease in a population, of which a fraction x is infected, is often governed by the equation

$$\frac{dx}{dt} = kx(1 - x) \tag{4}$$

Any such equation which connects the rates of change of variables with their values is called a *differential equation*. Equations (1), (3), and (4) are differential equations *of the first order* and equation (2) is *of the second order*, so called from the degree of their highest derivatives.

Now we can easily see that the differential equation

$$\frac{dy}{dx} = x^2$$

immediately leads to

$$y = \tfrac{1}{3}x^3 + c$$

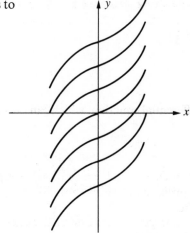

This is called its *general solution*; and by giving c a particular value we can obtain *particular solutions*, such as

$$y = \tfrac{1}{3}x^3 + 2 \quad \text{or} \quad y = \tfrac{1}{3}x^3 - 4$$

Graphically, we see that the general solution is represented by a family of curves, and that each particular solution is an individual member of this family.

Two further examples will illustrate this relationship.

Example 1

If $x^2 + y^2 = a^2$, form a differential equation independent of a.

As a varies, the equation

$$x^2 + y^2 = a^2$$

represents a family of circles, all with centre at the origin.

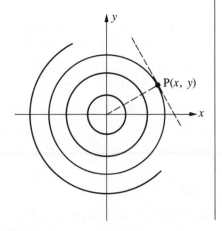

But $x^2 + y^2 = a^2$

$\Rightarrow \quad 2x + 2y\dfrac{dy}{dx} = 0$

$\Rightarrow \quad x + y\dfrac{dy}{dx} = 0$

This is the required differential equation, which can also be written as

$$\frac{y}{x}\frac{dy}{dx} = -1$$

and since the gradient of OP is y/x, this equation is simply expressing the fact that the tangent to each circle at every point is always perpendicular to the corresponding radius.

Example 2

A point P is moving along a straight line and after time t its displacement x is given by $x = A\sin 2t + B\cos 2t$. Form a differential equation which is independent of A and B.

As we here need to eliminate two arbitrary constants, we shall have to differentiate twice and form a second order differential equation.

Now $x = A\sin 2t + B\cos 2t$

$\Rightarrow \quad \dfrac{dx}{dt} = 2A\cos 2t - 2B\sin 2t$

$\Rightarrow \quad \dfrac{d^2x}{dt^2} = -4A\sin 2t - 4B\cos 2t$

So $\dfrac{d^2x}{dt^2} = -4x$

This, therefore, is the required equation, showing that for all values of A and B the acceleration is always $-4x$.

More frequently, however, we start with the differential equation and need to find appropriate solutions. In this case, for instance, the equation

$$\frac{d^2x}{dt^2} = -4x$$

might have arisen from investigation of a buoy bobbing up and down in the sea, and our task would generally be to show that this leads to a solution of the form

$$x = A \sin 2t + B \cos 2t$$

representing a periodic oscillation.

Exercise 13.1

1 Verify that the following differential equations have the solutions indicated (where A, B, ε are arbitrary constants):

a) $\dfrac{dx}{dt} = 3x, \quad x = Ae^{3t}$

b) $x\dfrac{dy}{dx} = 2y, \quad y = Ax^2$

c) $y\dfrac{dy}{dx} = x, \quad x^2 - y^2 = A$

d) $\dfrac{dx}{dt} = x(1 - x), \quad x = \dfrac{1}{1 + Ae^{-t}}$

e) $\dfrac{d^2x}{dt^2} + 3\dfrac{dx}{dt} + 2x = 0, \quad x = Ae^{-t} + Be^{-2t}$

f) $\dfrac{d^2x}{dt^2} + 2\dfrac{dx}{dt} + 5x = 1, \quad x = Ae^{-t}\sin(2t + \varepsilon) + \tfrac{1}{5}$

2 Find differential equations which are satisfied at all points of all members of the following families of curves (where A, B, a, c, etc. define the individual members of the families):

a) $xy = c$ **b)** $y = Ae^x$ **c)** $y^2 = ax$
d) $x^2 - y^2 = a^2$ **e)** $y = Ax^2 + Bx$ **f)** $x = Ae^t + Be^{-t}$

13.2 Step-by-step solutions

In the last section we started to see the relationship between a differential equation and its solutions. As with algebraic equations, the task of the mathematician is usually twofold:

a) to express a given situation—whether mechanical, electrical, chemical, biological, economic or whatever—in a suitable mathematical form (usually called a *model*), which very frequently leads to a differential equation; and

b) to find the appropriate solution of such a differential equation. Sometimes, as we shall see in the following sections, it is possible to obtain a general solution. More usually, however, we have to be satisfied with numerical solutions, and it is with simple examples of these that we shall now begin.

A liquid, which is initially at $10\,^\circ$C is placed in a heated container. Its temperature (θ) then rises at a rate given by the differential equation

$$\frac{d\theta}{dt} = 20 - \frac{\theta}{5}$$

where θ is measured in $^\circ$C and t is measured in minutes.

We wish to know the relationship between temperature and time, and in particular to estimate

a) the temperature after 5 minutes, and

b) the time taken for the temperature to rise to 90 $^\circ$C.

Let us consider what happens in the first minute. We know that the initial gradient of the solution curve (when $\theta = 10$) is $20 - \frac{10}{5} = 18$. So we can use the tangent at (0, 10) as an approximation to the curve for the first minute:

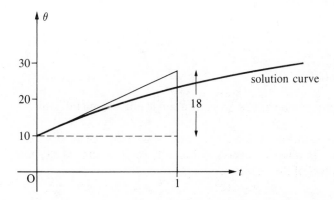

Therefore the approximate increase in temperature is 18°, giving a temperature of 28 $^\circ$C after one minute.

For the next one-minute interval, the initial gradient is $20 - \frac{28}{5} = 14.4$, giving a temperature rise of 14.4°. Therefore when $t = 2$, $\theta = 28 + 14.4 = 42.4\,^\circ$C.

This procedure can be repeated for the next three one-second intervals, giving:

t	0	1	2	3	4	5
θ	10	28	42.4	53.9	63.1	70.5

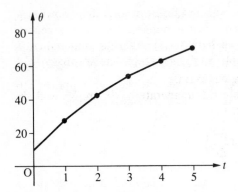

So the temperature after five minutes is approximately 70.5 °C.

Now the choice of one-second intervals was an arbitrary, though convenient one. Any other interval can be used, and clearly the smaller the interval, the better the approximation. (Indeed, the intervals do not have to be of the same length, though it is usual to make them so.)

With half-minute intervals, we have

$$\theta_{0.5} = 10 + 0.5 \times (20 - \tfrac{10}{5})$$
$$= 10 + 9$$
$$= 19\,°C$$

so that

$$\theta_1 = 19 + 0.5 \times (20 - \tfrac{19}{5})$$
$$= 19 + 8.1$$
$$= 27.1\,°C \quad (\text{compared with } 28\,°C \text{ using a single one-second interval})$$

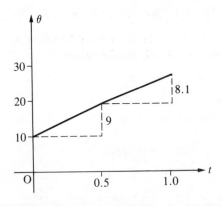

Continuing the process to $t = 5$, we have:

t	0	0.5	1.0	1.5	2.0	2.5	3.0	3.5	4.0	4.5	5.0
θ	10	19	27.1	34.4	41.0	46.9	52.2	57.0	61.3	65.2	68.7

So that a better approximation for the temperature after five minutes is 68.7 °C.

If, however, we now wish to estimate the time taken for the temperature to rise to 90 °C, it will be more appropriate to consider intervals of temperature, and estimate the corresponding increases in time.

From $\theta = 10$ to $\theta = 30$ we denote the temperature rise by $\delta\theta$, and the corresponding increase in time by δt.

Then $\quad \dfrac{\delta\theta}{\delta t} = 18 \quad \left(\text{the initial value of } \dfrac{d\theta}{dt}\right)$

$\Rightarrow \quad \delta t = \dfrac{\delta\theta}{18} = \dfrac{20}{18} \approx 1.11 \text{ minutes}$

So when $\theta = 30$, $t = 1.11$.

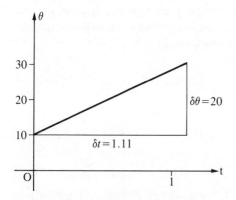

When $\theta = 30$, $\dfrac{d\theta}{dt} = 20 - \dfrac{30}{5} = 14$, so for the interval $30 \leqslant \theta \leqslant 50$, $\delta t = 20/14 \approx 1.43$. Therefore when $\theta = 50$, $t = 1.11 + 1.43 = 2.54$ minutes. Continuing the process for two more intervals, we obtain the values:

θ	10	30	50	70	90
t	0	1.11	2.54	4.54	7.87

and the graph:

So it will take approximately 7.9 minutes for the temperature to rise to 90 °C.

General procedure

For the differential equation $\dfrac{dy}{dx} = f(x, y)$, with initial point $P_0(x_0, y_0)$ we choose an x-interval of length δx.

Since $\dfrac{\delta y}{\delta x} \approx \dfrac{dy}{dx}$, the corresponding increment in y can be approximated by

$\delta y = \dfrac{dy}{dx} \times \delta x$. So for the next solution point, P_1, at (x_1, y_1)

$$x_1 = x_0 + \delta x \quad \text{and} \quad y_1 = y_0 + \frac{dy}{dx} \times \delta x$$

This process is then repeated to obtain P_2 from P_1, P_3 from P_2, etc.

The flow diagram summarises the procedure:

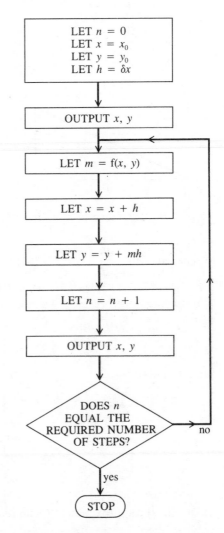

If, alternatively, we choose increments in y, and estimate the corresponding increments in x, then the instructions

$$\boxed{\text{LET } h = \delta x} \quad , \quad \boxed{\text{LET } x = x + h} \quad \text{and} \quad \boxed{\text{LET } y = y + mh}$$

are replaced by

$$\boxed{\text{LET } h = \delta y} \quad , \quad \boxed{\text{LET } y = y + h} \quad \text{and} \quad \boxed{\text{LET } x = x + h/m}$$

The following example illustrates how the detailed working can conveniently be set out.

Example 1

Given that $\dfrac{dy}{dx} = 2x + y$, and that $y = 1$ when $x = 0$, estimate the value of y when $x = 1$, using steps of $\delta x = 0.2$.

x-interval	δx	$\dfrac{dy}{dx}$	δy	x	y
				0	1
0–0.2	0.2	1	0.2		
0.2–0.4	0.2	1.6	0.32	0.2	1.2
0.4–0.6	0.2	2.32	0.464	0.4	1.52
0.6–0.8	0.2	3.184	0.637	0.6	1.984
0.8–1.0	0.2	4.221	0.844	0.8	2.621
				1.0	3.465

So when $x = 1$, y is approximately 3.5.

Example 2 overleaf compares step-by-step solutions with different step widths with the exact solution of a differential equation.

Example 2

Calculate the exact solution of the differential equation

$$\frac{dy}{dx} = \frac{1}{x+2}$$

given that $y = 0$ when $x = 2$. Compare this solution graphically over the interval $2 \leqslant x \leqslant 10$ with the step-by-step solutions obtained with steps of
a) $\delta x = 4$ **b)** $\delta x = 2$ **c)** $\delta x = 1$

If $\dfrac{dy}{dx} = \dfrac{1}{x+2}$ then $y = \ln(x+2) + c$, and since $y = 0$ when $x = 2$, it

follows that $c = -\ln 4$. So $y = \ln\left(\dfrac{x+2}{4}\right)$ is the exact solution of the equation.

For the step-by-step solutions:

a)

x-interval	δx	$\dfrac{dy}{dx}$	δy	x	y
				2	0
2–6	4	0.25	1	6	1
6–10	4	0.125	0.5	10	1.5

b)

x-interval	δx	$\dfrac{dy}{dx}$	δy	x	y
				2	0
2–4	2	0.25	0.5	4	0.5
4–6	2	0.167	0.333	6	0.833
6–8	2	0.125	0.25	8	1.083
8–10	2	0.1	0.2	10	1.283

c)

x-interval	δx	$\dfrac{dy}{dx}$	δy	x	y
				2	0
2–3	1	0.25	0.25	3	0.25
3–4	1	0.2	0.2	4	0.45
4–5	1	0.167	0.167	5	0.617
5–6	1	0.143	0.143	6	0.760
6–7	1	0.125	0.125	7	0.885
7–8	1	0.111	0.111	8	0.996
8–9	1	0.1	0.1	9	1.096
9–10	1	0.091	0.091	10	1.187

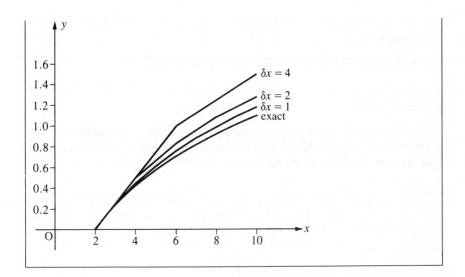

The choice of step length is a matter of judgement. The shorter the step used, the more accurate will be the solution, but balanced against this is the possible increase in time necessary to obtain the solution. However, as the simple flow diagram on p. 112 shows, the procedure can easily be programmed, and if computing facilities are available, the time factor will not be a problem.

Though the procedure in this section is the simplest step-by-step method, it is not the only one, and other more sophisticated methods can improve the accuracy of solutions. Examples of such improved methods can be found in exercise 13.2.

Exercise 13.2

1   Given that $\dfrac{\mathrm{d}y}{\mathrm{d}x} = x + y$, and that $y = 2$ when $x = 0$, estimate the value of y when $x = 1$, using steps of $\delta x = 0.2$.

2   Repeat no. **1**, but start with $y = 0$ when $x = 0$.

3   Repeat no. **1**, but start with $y = -2$ when $x = 0$.

4   In no. **1** estimate the value of x when $y = 4$, using steps of $\delta y = 0.5$.

5   $\dfrac{\mathrm{d}y}{\mathrm{d}x} = xy$, and $y = 1$ when $x = 0$. Estimate the value of y when $x = 2$, using steps of **a)** $\delta x = 0.4$ **b)** $\delta x = 0.2$.

6   $\dfrac{\mathrm{d}y}{\mathrm{d}x} = \dfrac{4}{x^2}$, and $y = 0$ when $x = 2$. Use a step-by-step method to estimate

the value of y when $x = 3$, using intervals of **a)** $\delta x = 0.5$ and **b)** $\delta x = 0.2$. Integrate the differential equation to find the exact value of y when $x = 3$. Find the percentage errors in your answers to **a)** and **b)**.

7 $\dfrac{dy}{dx} = y - 1$, and $y = 4$ when $x = 0$. Use a step-by-step method to estimate the value of x when $y = 10$, using intervals of **a)** $\delta y = 2$ and **b)** $\delta y = 1$. By writing the differential equation in the form $\dfrac{dx}{dy} = \dfrac{1}{y - 1}$ to obtain an exact solution, find the percentage errors in your answers to **a)** and **b)**.

8 A kettle is switched off when the water in it is boiling ($100\,^\circ$C). The rate at which the temperature (θ) drops is given by the differential equation $\dfrac{d\theta}{dt} = \dfrac{20 - \theta}{50}$ (where θ is measured in $^\circ$C and t is measured in minutes.)
a) Estimate the temperature after 10 minutes, using a step of $\delta t = 2$.
b) Estimate the time it takes for the temperature to fall to $60\,^\circ$C, using a step of $\delta\theta = 10$.

9 The acceleration of an object, with speed $v\,\mathrm{m\,s^{-1}}$, is given by

$$\frac{dv}{dt} = 10 - \frac{v^2}{1000} \quad (t \text{ being measured in seconds})$$

a) If the object starts from rest, estimate its speed after 10 seconds. (Use $\delta t = 2$.)
b) Using steps of $\delta v = 10$, estimate how long it takes to reach half of its terminal velocity. (That is the velocity at which its acceleration is zero.)

10 Obtain step-by-step solutions of the differential equation $\dfrac{dy}{dx} = \sqrt{(1 + xy)}$ over the interval $0 \leqslant x \leqslant 1$, with steps of $\delta x = 0.2$, for each of the initial points
a) $(0, 0)$ **b)** $(0, 1)$ **c)** $(0, 2)$ **d)** $(0, 3)$

Graph all these solutions on the same diagram.

11 Given that $\dfrac{dy}{dx} = \dfrac{1}{1 + x^2}$ and that $y = 0$ when $x = 0$,
a) integrate to find the exact value of y when $x = 5$;
b) use the step-by-step method with $\delta x = 1$, to estimate the value of y when $x = 5$.
c) use the step-by-step method with $\delta x = 1$, but this time use the value of $\dfrac{dy}{dx}$ at the mid-point of each interval (i.e. when $x = 0.5, 1.5$ etc.) instead of at its beginning.

12 If $\dfrac{dy}{dx} = x + y$ and $y = 1$ when $x = 0$, verify that with $\delta x = 0.1$, the point $(0.1, 1.1)$ is obtained. Evaluate $\dfrac{dy}{dx}$ at the 'mid-point' $(0.1, 1.1)$ of the interval $0 \leqslant x \leqslant 0.2$, and use this value of $\dfrac{dy}{dx}$ with $\delta x = 0.2$ to estimate the value of y when $x = 0.2$. Repeat this 'mid-point' procedure to estimate y when $x = 0.4, 0.6, 0.8$ and 1.0.

You will need to have access to computing facilities for nos **13–16**.

13 By successively reducing the step width by a factor 2, repeat the step-by-step solution of example 2 on p. 114, until the value of y, when $x = 10$, is correct to 3 significant figures.

14 Reduce the step width in example 1 on p. 113 until you have a value of y (when $x = 1$) which you can be confident is correct to 3 significant figures.

15 Repeat no. **1**, but reduce the step width until you have a solution which is correct to 3 significant figures.

16 Repeat no. **5**, but reduce the step width until you have a solution which is correct to 4 significant figures.

13.3 First order equations with separable variables

A typical first order equation is

$$\frac{dy}{dx} = f(x, y)$$

and if $f(x, y)$ is defined at every point of the x–y plane, it follows that at each point a corresponding gradient and direction are known, just as in a magnetic field plotted by a compass needle.

Example 1

$$\frac{dy}{dx} = \frac{x}{y}$$

The values of the gradient $\dfrac{dy}{dx}$ can be indicated by a table (see over):

		x						
		-3	-2	-1	0	1	2	3
	3	-1	$-\frac{2}{3}$	$-\frac{1}{3}$	0	$\frac{1}{3}$	$\frac{2}{3}$	1
	2	$-\frac{3}{2}$	-1	$-\frac{1}{2}$	0	$\frac{1}{2}$	1	$\frac{3}{2}$
	1	-3	-2	-1	0	1	2	3
y	0							
	-1	3	2	1	0	-1	-2	-3
	-2	$\frac{3}{2}$	1	$\frac{1}{2}$	0	$-\frac{1}{2}$	-1	$-\frac{3}{2}$
	-3	1	$\frac{2}{3}$	$\frac{1}{3}$	0	$-\frac{1}{3}$	$-\frac{2}{3}$	-1

These can then be plotted at the different points of the plane and already we
can begin to see the family of solutions, looking like the lines of force of a
magnetic field.

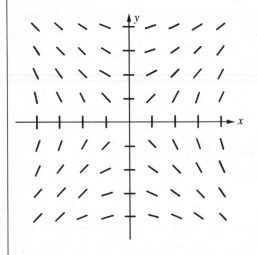

But we can also approach the problem quite differently. For the equation

$\dfrac{dy}{dx} = \dfrac{x}{y}$ can be written as $y\dfrac{dy}{dx} = x$.

$$\Rightarrow \quad \int y\frac{dy}{dx}\,dx = \int x\,dx$$

$$\Rightarrow \quad \tfrac{1}{2}y^2 = \tfrac{1}{2}x^2 + c$$

$$\Rightarrow \quad y^2 = x^2 + 2c$$

This, therefore, is the *general solution* of the given equation; and we can obtain different *particular solutions* by giving different values to the constant c.

Graphically, we see that each particular solution is a certain curve (in fact a rectangular hyperbola, each one having two branches), and the general solution represents a family of such hyperbolas. Finally, the sketch of these hyperbolas should be compared with our original 'compass-needle' figure.

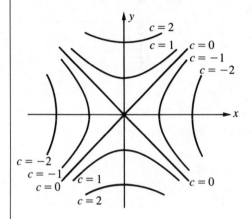

$$c = 0 \quad\Rightarrow\quad y^2 = x^2$$
$$c = 1 \quad\Rightarrow\quad y^2 = x^2 + 2$$
$$c = 2 \quad\Rightarrow\quad y^2 = x^2 + 4$$
$$c = -1 \quad\Rightarrow\quad y^2 = x^2 - 2$$
$$c = -2 \quad\Rightarrow\quad y^2 = x^2 - 4, \quad \text{etc.}$$

In the above solution we could have proceeded a little more rapidly by saying that

$$\frac{\mathrm{d}y}{\mathrm{d}x} = \frac{x}{y} \tag{1}$$

can be written $y\,\mathrm{d}y = x\,\mathrm{d}x$ (1a)

and then as $\displaystyle\int y\,\mathrm{d}y = \int x\,\mathrm{d}x$ (2)

$\Rightarrow$ $\tfrac{1}{2}y^2 = \tfrac{1}{2}x^2 + c$

$\Rightarrow$ $y^2 = x^2 + 2c$

Although the statement (1a) has not (as yet) any proper meaning, we have already justified the argument from (1) to (2), and (1a) clearly serves as a very useful intermediary. This is usually known as *separating the variables*: any first order equation which can similarly be expressed as

$$f(y)\,dy = g(x)\,dx$$

is said to have *separable variables*, and its general solution can be written as

$$\int f(y)\,dy = \int g(x)\,dx$$

Example 2

Solve the differential equation

$$\frac{dy}{dx} = -\frac{y}{x}$$

Here again, the variables are separable and we can write

$$\frac{dy}{y} = -\frac{dx}{x}$$

$$\Rightarrow \quad \ln y = -\ln x + c$$

$$\Rightarrow \quad \ln xy = c$$

$$\Rightarrow \quad xy = a \quad \text{(where } a = e^c)$$

The solution-curves are therefore:

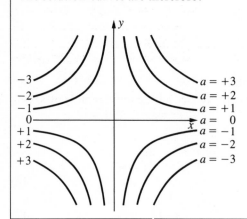

Lastly, we see that there is a relationship between the two examples we have chosen,

$$\frac{dy}{dx} = \frac{x}{y} \quad \text{and} \quad \frac{dy}{dx} = -\frac{y}{x}$$

Since $x/y \times y/x = -1$, it is clear that the directions at any point of the solution-curves of the two equations must be perpendicular, and this is very clearly seen when the figures are superimposed:

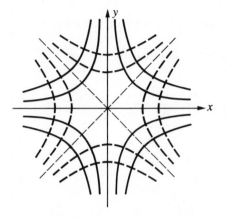

Such pairs of families of curves which are perpendicular everywhere they meet are called *orthogonal families* and are outstandingly important not only as *lines of forces* and their corresponding *equipotentials*, but throughout applied mathematics. There is indeed a particularly simple illustration of them on a map as the *contours* and the *lines of greatest slope* of a hillside, and this last diagram can be regarded as the map of a saddle-point with rainfall streaming from the hillsides of the North-East and South-West into the valleys of the North-West and South-East:

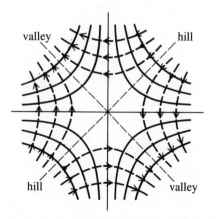

Example 3

A stone falls vertically from rest against an air resistance which is proportional to its velocity v, so that its acceleration is $g - kv$, where g and k are constants.

a) Write down differential equations connecting v with t, the time; and v with x, the distance fallen.

b) Find expressions for v in terms of t and x in terms of t.

c) Find the terminal velocity, as $t \to \infty$.

a) The downwards acceleration $= \dfrac{dv}{dt} = \dfrac{dx}{dt} \times \dfrac{dv}{dx} = v\dfrac{dv}{dx}.$

So $\qquad \dfrac{dv}{dt} = g - kv$ $\qquad\qquad\qquad\qquad\qquad$ (1)

and $\quad v\dfrac{dv}{dx} = g - kv$ $\qquad\qquad\qquad\qquad\qquad$ (2)

b) From (1), we see that

$$\frac{dv}{g - kv} = dt \quad \Rightarrow \quad -\frac{1}{k}\ln(g - kv) = t + c$$

But when $t = 0, v = 0$, so $\quad c = -\dfrac{1}{k}\ln g$

Hence $\qquad -t = \dfrac{1}{k}\ln(g - kv) - \dfrac{1}{k}\ln g$

$$= \frac{1}{k}\ln\frac{g - kv}{g}$$

$\Rightarrow \quad \ln\left(1 - \dfrac{kv}{g}\right) = -kt$

$\Rightarrow \qquad 1 - \dfrac{kv}{g} = e^{-kt}$

$\Rightarrow \qquad\qquad v = \dfrac{g}{k}(1 - e^{-kt})$ $\qquad\qquad\qquad\qquad$ (3)

Hence $\qquad \dfrac{dx}{dt} = \dfrac{g}{k}(1 - e^{-kt})$

$\Rightarrow \qquad\qquad x = \dfrac{g}{k}\left(t + \dfrac{1}{k}e^{-kt}\right) + d$

But when $t = 0$, $x = 0$, so $\quad d = -\dfrac{g}{k^2}$

Hence $\quad x = \dfrac{gt}{k} + \dfrac{g}{k^2}(e^{-kt} - 1)$ (4)

c) Finally, we see from (3) that as $t \to \infty$, $v \to g/k$, which is therefore called its terminal velocity. (And the same value can be found more simply from equation (1) if we suppose that the acceleration eventually tends to zero.)

Example 4

The rate at which a tree takes in its 'food' is proportional to the area of its rooting surface. The food is used
a) to maintain the tree (this requires an amount proportional to its volume) and
b) for it to grow (this requires an amount proportional to the rate of increase of volume).

If the complete tree, including roots, retains a constant shape show that its height h satisfies a differential equation of the form

$$\frac{dh}{dt} + Ah = B$$

and that the tree will tend to a maximum height.

A tree is planted when it is 2 m high and its initial rate of growth in height is 0.3 m per year. Its final height is 100 m. How long does it take to reach a height of 50 m?

This example shows how a so-called 'mathematical model' can be set up to describe a given situation; whether or not the model is valid can then be tested from the accuracy of its predictions.

Suppose that after time t the volume of the tree is V and the area of its rooting surface is S. Then, since the tree retains a constant shape,

$$S = ah^2, \quad V = bh^3 \quad \text{(where } a, b \text{ are constants)}$$

Also the rate at which it takes in its food is kS.

So $\qquad kS = cV + d\dfrac{dV}{dt}$ (where k, c, d are constants)

$\Rightarrow \qquad kah^2 = cbh^3 + d \times 3bh^2\dfrac{dh}{dt}$

$\Rightarrow \qquad ka = bch + 3bd\dfrac{dh}{dt}$

$\Rightarrow \quad \dfrac{dh}{dt} + Ah = B$ (1)

When $t = 0$, $h = 2$ and $\dfrac{dh}{dt} = 0.3$, so

$$0.3 + 2A = B \tag{2}$$

Also, as the tree nears its final height, $h \to 100$ and $\dfrac{dh}{dt} \to 0$; so

$$100A = B \tag{3}$$

From (2) and (3), $A = 0.3/98$ and $B = 30/98$.

Hence $$\dfrac{dh}{dt} + \dfrac{0.3}{98}h = \dfrac{30}{98}$$

$\Rightarrow$ $$98\dfrac{dh}{dt} = 0.3\,(100 - h)$$

$\Rightarrow$ $$\dfrac{98\,dh}{100 - h} = 0.3\,dt$$

$\Rightarrow$ $-98\ln\,(100 - h) = 0.3t + A$

But when $t = 0$, $h = 2$; so

$-98\ln 98 = A$

$\Rightarrow$ $$0.3t = 98\ln\dfrac{98}{100 - h} \tag{4}$$

Putting $h = 50$, $0.3t = 98\ln\frac{98}{50}$

$\Rightarrow$ $$t = \dfrac{98}{0.3}\ln 1.96$$

$$= \dfrac{98 \times 0.6729}{0.3}$$

$$\approx 220$$

So the tree would take just over 220 years to grow to a height of 50 m.

   Moreover, from (4) we can obtain a more general formula for its height h after t years.

For $\ln\dfrac{100 - h}{98} = -\dfrac{0.3t}{98}$

$\Rightarrow$ $$\dfrac{100 - h}{98} = e^{-0.3t/98}$$

$\Rightarrow$ $$h = 100 - 98\,e^{-0.3t/98}$$

$\Rightarrow$ $$h = 100 - 98\,e^{-0.00306t}$$

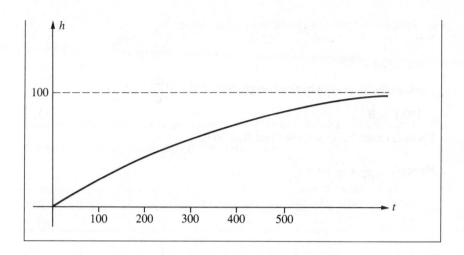

Exercise 13.3

1 For each of the following equations,
  i) illustrate the equation by means of a compass-needle diagram;
  ii) find and illustrate its general solution:

a) $\dfrac{dy}{dx} = y$
b) $\dfrac{dy}{dx} = -xy$
c) $\dfrac{dy}{dx} = 2x$

d) $\dfrac{dy}{dx} = -y^2$
e) $\dfrac{dy}{dx} = \dfrac{y}{x}$
f) $\dfrac{dy}{dx} = -\dfrac{2x}{y}$

2 For each of the following families of curves,
  i) eliminate the parameter c to form a differential equation which represents the family (i.e. is true at every point of every curve);
  ii) illustrate by means of a compass-needle diagram;
  iii) find the differential equation representing the orthogonal family;
  iv) find and illustrate the general solution of this equation:

a) $y = cx$
b) $y = x + c$
c) $y = cx^2$

d) $y^2 = x + c$,
e) $y = ce^{-x}$
f) $y = \dfrac{c}{x^2}$

3 Solve the differential equation

$$\frac{dy}{dx} = \tan y \cot x$$

given that, when $x = \tfrac{1}{2}\pi$, $y = \tfrac{1}{6}\pi$.

(OC)

4 Find the solution of the differential equation

$$x\frac{dy}{dx} = \sin y$$

(for $x > 0$) which passes through the point $(2, \frac{1}{2}\pi)$.   (SMP)

5 The size S of a population at time t satisfies approximately the differential equation $\frac{dS}{dt} = kS$, where k is a constant. Integrate this equation to find S as a function of t.

   The population numbered 32000 in the year 1900 and had increased to 48000 by 1970. Estimate what its size will be (correct to the nearest 1000) in the year 2000.   (SMP)

6 If a ship's mooring rope is wrapped round a rough bollard, with coefficient of friction μ, and is on the point of slipping, it can be shown that its tension T at angle θ from its taut end is such that

$$\frac{dT}{d\theta} = -\mu T$$

Deduce that

$$T = T_0 e^{-\mu\theta}$$

where T_0 is the tension from the ship. If $\mu = \frac{1}{2}$, show that one complete turn will reduce this tension in the ratio $1:e^\pi$ (or approximately $1:23$).

7 If the rate of erosion of mass of a spherical pebble is proportional to its surface area and it is half-eroded after 100 years, how much longer will it last?

8 Show that the acceleration $\frac{dv}{dt}$ of a particle moving in a straight line can be written, in terms of its velocity v and its displacement x from a point of the line, in the form $v\frac{dv}{dx}$.

   At a distance x km from the centre of the Earth the gravitational acceleration in km s^{-2} is given by the formula c/x^2 where $c = 4 \times 10^5$. If a lunar vehicle 10000 km from the centre of the Earth is moving directly away from it at a speed of 10 km s^{-1}, at what distance will its speed be half that value?   (SMP)

9 A rumour is spreading through a large city at a rate which is proportional to the product of the numbers of those who have heard it and of those who have not heard it, so that if x is the fraction who have heard it after time t,

$$\frac{dx}{dt} = kx(1 - x)$$

If initially a fraction c of the population has heard the rumour, deduce that

$$x = \frac{c}{c + (1 - c)e^{-kt}}$$

If 10% have heard the rumour at noon and another 10% by 3.00 p.m., find x as a function of t. What further proportion would you expect to have heard it by 6.00 p.m.?

10   The inhabitants of a country are in two racial groups which do not mix reproductively. Each group's population increases at a rate proportional to the population. At a certain stage 80% are of type A, type B will double in 50 years, but the whole population will take 100 years to double. How long will it take group A to double? How long will it take before the two groups are equal in numbers?   (MEI)

11   Heat is supplied to an object at a constant rate Wa joules s^{-1}; heat is lost to its surroundings at a rate of $bW\theta$ joules s^{-1}, where W is the number of joules needed to raise the temperature of the object by 1 °C, θ °C is the difference in temperature between the object and its surroundings, and a, b are constants. Obtain a differential equation to connect θ with the time t.

     Use this mathematical model to solve the following problem. An electric kettle, open to a room temperature of 20 °C, reaches 70 °C in 3 min after being switched on, and reaches boiling-point in a further 3 min. Find the time taken to reach boiling-point from 20 °C if it is completely heat insulated.

     What deficiencies has the mathematical model in representing the actual physical situation?   (MEI)

12   The food energy taken in by a human body goes partly to increase the mass and partly to fulfil the requirements of the body; these daily requirements are taken to be proportional to the mass M. The rate of increase of mass is proportional to the number of joules available for this. Write down a differential equation connecting M, the time t and the daily intake of joules $f(t)$.

     A man's mass is 100 kg; if he took in no energy he would reduce his weight by 10% in 10 days. How long would it take him to reduce by this amount if, instead, he took in exactly half the number of joules needed to keep his mass constant at 100 kg?   (MEI)

13 A stone is thrown vertically upwards with velocity u, and air resistance is proportional to the square of its speed ($= kv^2$ per unit mass).

Find:

a) the differential equation connecting its velocity v and the height x it has then reached;

b) x in terms of v;

c) its maximum height;

d) its terminal (or limiting) velocity.

14 A flask contains $10\,\text{dm}^3$ of a solution of water and a thoroughly dissolved chemical. Each minute, $2\,\text{dm}^3$ of water flow into the flask and $3\,\text{dm}^3$ of solution flow out of the flask. Write down an expression for the number of kilograms of chemical per cubic decimetre in the solution after t min. If there are x kg of chemical in the flask at this time, show that

$$\frac{\mathrm{d}x}{\mathrm{d}t} = \frac{-3x}{10 - t}$$

Find the general solution of this equation. If there were 1 kg of chemical in the flask at time $t = 0$, find the mass of chemical in the flask after 5 min.

(AEB)

15 The rate of increase of thickness of ice on a pond is inversely proportional to the thickness of ice already present. It is known that, when the thickness of the ice is x cm and the temperature of the air above the ice is $-\alpha\,°\text{C}$, the rate is $(\alpha/14400\,x)\,\text{cm}\,\text{s}^{-1}$. Form an appropriate differential equation, and hence show that, if the air temperature is $-10\,°\text{C}$, the time taken for the thickness to increase from 5 cm to 6 cm is a little more than 2 h.

(C)

* 13.4 Linear equations of first order

A particularly important class of differential equations consists of those which are *linear*, where y and its derivatives occur only in linear combinations even though their coefficients may be functions of x. So a linear equation of the first order can be written as

$$f(x)\frac{\mathrm{d}y}{\mathrm{d}x} + g(x)y = h(x) \tag{1}$$

Example 1

$$x^2\frac{\mathrm{d}y}{\mathrm{d}x} + 2xy = 1$$

We quickly notice that this is a particularly convenient equation since it can be written as

$$\frac{d}{dx}(x^2y) = 1$$

and, as the left-hand side is an exact derivative, the original equation is also called *exact*.

Proceeding with the solution, we see that

$$x^2y = x + A$$

$$\Rightarrow \quad y = \frac{1}{x} + \frac{A}{x^2}$$

If we now return to the more general linear equation (1) we see that it can be written as

$$\frac{dy}{dx} + \frac{g(x)}{f(x)}y = \frac{h(x)}{f(x)}$$

or $\quad \dfrac{dy}{dx} + Py = Q$ $\hspace{3cm}$ (2)

where P and Q are functions of x.

This is usually taken as the standard form of the first order linear equation. Unfortunately (unlike example 1) it is not usually exact, so we now investigate whether there is any way in which it can be made into an exact equation.

If we multiply each side of (2) by another function, denoted by R, it becomes

$$R\frac{dy}{dx} + PRy = QR$$

and we see that the left-hand side would be our exact derivative if only it could be written as $\dfrac{d}{dx}(Ry)$.

But $\quad \dfrac{d}{dx}(Ry) = R\dfrac{dy}{dx} + \dfrac{dR}{dx}y$

So $R\dfrac{dy}{dx} + PRy$ is an exact derivative provided that

$$\frac{dR}{dx} = PR$$

$$\Leftrightarrow \quad \frac{dR}{R} = P\,dx$$

$$\Leftrightarrow \quad \ln R = \int P \, dx$$

$$\Leftrightarrow \quad R = e^{\int P \, dx}$$

As this function enables us to transform (2) into an exact equation, $e^{\int P \, dx}$ is usually known as an *integrating factor*. By this means the equation becomes

$$e^{\int P \, dx}\frac{dy}{dx} + P e^{\int P \, dx} y = Q e^{\int P \, dx}$$

$$\Rightarrow \quad \frac{d}{dx}(e^{\int P \, dx}y) = Q e^{\int P \, dx}$$

and the problem is reduced to one of ordinary integration.

Two further examples will make this clear:

Example 2

$$\frac{dy}{dx} = x + y$$

Unlike example 1, this is not an exact equation. But it can be written as

$$\frac{dy}{dx} - y = x$$

---

so that a possible integrating factor is

$$e^{\int P\,dx} = e^{\int -1\,dx} = e^{-x}$$

Hence $\quad e^{-x}\dfrac{dy}{dx} - e^{-x}y = xe^{-x}$

$\Rightarrow \qquad\qquad \dfrac{d}{dx}(e^{-x}y) = xe^{-x}$

$\Rightarrow \qquad\qquad e^{-x}y = \displaystyle\int xe^{-x}\,dx$

$$\qquad\qquad\qquad = -xe^{-x} - e^{-x} + A$$

$\Rightarrow \qquad\qquad y = Ae^x - x - 1$

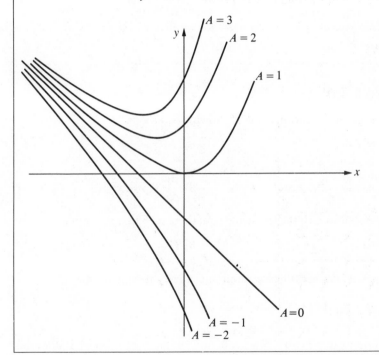

Example 3

$$\sin x\,\dfrac{dy}{dx} + 2y\cos x = 1$$

In standard form, this becomes

$$\frac{dy}{dx} + (2 \cot x)y = \operatorname{cosec} x$$

and the necessary integrating factor is

$$e^{\int 2 \cot x \, dx} = e^{2 \ln \sin x} = e^{\ln \sin^2 x} = \sin^2 x$$

So the equation becomes

$$(\sin^2 x)\frac{dy}{dx} + (2 \sin x \cos x)y = \sin x$$

$$\Rightarrow \qquad \frac{d}{dx}(y \sin^2 x) = \sin x$$

$$\Rightarrow \qquad y \sin^2 x = -\cos x + A$$

$$\Rightarrow \qquad y = \frac{A - \cos x}{\sin^2 x}$$

Exercise 13.4

1 Find the general solutions of the following equations:

a) $\dfrac{dy}{dx} = x - y$

b) $x\dfrac{dy}{dx} = x + y$

c) $x\dfrac{dy}{dx} - 2y = (x - 2)e^x$

d) $\dfrac{dy}{dx} + y = x^2$

2 Find the solutions of the following equations such that $y = 1$ when $x = 0$:

a) $(1 - x^2)\dfrac{dy}{dx} - xy = x$

b) $(1 + x)\dfrac{dy}{dx} = x(y + 1)$

3 The current i in a certain electric circuit satisfies the equation

$$\frac{di}{dt} + i = \sin t.$$

If it is known that $i = 0$ when $t = 0$, find i in terms of t. To what does this approximate as $t \rightarrow \infty$?

4 A radioactive substance P decays and changes (without loss of mass) into a substance Q, which itself similarly changes into a third substance R. R suffers no further change. The masses of P, Q, and R present at time t are given by p, q, and r respectively. The rates of change are such that

$$\frac{dp}{dt} = -2p \quad \text{and} \quad \frac{dr}{dt} = q \qquad\qquad (1)$$

Show that

$$\frac{dq}{dt} = 2p - q$$

Initially (at time $t = 0$) there is 1 g of substance P and none of substance Q. Integrate equation (1) and hence show that q satisfies the differential equation

$$\frac{dq}{dt} + q = 2e^{-2t} \qquad (2)$$

Show that (2) may be written in the form

$$\frac{d}{dt}(q e^t) = 2e^{-t}$$

and integrate to find q as a function of t. Hence prove that, at any subsequent time, there is never more than $\frac{1}{2}$ g of Q present.   (SMP)

5

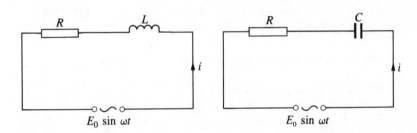

$E_0 \sin \omega t$ $E_0 \sin \omega t$

In each of the above circuits an alternating potential $E_0 \sin \omega t$ is applied and the current i satisfies the following equations (respectively), where R, L, C are constants, known as resistance, inductance, and capacitance.

a) $L\dfrac{di}{dt} + Ri = E_0 \sin \omega t$ **b)** $R\dfrac{di}{dt} + \dfrac{1}{C}i = \omega E_0 \cos \omega t$

In each case, find:
  i) the appropriate integrating factor;
  ii) the general solution for i, as a function of t;
  iii) the ultimate 'steady' alternating current (when $t \to \infty$ and any 'transient' terms have faded away).
    [It may be assumed—or proved by repeated integration by parts—that

$$\int e^{kt} \sin \omega t \, dt = \frac{e^{kt}}{k^2 + \omega^2}(k \sin \omega t + \omega \cos \omega t)$$

$$\text{and} \int e^{kt} \cos \omega t \, dt = \frac{e^{kt}}{k^2 + \omega^2}(\omega \sin \omega t - k \cos \omega t).]$$

*13.5 Linear equations with constant coefficients

A particularly important (and fortunately simple) linear differential equation is one in which the coefficients of y and its derivatives are all constants. We shall consider the second-order equation

$$a\frac{d^2y}{dx^2} + b\frac{dy}{dx} + cy = f(x) \tag{1}$$

where a, b, c are constants.

The solution of such equations can be divided into two parts:

a) finding a *general* solution of the *associated equation*,

$$a\frac{d^2y}{dx^2} + b\frac{dy}{dx} + cy = 0 \tag{2}$$

in which $f(x)$ has been replaced by zero. This solution is known as the *complementary function (c.f.)*.

b) finding a *particular* solution of the original equation (1). This is called a *particular integral (p.i.)*.

These processes will be clear when we take particular examples, but the general method can be summarised as follows:

a) The complementary function
To find a general solution of the associated equation

$$a\frac{d^2y}{dx^2} + b\frac{dy}{dx} + cy = 0$$

we put $y = e^{mx}$ and observe that this is a solution provided that

$$am^2 e^{mx} + bm e^{mx} + c e^{mx} \equiv 0 \quad \Leftrightarrow \quad am^2 + bm + c = 0$$

So if this equation† has roots α and β, it follows that $e^{\alpha x}$ and $e^{\beta x}$ are both solutions of equation (2).

Moreover, if $\quad u = A e^{\alpha x} + B e^{\beta x}\quad$ (where A, B are any constants)

---

† The very important case when $am^2 + bm + c = 0$ does not have any real roots will be considered in examples 4 and 5. Meanwhile, we see that if it has identical roots α, α, then

$$u = A e^{\alpha x} + B e^{\alpha x} = (A + B) e^{\alpha x}$$

which effectively contains only one arbitrary constant $A + B$, rather than two independent constants A and B. But in this case it can quite easily be shown that $x e^{\alpha x}$ is another solution, so that

$$u = A e^{\alpha x} + Bx e^{\alpha x}$$

is also a solution, thus restoring two independent constants.

then $\dfrac{\mathrm{d}u}{\mathrm{d}x} = A\alpha\,\mathrm{e}^{\alpha x} + B\beta\,\mathrm{e}^{\beta x}$

and $\dfrac{\mathrm{d}^2 u}{\mathrm{d}x^2} = A\alpha^2\,\mathrm{e}^{\alpha x} + B\beta^2\,\mathrm{e}^{\beta x}$

Hence $a\dfrac{\mathrm{d}^2 u}{\mathrm{d}x^2} + b\dfrac{\mathrm{d}u}{\mathrm{d}x} + cu$

$$= a(A\alpha^2\,\mathrm{e}^{\alpha x} + B\beta^2\,\mathrm{e}^{\beta x}) + b(A\alpha\,\mathrm{e}^{\alpha x} + B\beta\,\mathrm{e}^{\beta x}) + c(A\,\mathrm{e}^{\alpha x} + B\,\mathrm{e}^{\beta x})$$
$$= A(a\alpha^2 + b\alpha + c)\,\mathrm{e}^{\alpha x} + B(a\beta^2 + b\beta + c)\,\mathrm{e}^{\beta x}$$
$$= 0 + 0 = 0$$

So $u = A\,\mathrm{e}^{\alpha x} + B\,\mathrm{e}^{\beta x}$ is also a solution of (2). This is therefore the complementary function.

b) A particular integral

The discovery of a *particular* solution of the original equation is frequently a matter of trial and error, and we shall restrict ourselves to the simplest cases. But once we have found such a p.i. we can immediately combine it with the c.f. to provide a *general* solution of the original equation.

For suppose that $v(x)$ is a particular integral.

Then $a\dfrac{\mathrm{d}^2 v}{\mathrm{d}x^2} + b\dfrac{\mathrm{d}v}{\mathrm{d}x} + cv = \mathrm{f}(x)$

But the complementary function $u(x)$ was such that

$$a\dfrac{\mathrm{d}^2 u}{\mathrm{d}x^2} + b\dfrac{\mathrm{d}u}{\mathrm{d}x} + cu = 0$$

Hence $a\dfrac{\mathrm{d}^2}{\mathrm{d}x^2}(u + v) + b\dfrac{\mathrm{d}}{\mathrm{d}x}(u + v) + c(u + v) = \mathrm{f}(x)$

and we see that $u(x) + v(x)$ is a general solution of the original equation. So for such a linear equation

general solution = complementary function + particular integral

Example 1

Find a general solution of the differential equation

$$\dfrac{\mathrm{d}^2 y}{\mathrm{d}x^2} + \dfrac{\mathrm{d}y}{\mathrm{d}x} - 6y = x + 1$$

a) Complementary function

The associated equation is

$$\frac{d^2y}{dx^2} + \frac{dy}{dx} - 6y = 0$$

and $y = e^{mx}$ will be a solution provided that

$$m^2 e^{mx} + m e^{mx} - 6 e^{mx} \equiv 0$$
$$\Leftrightarrow \qquad m^2 + m - 6 = 0$$
$$\Leftrightarrow \qquad (m - 2)(m + 3) = 0$$
$$\Leftrightarrow \qquad m = 2 \quad \text{or} \quad -3$$

So the c.f. is $y = A e^{2x} + B e^{-3x}$

b) Particular integral

If we seek for y as a function of x such that

$$\frac{d^2y}{dx^2} + \frac{dy}{dx} - 6y \equiv x + 1$$

we naturally first look amongst simple linear functions of the form

$$y = ax + b$$

Then $\dfrac{dy}{dx} = a$

and $\dfrac{d^2y}{dx^2} = 0$

So $\qquad \dfrac{d^2y}{dx^2} + \dfrac{dy}{dx} - 6y \equiv x + 1$

$$\Leftrightarrow \qquad 0 + a - 6(ax + b) \equiv x + 1$$
$$\Leftrightarrow \qquad a = -\tfrac{1}{6} \quad \text{and} \quad a - 6b = 1 \quad \Rightarrow \quad b = -\tfrac{7}{36}$$

So a p.i. is $\quad y = -\tfrac{1}{6}x - \tfrac{7}{36}$

and a general solution of the original equation is

$$y = A e^{2x} + B e^{-3x} - \tfrac{1}{6}x - \tfrac{7}{36}$$

We can clearly also apply this method to similar first-order equations. For instance, the equation of example 2 (p. 130) can be reconsidered:

Example 2

$$\frac{dy}{dx} - y = x$$

Here we can immediately obtain the necessary complementary function and particular integral:

a) Complementary function
Put $y = e^{mx}$ in the associated equation

$$\frac{dy}{dx} - y = 0$$

Then $m e^{mx} - e^{mx} \equiv 0$
$\Rightarrow \qquad\qquad m = 1$
so that the c.f. is $y = A e^x$

b) Particular integral
First we try $y = ax + b$. This would be a p.i. provided that

$$a - (ax + b) \equiv x$$
$$\Rightarrow \quad -ax + (a - b) \equiv x$$
$$\Rightarrow \quad a = -1, \quad a - b = 0$$
$$\Rightarrow \qquad\qquad a = b = -1$$

So a p.i. is $\qquad\qquad y = -x - 1$

Hence the general solution of the given equation is

$$y = A e^x - x - 1$$

Example 3

If an alternating potential is applied to a certain electric circuit then (using suitable units) the current x after time t is such that

$$\frac{d^2x}{dt^2} + 4\frac{dx}{dt} + 3x = \sin t$$

Find the general solution.

a) Complementary function
The associated equation is

$$\frac{d^2x}{dt^2} + 4\frac{dx}{dt} + 3x = 0$$

and $x = e^{mt}$ is a solution provided that

$$m^2 e^{mt} + 4m e^{mt} + 3 e^{mt} \equiv 0$$
$$\Leftrightarrow \qquad\quad m^2 + 4m + 3 = 0$$
$$\Leftrightarrow \qquad\qquad m = -1 \quad \text{or} \quad -3$$

So the c.f. is $\qquad\qquad x = A e^{-t} + B e^{-3t}$

b) Particular integral

Perhaps we would first try to find a function of the type

$$x = a \sin t$$

which satisfies the original equation for all values of t.

But this would entail

$$-a \sin t + 4a \cos t + 3a \sin t \equiv \sin t$$
$$\Rightarrow \qquad 2a \sin t + 4a \cos t \equiv \sin t$$

This identity is clearly impossible, so we now try

$$x = a \sin t + b \cos t$$

In this case $\quad \dfrac{\mathrm{d}x}{\mathrm{d}t} = a \cos t - b \sin t$

and $\qquad \dfrac{\mathrm{d}^2 x}{\mathrm{d}t^2} = -a \sin t - b \cos t$

So $\quad (-a \sin t - b \cos t) + 4(a \cos t - b \sin t) + 3(a \sin t + b \cos t) \equiv \sin t$
$\Rightarrow \quad (2a - 4b) \sin t + (4a + 2b) \cos t \equiv \sin t$
$\Rightarrow \quad \left. \begin{array}{r} 2a - 4b = 1 \\ 4a + 2b = 0 \end{array} \right\} \quad \Rightarrow \quad a = \tfrac{1}{10}, \quad b = -\tfrac{1}{5}$

and we see that

$$x = \tfrac{1}{10} \sin t - \tfrac{1}{5} \cos t \quad \text{is a p.i.}$$

Hence a general solution is

$$x = A \mathrm{e}^{-t} + B \mathrm{e}^{-3t} + \tfrac{1}{10}(\sin t - 2 \cos t)$$

In a particular case, the constants A and B would usually be determined by the initial conditions, e.g. from one's knowledge of the current x at $t = 0$ and how it was changing. But it is noticeable that both terms of the complementary function $A \mathrm{e}^{-t} + B \mathrm{e}^{-2t}$ tend to zero as $t \to \infty$; and this happens very rapidly, so that in every case the steady-state current after the initial surge (or other transient effect) is

$$x \approx \tfrac{1}{10}(\sin t - 2 \cos t)$$

Example 4

If $x = 4$ and $\dfrac{dx}{dt} = 6$ when $t = 0$, solve the equation

$$\frac{d^2x}{dt^2} + 9x = 0$$

which represents the vibrations of a certain spring.

Putting $x = e^{mt}$, we obtain

$$
\begin{aligned}
& m^2 e^{mt} + 9 e^{mt} = 0 \\
\Rightarrow\quad & m^2 + 9 = 0 \\
\Rightarrow\quad & m = \pm 3\mathrm{i}
\end{aligned}
$$

So the general solution is

$$x = A e^{3\mathrm{i}t} + B e^{-3\mathrm{i}t}$$

At first this seems hopelessly dependent on $\sqrt{-1}$. However, recalling (from section 12.5) that $e^{\mathrm{i}\theta} = \operatorname{cis}\theta$, we quickly see that this can be written as

$$
\begin{aligned}
x &= A(\cos 3t + \mathrm{i}\sin 3t) + B(\cos 3t - \mathrm{i}\sin 3t) \\
&= (A + B)\cos 3t + \mathrm{i}(A - B)\sin 3t
\end{aligned}
$$

Furthermore, A and B can perfectly well be complex numbers and we can write

$$A = \tfrac{1}{2}(a - \mathrm{i}b), \quad B = \tfrac{1}{2}(a + \mathrm{i}b)$$

where a and b are real;

so that $A + B = a, \quad \mathrm{i}(A - B) = b$

and $x = a\cos 3t + b\sin 3t$

Hence $\dfrac{dx}{dt} = -3a\sin 3t + 3b\cos 3t$

and when $t = 0$, $x = a = 4$

$$\frac{dx}{dt} = 3b = 6$$

Hence $a = 4$, $b = 2$, and the required solution which describes the vibrations of the given spring is

$$x = 4\cos 3t + 2\sin 3t$$

This is known as *simple harmonic motion*.

Example 5

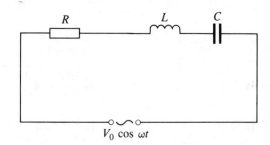

$V_0 \cos \omega t$

If an alternating potential $V_0 \cos \omega t$ is applied to a circuit containing resistance R, inductance L and capacitance C, all in series, the differential equation satisfied by Q, the quantity of electricity which has passed at time t, can be shown to be

$$L\ddot{Q} + R\dot{Q} + Q/C = V_0 \cos \omega t$$

Suppose now that for a particular circuit, the differential equation arising is

$$\ddot{Q} + 6\dot{Q} + 25Q = 60 \cos 5t$$

and that we wish to find the current $I(=\dot{Q})$ flowing at time t.

Seeking first the complementary function, we obtain the quadratic equation $k^2 + 6k + 25 = 0$, so that $k = -3 \pm 4i$.

Thus the c.f. is
$$\begin{aligned}
Q &= A\,e^{(-3+4i)t} + B\,e^{(-3-4i)t} \\
&= e^{-3t}(A\,e^{4it} + B\,e^{-4it}) \\
&= e^{-3t}\{A(\cos 4t + i \sin 4t) + B(\cos 4t - i \sin 4t)\} \\
&= e^{-3t}(a\cos 4t + b\sin 4t)
\end{aligned}$$

where $a = A + B$ and $b = i(A - B)$ are a fresh pair of arbitrary constants to replace the pair A, B.

The particular integral will be of the form $Q = L\cos 5t + M\sin 5t$. Direct substitution back into the differential equation enables us to find the values of L, M, namely $L = 0$, $M = 2$, so the complete solution of the differential equation is

$$Q = e^{-3t}(a\cos 4t + b\sin 4t) + 2\sin 5t$$

Hence the current $I = \dot{Q} = e^{-3t}(E\cos 4t + F\sin 4t) + 10\cos 5t$, where E, F are yet another replacement pair of arbitrary constants. Their values could be found from two pieces of information, such as the initial charge and current. However, since the factor e^{-3t} diminishes rapidly with time, the complementary function (known as the *transient*) represents a heavily damped oscillation which soon ceases to be of any consequence, and the term $10\cos 5t$ gives the value of the permanent current resulting from the application of the given alternating input.

Exercise 13.5

1 Find the general solutions of:

a) $\dfrac{d^2y}{dx^2} - 4\dfrac{dy}{dx} + 3y = 0$

b) $\dfrac{d^2y}{dx^2} - 4\dfrac{dy}{dx} - 5y = 0$

c) $\dfrac{d^2y}{dx^2} - 4y = 0$

d) $\dfrac{d^2y}{dx^2} - 4\dfrac{dy}{dx} = 0$

e) $\dfrac{d^2x}{dt^2} + 4x = 0$

f) $\dfrac{d^2x}{dt^2} - 2\dfrac{dx}{dt} + 2x = 0$

g) $\dfrac{d^2x}{dt^2} + 2\dfrac{dx}{dt} + 5x = 0$

h) $\dfrac{d^2x}{dt^2} + 4\dfrac{dx}{dt} + 4x = 0$

2 Verify that the given expressions are particular integrals of the corresponding equations, and hence find their general solutions:

a) $\dfrac{d^2y}{dx^2} - 3\dfrac{dy}{dx} + 2y = 6, \quad y = 3$

b) $\dfrac{d^2y}{dx^2} - 3\dfrac{dy}{dx} + 2y = x^2, \quad y = \tfrac{1}{4}(2x^2 + 6x + 7)$

c) $\dfrac{d^2x}{dt^2} + 3\dfrac{dx}{dt} + 2x = e^t, \quad x = \tfrac{1}{6}e^t$

d) $\dfrac{d^2x}{dt^2} + 2\dfrac{dx}{dt} + 5x = \sin t, \quad x = \tfrac{1}{10}(2\sin t - \cos t)$

3 For each of the following equations find the complementary function, a particular integral and the general solution:

a) $\dfrac{d^2y}{dx^2} - 2\dfrac{dy}{dx} - 3y = x$

b) $\dfrac{d^2y}{dx^2} + 2\dfrac{dy}{dx} - 3y = e^{-x}$

c) $\dfrac{d^2x}{dt^2} - 3\dfrac{dx}{dt} + 2x = \sin t$

d) $\dfrac{d^2\theta}{dt^2} + 2\dfrac{d\theta}{dt} + 2\theta = \cos t$

4 Find the solutions of the following equations which satisfy the given initial conditions:

a) $\dfrac{d^2y}{dx^2} - 3\dfrac{dy}{dx} + 2y = 0;$ $y = 3$ and $\dfrac{dy}{dx} = 4$ when $x = 0$

b) $\dfrac{d^2x}{dt^2} = 4x;$ $x = 0$ and $\dfrac{dx}{dt} = 4$ when $t = 0$

c) $\dfrac{d^2\theta}{dt^2} + \dfrac{d\theta}{dt} - 2\theta = 4;$ $\theta = 0$ and $\dfrac{d\theta}{dt} = -1$ when $t = 0$

5 The needle of a certain heavily damped instrument makes an angle θ with its zero position after time t, and it can be shown that

$$\frac{d^2\theta}{dt^2} + 5\frac{d\theta}{dt} + 6\theta = 0$$

Find an expression for θ in terms of t if initially (i.e. when $t = 0$) the needle was at the zero position but moving at 2 rad s^{-1}. Hence find the maximum deflection of the needle and sketch the graph of θ against t.

6

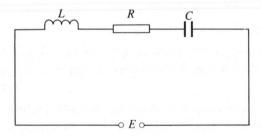

If a potential E is applied to a simple circuit with inductance L, resistance R and capacitance C, it can be shown that the charge q (coulomb) on the capacitor after time t must satisfy the equation

$$L\frac{d^2q}{dt^2} + R\frac{dq}{dt} + \frac{q}{C} = E$$

Find the general solution of this equation for the circuit in which $L = 5 \times 10^{-3}$ (henry), $R = 6$ (ohm), $C = 10^{-3}$ (farad) when the applied potential is 100 (volt), and show that the charge on the capacitor very soon becomes $\frac{1}{10}$ C.

7 In exercise 13.4 no. **5a)** and **b)**, i) find the corresponding complementary functions and appropriate particular integrals.
    Hence proceed to parts ii) and iii).

Miscellaneous problems

1   In a purification system, the concentration of impurity, x milligrams per litre, t seconds after the system is switched on, satisfies the differential equation

$$\frac{\mathrm{d}x}{\mathrm{d}t} = -\frac{x}{500}$$

Given that the initial concentration of impurity is 20 milligrams per litre, calculate, to the nearest second, the time taken to reduce the concentration to 4 milligrams per litre.   (c)

2   A submarine of mass m, when moving in a horizontal line Ox with speed v, experiences a resistance kv^2, where k is a constant. The power P of the engines is constant and is just sufficient to maintain a speed u. Show that

$$mv^2 \frac{\mathrm{d}v}{\mathrm{d}x} = \frac{P}{u^3}(u^3 - v^3)$$

Show also that the distance moved while the submarine accelerates from a speed $\frac{1}{4}u$ to a speed $\frac{3}{4}u$ is

$$\frac{mu^3}{3P}\ln\frac{63}{37}$$

Find the time taken for the speed of the submarine to reduce from $\frac{3}{4}u$ to $\frac{1}{4}u$ when moving against the resistance with no power supplied by the engines.   (JMB)

3   An elementary particle of mass m and charge e moves under the action of uniform and constant electric and magnetic fields $E\mathbf{i}$ and $B\mathbf{i}$ respectively. At time t, the position vector of the particle is $\mathbf{r} = x\mathbf{i} + y\mathbf{j} + z\mathbf{k}$ with the usual notation, where (as a consequence of Newton's second law and the laws of electromagnetism)

$$m\ddot{x} = eE, \quad m\ddot{y} = eB\dot{z}, \quad m\ddot{z} = -eB\dot{y}$$

Initially the particle is at the origin with velocity $W\mathbf{k}$.
a) Find x in terms of t.
b) By integrating the equations involving y and z, show that, for all t,

$$\dot{y} = \Omega z, \quad \dot{z} = -\Omega y + W$$

where $\Omega = eB/m$. Deduce that $\ddot{z} = -\Omega^2 z$, and hence find y and z in terms of t.
c) Give a brief description of the path of the particle.   (MEI)

4   A cylindrical container of radius a is mounted with its axis vertical and filled to a depth h with liquid. The liquid can run out through a hole in the

base of radius b $(b \ll a)$ under the influence of a vertical gravitational field g.

Assuming ideal conditions and no resistance, use conservation of mass and conservation of energy to show that the rate of outflow is proportional to the square root of the depth of the liquid remaining.

Determine the time it takes for the container to empty.   (MEI)

5 A ball of mass m is thrown into the air with initial velocity **u**, and during its flight it experiences a resistance from the air given by the vector $-mc\mathbf{v}$, where **v** is the velocity of the ball at that instant. Write down a differential equation for the velocity as a function of the time, and verify that this equation and the initial conditions are both satisfied if

$$\mathbf{v} = \frac{1 - e^{-ct}}{c}\mathbf{g} + e^{-ct}\mathbf{u}$$

where **g** is the vector acceleration due to gravity.

Find an expression for **r**, the displacement of the ball from its initial position at time t; and show that, if c is small, then this is approximately equal to

$$\mathbf{r^*} - c(\tfrac{1}{6}t^3\mathbf{g} + \tfrac{1}{2}t^2\mathbf{u})$$

where **r*** is the position that it would occupy in the absence of air resistance.   (MEI)

6 A particle of unit mass is attracted to a point O with a force $-9\mathbf{r}$, where **r** is the position vector of the particle relative to O. Initially the particle is at a point with position vector 16**i** and moving with velocity 24**j**, where **i** and **j** are fixed unit vectors in perpendicular directions. Verify that the subsequent motion is described by the equation

$$\mathbf{r} = 16\mathbf{i}\cos 3t + 8\mathbf{j}\sin 3t$$

and interpret this equation geometrically.

If there were also a resistance to motion of magnitude ten times the speed, obtain a differential equation for the motion and show that it has a solution in the form

$$\mathbf{r} = \mathbf{A}e^{-t} + \mathbf{B}e^{-9t}$$

With the same initial conditions as before, find the values of **A** and **B**, and investigate the nature of the motion for $t > 0$.   (SMP)

7 A particle of mass m is projected with velocity V at an angle α to the horizontal. The air resistance is mk times the speed of the particle, where k is a constant, and opposes the direction of motion at all times. Prove that the equations of motion of the particle can be put in the form

$$\ddot{x} + k\dot{x} = 0$$
$$\ddot{y} + k\dot{y} + g = 0$$

where x and y are respectively the horizontal and vertical distances from the point of projection. Show that

$$x = \frac{V}{k}\cos\alpha(1 - e^{-kt})$$

and find a similar expression for y.

Sketch the path of the projectile.   (MEI)

8   In a population of N rats, n initially have a disease, the rest have not yet caught it. If a rat recovers from the disease it becomes immune. At a later time, t, the population is therefore composed of $I(t)$ rats ill with the disease, $D(t)$ dead rats, $R(t)$ rats who have already recovered and $S(t)$ who are still susceptible.

It is observed that at any stage

$$\frac{dR}{dt} = aI, \quad \frac{dD}{dt} = bI, \quad \frac{dS}{dt} = -cSI$$

where a, b, and c are constants, and $N - n > (a + b)/c$.

Prove that I increases until S falls to $(a + b)/c$, and then decreases. Prove also that the maximum value of I is

$$N - \frac{a + b}{c}\ln\frac{ec(N - n)}{a + b}$$ (OS)

9   Two tribes live on a desert island. Each tribe breeds only among itself and, in isolation, the population of each would increase at the same constant rate. But each tribe is eaten (by the other) at a rate equal to a constant, b, times the population of the other tribe. At a certain time the total population of both tribes together is N_0. Investigate how the total population varies with time subsequently.   (OS)

10   The warm-up temperature $T(°C)$ of an electric iron is described as a function of time t (min) by the differential equation,

$$\frac{d^2T}{dt^2} + 2\frac{dT}{dt} + 5T = 1000$$

where initially the temperature is $20°C$ and increasing at the rate of $100°C$ per minute.

State the equilibrium temperature.

Find the temperature as a function of time and sketch the relationship.

Determine the time at which the equilibrium temperature is first reached, the maximum overshoot temperature and the time at which this maximum occurs.   (MEI)

14 Further mechanics

14.1 Elasticity: Hooke's law and elastic energy

Hooke's law

It was first shown by Robert Hooke (1635–1703) that if a spring is extended from its natural length but remains within its elastic limits, the tension T is proportional to the extension x:

$$T = kx$$

where k is a constant of the particular spring, known as its *stiffness*.†

(If, in the case of a spring, x is negative, then so is T: the spring is under compression and is exerting a negative tension, or *thrust*.)

Elastic energy

When a string or a spring is extended, we see that T and x are in opposite directions, so the work done *by the tension T* during this extension is

$$\int_0^x -kx\,\mathrm{d}x = -\left[\tfrac{1}{2}kx^2\right]_0^x = -\tfrac{1}{2}kx^2$$

whose magnitude can also be found from the force–distance graph:

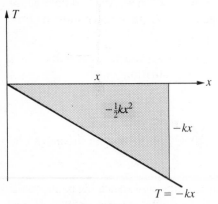

$$T = -kx$$

---

† If the natural length of the spring (or string) is l, then k is sometimes written as λ/l, where λ is called its *modulus of elasticity*. Hence $T = \lambda x/l$, and the corresponding elastic energy is $\lambda x^2/2l$.

We have already seen (section 9.6) that when a mass m is at a height h above a given base-level, its weight mg would do an amount of work mgh if it were to return to its original level. This, therefore, was called its *gravitational potential energy* (referred to the given base-level) when at height h. Similarly we now note that if our loaded elastic spring with extension x were to return to its natural length, the tension would clearly do a positive amount of work $\frac{1}{2}kx^2$. This, therefore, is called the *elastic potential energy* of the spring (or, more briefly, its *elastic energy*), being its stored capacity for doing work.

Example

An elastic spring of unstretched length 1 m hangs from the ceiling of a room whose height is 3 m. When a 1 kg mass is attached, the spring extends to a length 2 m, and it is then pulled down further, so that the mass touches the floor. If it is then released, find its velocity at heights 1 m, 2 m above the floor, and calculate the kinetic energy, gravitational potential energy, and elastic potential energy in all three positions.

Let the stiffness of the spring be k. Then, taking $g = 10\,\mathrm{m\,s^{-2}}$, the weight of the mass will be 10 N

and $\quad k \times 1 = 10 \quad \Rightarrow \quad k = 10$

If we measure depth x from the unstretched position, we can now investigate:

a) The displacement from $x = 2$ to $x = 1$
Work done by weight $\quad = -10 \times 1 = -10\,\mathrm{J}$
Also tension in spring $\quad = 10x$

So work done by spring $= \displaystyle\int_{2}^{1} -10x\,\mathrm{d}x$

$$= \left[5x^2 \right]_{1}^{2}$$

$$= 15\,\mathrm{J}$$

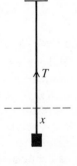

Furthermore, if velocity is v when $x = 1$, gain in k.e. $= \frac{1}{2} \times 1 \times v^2$

So $\quad -10 + 15 = \frac{1}{2}v^2$
$\Rightarrow \qquad\qquad v^2 = 10$
$\Rightarrow \qquad\qquad v = \sqrt{10} = 3.2\,\mathrm{m\,s^{-1}}$

b) The displacement from $x = 2$ to $x = 0$
Work done by weight $= -10 \times 2 = -20\,\mathrm{J}$

Work done by spring $= \int_{2}^{0} -10x \, dx$

$$= \left[5x^2 \right]_{0}^{2}$$

$$= 20 \, J$$

So if velocity is V when $x = 1$, gain in k.e. $= \frac{1}{2} \times 1 \times V^2 = \frac{1}{2}V^2$

So $\quad -20 + 20 = \frac{1}{2}V^2$

$\Rightarrow \qquad V = 0$

Finally, if we measure gravitational energy from floor level and elastic energy from the unstretched position (so that elastic p.e. $= \frac{1}{2} \times kx^2 = 5x^2$), we can express the three forms of energy in the following table:

depth from ceiling/m	gravitational energy/J	elastic energy/J	kinetic energy/J
1 ($x = 0$)	20	0	0
2 ($x = 1$)	10	5	5
3 ($x = 2$)	0	20	0

So we note that at each level the sum of the three forms of energy is 20 J.

Hence the total energy of the system remains constant. More generally, this is true in any system, *provided that the work done by its forces to achieve any particular position is independent of the route by which this position is achieved.*

The general result is then known as the *principle of conservation of energy.*

It is, however, necessary to issue a warning that not all systems are so well-behaved, or *conservative*, as this. If, for instance, we consider a mass being moved across a rough table from a fixed starting point to a particular position, it is clear that the work done by friction is entirely dependent on the route taken. So friction cannot have a potential and is a *non-conservative* force, so that it is impossible to have a principle of conservation of mechanical energy in this case. We shall meet many similar instances, such as the impact of particles which are not perfectly elastic, when the total mechanical energy of the system is seen to diminish, though this is accompanied by the production of other forms of energy, such as heat or sound, which conserve the overall energy of the system.

Exercise 14.1

1 If a spring has stiffness $200\,\text{Nm}^{-1}$, find:
   a) the tension needed to extend it by $4\,\text{cm}$;
   b) the force required to compress it by $5\,\text{mm}$;
   c) its extension when pulled out by a force of $40\,\text{N}$;
   d) its compression when pushed in by a force of $10\,\text{N}$.

2 In each case of no. **1**, find the elastic energy stored in the spring.

3 An elastic string is extended $2\,\text{mm}$ by a tension of $4\,\text{N}$. Find:
   a) its stiffness;
   b) its elastic energy.

4 A spring is compressed a distance of $1\,\text{cm}$ by a force of $500\,\text{N}$. Find:
   a) its stiffness;
   b) its elastic energy.

5 An elastic string has natural length $2\,\text{m}$ and stiffness $50\,\text{N}\,\text{m}^{-1}$. Find:
   a) how far it stretches when a mass $20\,\text{kg}$ hangs from it;
   b) its elastic energy in this position;
   c) the work done by its weight if the $20\,\text{kg}$ was attached to the end of the string when unstretched and gently lowered into its equilibrium position. (Take $g \approx 10\,\text{m}\,\text{s}^{-2}$.)

6 The particle of mass $20\,\text{kg}$ in no. **5** is attached to the string and dropped vertically from its point of suspension. When the string has become taut and is extended by a further amount x, find:
   a) the loss in p.e. of the weight;
   b) the elastic energy of the string;
   c) the k.e. of the particle.
   Hence find the maximum extension of the string. (Take $g \approx 10\,\text{m}\,\text{s}^{-2}$.)

7 A nylon climbing rope of length $40\,\text{m}$ is stretched by a force F. The relation between the force F and the corresponding extension x is given in the following table:

x/m	0	1	2	3	4	5	6	7	8	9	10
F/N	0	1600	2400	3000	3600	4100	4600	5200	5700	6400	7900

Use Simpson's rule to estimate the energy stored in the rope when it is extended by $10\,\text{m}$.

   A climber of mass $64\,\text{kg}$ is climbing on a vertical rock-face h m above the point to which he is securely attached by the rope of length $40\,\text{m}$ described above. The climber slips and falls freely, being brought to rest for the first time by the rope when it has been extended by $10\,\text{m}$.

   Calculate the value of h to the nearest metre.

Determine also the velocity of the climber at the moment the rope becomes taut, and hence, or otherwise, calculate the average deceleration of the climber from that moment until he comes to rest. (Take $g = 9.8\,\text{m s}^{-2}$.)

(JMB)

14.2 Oscillations: simple harmonic motion

Amplitude, period and frequency

Before we use Hooke's law to investigate simple oscillations we first need to remind ourselves about their amplitude and period and to introduce the idea of *frequency*.

Suppose that two objects, A and B, are bobbing up and down in water and after t seconds their heights x m above their equilibrium positions are

a) $x = 2\sin t$ **b)** $x = 3\sin 5\pi t$

a) Object A $x = 2\sin t$

In this case the graph of x against t is

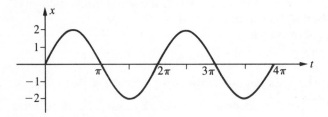

and

i) the maximum displacement, or *amplitude*, of the oscillation is $2\,\text{m}$;

ii) the *period* of the oscillation is clearly $2\pi \approx 6.28$ s. So the oscillations of this object are very slow (presumably because its density is nearly that of the water, or because it is badly water-logged);

iii) the *frequency* of the oscillation is the number of oscillations (or *cycles*) per second, which is

$$\frac{1}{2\pi} \approx \frac{1}{6.28} \approx 0.159\,\text{s}^{-1}$$

(and 1 cycle per second is usually called 1 Hertz: $1\,\text{s}^{-1} = 1\,\text{Hz}$)

Furthermore, the equation $x = 2\sin t$ enables us to find the displacement at any moment of time: after 5 seconds, for instance,

$$x = 2 \sin 5$$
$$= 2 \sin 286.5°$$
$$= 2 \times -0.9588 = -1.92$$

So the object is 1.92 m below its equilibrium position (and so only 0.08 m above its lowest point).

b) Object B $x = 3 \sin 5\pi t$

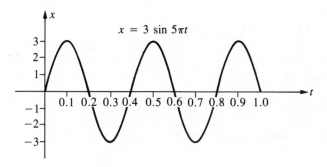

In this case the graph has
 i) a larger amplitude, 3 m;
ii) successive cycles beginning when

$$5\pi t = 0, 2\pi, 4\pi, 6\pi$$

i.e. when $t = 0, 0.4, 0.8, 1.2, \ldots$
Hence the period $= 0.4\,\text{s}$

and   frequency $= \dfrac{1}{0.4} = 2.5\,\text{s}^{-1}$

More generally, $x = a \sin nt$ represents an oscillation whose successive cycles begin at

$$nt = 0, 2\pi, 4\pi, 6\pi \ldots$$

i.e. $t = 0, \dfrac{2\pi}{n}, \dfrac{4\pi}{n}, \dfrac{6\pi}{n}, \ldots$

So the oscillation $x = a \sin nt$

has   period $= \dfrac{2\pi}{n}$

           frequency $= \dfrac{n}{2\pi}$

           amplitude $= a$

Exercise 14.2a

1   For each of the following oscillations find: i) the amplitude,   ii) the period,   iii) the frequency,   iv) the displacement after 2 seconds. (Assume that t is measured in seconds.)
   a) $x = 4 \cos t$ **b)** $x = 2 \sin 10t$
   c) $x = 5 \sin 100t$ **d)** $x = a \sin 2\pi nt$

2   Write in the form $x = a \sin nt$ an equation for a wave:
   a) with amplitude 10 and period 2π s;
   b) with amplitude 3 and period 2 s;
   c) with amplitude 0.5 and frequency $100 \, \text{s}^{-1}$.

Springs, buoys and pendulums

Example 1

If a spring of stiffness k hangs vertically from a fixed point, with its other end attached to a mass m, investigate its equation of motion at a point when it is pulled down a further distance x.

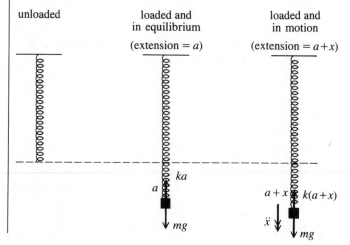

unloaded   loaded and   loaded and
                        in equilibrium   in motion

                        (extension $= a$)   (extension $= a+x$)

                        ka

                a

                        mg $a+x$ $k(a+x)$

                                       $\ddot{x}$

                                              mg

By Hooke's law, when extension $= a$, the tension $= ka$. So, since the mass is then in equilibrium, $ka = mg$.

Also, when extension $= a + x$, tension $= k(a + x)$.

So the general equation of motion is

$$mg - k(a + x) = m\ddot{x}$$

But $\qquad\qquad ka = mg$

So $\qquad\qquad -kx = m\ddot{x}$

$\Rightarrow \qquad\qquad \ddot{x} = -\dfrac{k}{m}x$

Hence the acceleration of the mass is proportional to its extra displacement and, of course, in the opposite direction: the tension in the spring acts as a *restoring force*, attempting to re-establish equilibrium.

Example 2

Investigate the equation of motion of a cylindrical canister buoy, bobbing up and down at sea.

We shall suppose that the buoy has constant cross-sectional area A and height h, and that in equilibrium it is submerged to a depth d. If the density of the buoy is ρ and of the sea is σ, we can investigate its motion when it is submerged by a further amount x.

In each case, both of equilibrium and of submersion by a further amount x, there are just two forces acting:

a) the *weight* of the buoy;

Now   its volume $= Ah$
so   its mass $= Ah\rho$
and   its weight $= Ah\rho g$

b) the *buoyancy*, or *upthrust* of water on the buoy, which Archimedes showed to be equal to the weight of the water displaced.

Now, in the equilibrium position,

$\qquad$ volume of water displaced $= Ad$
$\Rightarrow \qquad$ mass of water displaced $= Ad\sigma$
$\Rightarrow \qquad$ weight of water displaced $= Ad\sigma g$
$\Rightarrow \qquad\qquad$ buoyancy $= Ad\sigma g$

and after submersion by a further amount x, the buoyancy is $A(d + x)\sigma g$.

At the equilibrium position, we see that

$$Ad\sigma g = Ah\rho g$$
$\Rightarrow \quad d\sigma = h\rho$

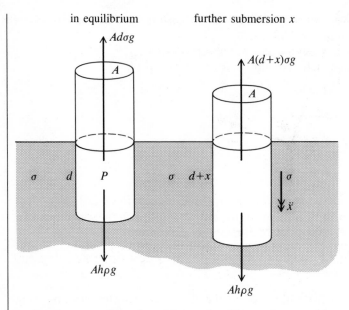

in equilibrium   further submersion x

and at the more general position, using Newton's second law,

$$Ah\rho g - A(d + x)\sigma g = Ah\rho\ddot{x}$$
$$\Rightarrow \qquad -Ax\sigma g = Ah\rho\ddot{x}$$

$$\Rightarrow \qquad \ddot{x} = -\frac{\sigma g}{\rho h}x$$

So, again, we see that the acceleration of the buoy is proportional to its displacement from the equilibrium position and in the opposite direction.

As our final example, we now investiate an oscillation which, unlike the previous cases, is not along a straight line.

Example 3 Simple pendulum

A mass m is suspended from a fixed point by means of a light inextensible string of length l and swings to and fro in a vertical plane.

When the string makes an angle θ with the vertical, the acceleration of the mass has two components:

   $l\ddot{\theta}$ perpendicular to the string, and
   $l\dot{\theta}^2$ along the string.

Furthermore, there are just two forces acting:
a) the weight mg;

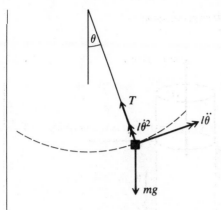

b) the tension T in the string.

Hence, by Newton's second law and resolving perpendicular to the string:

$$-mg \sin \theta = ml\ddot{\theta}$$

$$\Rightarrow \qquad \ddot{\theta} = -\frac{g}{l}\sin \theta$$

In this case, therefore, $\ddot{\theta}$ is proportional not to θ, but to $\sin \theta$. However, if the string is swinging through a small angle, we know that $\sin \theta \approx \theta$.

Hence $\ddot{\theta} \approx -\dfrac{g}{l}\theta$

Simple harmonic motion

In all three examples the final equation of motion was of the form

$$\ddot{x} = -kx \quad (k > 0)$$

which is evidently of considerable importance. Any motion which can be modelled by such an equation is called *simple harmonic motion* (or *s.h.m.*).

As $k > 0$, we can let $k = \omega^2$ (so that ω is another constant) and obtain the equation:

$$\ddot{x} = -\omega^2 x$$

This proves to be a more convenient form of the equation and is therefore adopted as standard.

We can now proceed to find its solutions by either of two methods:

Method 1

$$\ddot{x} = -\omega^2 x$$

$$\Rightarrow \quad 2\dot{x}\ddot{x} = -2\omega^2 x\dot{x}$$

If we now integrate with respect to t, it follows that

$$\dot{x}^2 = -\omega^2 x^2 + A$$

Supposing that when $\dot{x} = 0$ the value of x is a (i.e. that a is the *amplitude* of the oscillation), it follows that

$$0^2 = -\omega^2 a^2 + A$$

$$\Rightarrow \quad A = \omega^2 a^2$$

So $\dot{x}^2 = \omega^2(a^2 - x^2)$

Hence $$\frac{dx}{dt} = \pm\omega\sqrt{(a^2 - x^2)}$$

$$\Rightarrow \quad \pm\frac{dx}{\sqrt{(a^2 - x^2)}} = \omega\, dt$$

$$\Rightarrow \quad \pm\sin^{-1}\frac{x}{a} = \omega t + \varepsilon \quad \text{(where } \varepsilon \text{ is a constant angle)}$$

$$\Rightarrow \quad \frac{x}{a} = \pm\sin(\omega t + \varepsilon)$$

$$\Rightarrow \quad x = \pm a\sin(\omega t + \varepsilon)$$

Now $-\sin(\omega t + \varepsilon) = \sin(\omega t + \varepsilon + \pi)$,

so $x = a\sin(\omega t + \varepsilon)$ or $a\sin(\omega t + \varepsilon + \pi)$

But ε is itself an arbitrary constant, so the second of these general solutions is the same as the first.

Hence $x = a\sin(\omega t + \varepsilon)$ and any simple harmonic motion can be described by an ordinary sine oscillation. The constant ε is called the *angle of phase*, or simply the *phase*.

Moreover, $\sin(\omega t + \varepsilon)$ takes the same value when ωt is increased by 2π, i.e., when t is increased by $2\pi/\omega$.

So the *period* of the s.h.m. is $2\pi/\omega$ (and its *frequency* is $\omega/2\pi$).

Summarising, we see that the s.h.m. has the properties:

$$\ddot{x} = -\omega^2 x$$
$$\dot{x}^2 = \omega^2(a^2 - x^2)$$
$$x = a\sin(\omega t + \varepsilon)$$

amplitude $= a$ period $= \dfrac{2\pi}{\omega}$ phase angle $= \varepsilon$

Method 2

Alternatively, $\ddot{x} = -\omega^2 x$ can be written

$$\frac{d^2x}{dt^2} + \omega^2 x = 0$$

Now $x = A\,e^{mt}$ is a solution, provided that

$$m^2 + \omega^2 = 0 \quad \Rightarrow \quad m = \pm i\omega$$

Hence the complementary function is

$$x = A\,e^{i\omega t} + B\,e^{-i\omega t}$$
$$= A(\cos\omega t + i\sin\omega t) + B(\cos\omega t - i\sin\omega t)$$
$$= C\sin\omega t + D\cos\omega t$$

where C, D (like A, B) are arbitrary constants.
If we now choose a, ε so that

$$a\cos\varepsilon = C$$
$$a\sin\varepsilon = D$$

it follows that

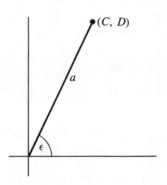

$$x = a(\sin\omega t\cos\varepsilon + \cos\omega t\sin\varepsilon)$$
$$\Rightarrow \qquad x = a\sin(\omega t + \varepsilon)$$

Hence $\quad \dot{x} = a\omega\cos(\omega t + \varepsilon)$

and it follows immediately that

$$\dot{x}^2 = \omega^2(a^2 - x^2)$$

In particular, we can now calculate the periods of the oscillations in the previous examples:

Spring of stiffness k supporting particle of mass m (example 1):

$$\omega = \sqrt{\frac{k}{m}} \quad \Rightarrow \quad \text{period} = 2\pi\sqrt{\frac{m}{k}}$$

Buoy of density ρ in sea of density σ (example 2):

$$\omega = \sqrt{\frac{\sigma g}{\rho h}} \quad \Rightarrow \quad \text{period} = 2\pi\sqrt{\frac{\rho h}{\sigma g}}$$

Simple pendulum of length l (example 3):

$$\omega = \sqrt{\frac{g}{l}} \quad \Rightarrow \quad \text{period} = 2\pi\sqrt{\frac{l}{g}}$$

Example 4

A particle is describing simple harmonic motion in a straight line. When its distance from O, the centre of its path, is 3 m, its velocity is $16 \, \text{m s}^{-1}$ towards O and its acceleration is $48 \, \text{m s}^{-2}$ towards O. Find:

a) the period of the motion;
b) the amplitude of the motion;
c) the time taken by the particle to reach O;
d) the velocity of the particle as it passes through O.

a) With the usual notation, $\ddot{x} = -\omega^2 x$

But when $x = 3$, $\ddot{x} = -48$

So $-48 = -3\omega^2$ $\Rightarrow$ $\omega = 4$

Hence period $= \dfrac{2\pi}{\omega} = \dfrac{\pi}{2} \approx 1.57 \, \text{s}$

b) Furthermore, $\dot{x}^2 = \omega^2(a^2 - x^2) = 16(a^2 - x^2)$

But when $x = 3$, $\dot{x} = -16$.

So $256 = 16(a^2 - 9)$
$\Rightarrow$ $a^2 - 9 = 16$ $\Rightarrow$ $a = 5$

So   amplitude $= 5 \, \text{m}$

c) Moreover, measuring t from the instant when $x = 0$, we know that

   $x = a \sin \omega t = 5 \sin 4t$

So, letting $x = 3$, we see that

   $3 = 5 \sin 4t$ $\Rightarrow$ $4t = \sin^{-1} 0.6 = 0.6434$
$\Rightarrow$ $t = 0.16 \, \text{s}$

So   time taken $= 0.16 \, \text{s}$

d) Finally $\dot{x}^2 = 16(25 - x^2)$

and putting $x = 0$

we obtain $\dot{x}^2 = 16 \times 25$
$\Rightarrow$ $\dot{x} = 4 \times 5 = 20$

So velocity of particle as it passes through O is $20 \, \text{m s}^{-1}$.

Example 5

A particle of mass m hangs at rest from the end of a vertical string which is attached to a fixed point and stretched by an amount e. If it is then given a further extension a and then released, find the period of its subsequent oscillations if **a)** $a < e$ **b)** $a > e$

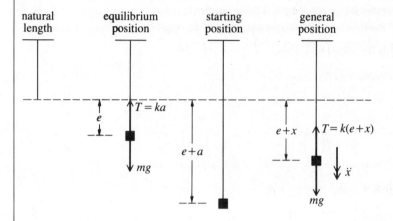

Let the stiffness of the string be k. Then, by considering the position of equilibrium,

$$mg = ke \quad \Rightarrow \quad k = \frac{mg}{e}$$

At the general position, applying Newton's second law:

$$mg - k(e + x) = m\ddot{x}$$
$$\Rightarrow \qquad\qquad -kx = m\ddot{x}$$

$$\Rightarrow \qquad\qquad \ddot{x} = -\frac{k}{m}x = -\frac{g}{e}x$$

So the particle executes simple harmonic motion with $\omega^2 = g/e$, provided only that the string remains taut. It is therefore now necessary to distinguish two cases:

a) $a < e$
In this case, the string remains taut, so the period of the oscillation is

$$\frac{2\pi}{\omega} = \frac{2\pi}{\sqrt{g/e}} = 2\pi\sqrt{\frac{e}{g}}$$

b) $a > e$

In this case the string becomes slack as when $x = -e$ the particle is still travelling upwards. So the full cycle consists of three parts:

i) Particle below equilibrium position ($x > 0$)

$$\text{Time taken} = \frac{\pi}{\omega} = \pi\sqrt{\frac{e}{g}}$$

ii) Particle above equilibrium position but string still taut ($-e < x < 0$)

As the motion is simple harmonic, and if we measure time from the instant of passing through the equilibrium position,

$$x = -a\sin\omega t = -a\sin\sqrt{\left(\frac{g}{e}\right)}t$$

Putting $x = -e$,

$$-e = -a\sin\sqrt{\left(\frac{g}{e}\right)}t \quad\Rightarrow\quad t = \sqrt{\left(\frac{e}{g}\right)}\sin^{-1}\frac{e}{a}$$

So   time taken $= 2\sqrt{\left(\frac{e}{g}\right)}\sin^{-1}\frac{e}{a}$

iii) String slack

The string becomes slack when $x = -e$.

But $\dot{x}^2 = \omega^2(a^2 - x^2)$

So at this point $\dot{x} = \sqrt{\frac{g(a^2 - e^2)}{e}}$

Hence time taken with string slack

$$= \frac{2\dot{x}}{g} = 2\sqrt{\frac{a^2 - e^2}{ge}}$$

So total time taken

$$= \pi\sqrt{\frac{e}{g}} + 2\sqrt{\frac{e}{g}}\sin^{-1}\frac{e}{a} + 2\sqrt{\frac{a^2 - e^2}{ge}}$$

$$= \sqrt{\frac{e}{g}}\left\{\pi + 2\sin^{-1}\frac{e}{a} + 2\sqrt{\frac{a^2 - e^2}{e^2}}\right\}$$

Exercise 14.2b

1 A spring has natural length $4\,\text{m}$ and stiffness $20\,\text{N}\,\text{m}^{-1}$. One end is attached to a fixed point O of a smooth horizontal plane and the other end to a particle of mass $5\,\text{kg}$ which is extended by a further distance of $3\,\text{m}$ and then released to oscillate in a straight line. After time t the extension of the spring is x. Taking $g = 10\,\text{m}\,\text{s}^{-2}$, find:
   a) the equation of motion of the particle;
   b) the period of its oscillation;
   c) x in terms of t;
   d) the maximum speed of the particle.

2 A vertical light spring has two equal masses fixed to it at its ends, as shown. When the lower mass lies on a table, the spring is compressed a distance d by the weight of the upper mass. If the upper mass is now pushed down a further distance $2.5d$ and then released, examine whether the lower mass will at some time leave the table.

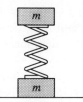

  (You may assume that if the lower mass were fixed, the upper mass would oscillate symmetrically about its equilibrium position.)

3 A boat is tossing (moving vertically up and down) in s.h.m. so that at one extreme position the weight of a passenger appears to be halved.
   a) What is the ratio of apparent weight to actual weight at the other extreme?
   b) If the total range of motion is $10\,\text{m}$, what is the period of a complete oscillation?
   c) What is the ratio of the volumes of water displaced by the boat at the two extremes?
   d) During what fraction of the time is the apparent weight within 10 per cent of the real weight?   (MEI)

4 A particle of mass $2\,\text{kg}$ is hung from the end of a vertical string and stretches it by $10\,\text{cm}$. Find the period of the oscillation if it is then pulled down a further distance of **a)** $5\,\text{cm}$, **b)** $15\,\text{cm}$, and then released. (Take $g \approx 10\,\text{m}\,\text{s}^{-2}$.)

5 One end O of a light elastic string OA, of natural length $4a$ and modulus $3mg$, is attached to a fixed point of a smooth horizontal table. A particle of mass m is attached to the other end A. The particle is pulled along the table until it is at a distance $5a$ from O and is then released with the string straight. Show that the particle first reaches O after a time

$$(\pi + 8)\sqrt{\frac{a}{3g}} \qquad \text{(OC)}$$

6 A woman stands on a horizontal rough plank which performs a horizontal simple harmonic motion of period T and amplitude a. The coefficient of friction between the woman and the plank is μ. Show that she will not slip on the plank if

$$\mu g T^2 \geqslant 4\pi^2 a \qquad \text{(OC)}$$

7 Two particles of equal mass on a smooth horizontal table are connected by an elastic string of natural length a and modulus of elasticity equal to the weight of one particle. If the particles are held at rest at a distance $3a$ apart and are then released simultaneously, find the time which elapses before they collide.

8 A particle P, of mass m, is fastened to one end of a light string, of natural length a and modulus of elasticity λ, whose other end is fixed at O. A particle Q, also of mass m, is fastened to one end of a light inextensible string of length a whose other end is fastened to P. Initially the system is at rest with OPQ in a vertical straight line and OP of length a, so that P is at the middle point of OQ. The system is now released. Prove, by considerations of energy or otherwise, that the length of OP is never less than a.

The length of OP at time t is $a + x$. Prove that,

$$\text{if} \quad n^2 = \frac{\lambda}{2ma}, \quad \text{then} \quad \frac{d^2x}{dt^2} + n^2 x = g$$

Show that

$$x = \frac{g}{n^2}(1 - \cos nt)$$

and deduce the tension at time t in the string PQ. (C)

9 A particle of mass m hangs at rest from a light inextensible string of length l. The particle is suddenly struck a small blow of impulse mv, perpendicular to the string. Find the amplitude of the subsequent motion and the time taken for the particle to rise to a height equal to one-half of its greatest height, measured from the initial position. (MEI)

10 A particle of mass m is attached to the mid-point of a stretched light string whose tension is T and whose ends are fixed to two points distance l apart on a smooth horizontal plane. Show that if the particle is moved a distance x perpendicular to the string, the restoring force is $2T \sin \theta$, where θ is the angle which the parts of the string make with their original line.

Hence find the equation of motion when θ is small, and the period of oscillation.

14.3 Elasticity: Newton's law of impact

Suppose that two bodies of mass m_1, m_2 collide and that the components of their velocities along the line of impact (i.e. along the common normal at the point of contact)

before the impact are u_1, u_2:

and   after the impact are v_1, v_2:

Then *Newton's experimental law of impact* states that

$$v_1 - v_2 = -e(u_1 - u_2)$$

where e is a constant for the pair of bodies known as their *coefficient of restitution*.

Alternatively, this can be expressed as

$$v_2 - v_1 = e(u_1 - u_2)$$

i.e. that their relative velocity of separation is proportional to their relative velocity of approach.

If the second body is at rest, for instance when it is a fixed hard floor on to which the first body is dropped, it follows that $v_1 = -eu_1$; and it is quickly seen that e can range from 0 (in the case of a piece of putty dropped on to the floor) to nearly 1 (for a ball with a very good bounce, which is said to be almost *perfectly elastic*).

So $0 \leqslant e \leqslant 1$

Example 1

A ball is dropped from a height of 5 m on to a hard horizontal pavement. If $e = \frac{1}{2}$, calculate:
a) the height to which it returns after its first bounce;
b) the total distance travelled after repeated bounces;
c) the total time taken.
(Take $g = 10\,\mathrm{m\,s^{-2}}$.)

a) After falling from a height of 5 m the speed of the ball is

$$\sqrt{(2 \times 10 \times 5)} = 10\,\mathrm{m\,s^{-1}}$$

and since $e = \frac{1}{2}$, it rebounds with a speed of $\frac{1}{2} \times 10 = 5\,\mathrm{m\,s^{-1}}$.
Hence it will rebound to a height h, where

$$5^2 = 2 \times 10 \times h \quad \Rightarrow \quad h = 1.25\,\mathrm{m}$$

b) Just as the height of the first rebound is $\frac{1}{4} \times 5\,\mathrm{m}$, so, similarly, the height of its second rebound is

$$\tfrac{1}{4} \times (\tfrac{1}{4} \times 5) = (\tfrac{1}{4})^2 \times 5\,\mathrm{m}$$

and the height of its third rebound is

$$\tfrac{1}{4} \times (\tfrac{1}{4})^2 \times 5 = (\tfrac{1}{4})^3 \times 5\,\mathrm{m}, \quad \text{etc.}$$

Hence its total distance travelled (up and down) in making an infinite number of bounces

$$= 5 + 2\{\tfrac{1}{4} \times 5 + (\tfrac{1}{4})^2 \times 5 + (\tfrac{1}{4})^3 \times 5 + \cdots\}$$
$$= 5 + 2 \times \tfrac{1}{4} \times 5\{1 + \tfrac{1}{4} + (\tfrac{1}{4})^2 + \cdots\}$$
$$= 5 + \tfrac{5}{2}\frac{1}{1 - \tfrac{1}{4}} = 5 + \tfrac{5}{2} \times \tfrac{4}{3} = 8\tfrac{1}{3}\,\mathrm{m}$$

c) The time taken to drop from 5 m $= 1\,\mathrm{s}$;
the time taken between first and second bounces $= 2 \times \frac{5}{10} = 1\,\mathrm{s}$;

the time taken between second and third bounces $= 2 \times \dfrac{\tfrac{5}{2}}{10} = \tfrac{1}{2}\,\mathrm{s}$; etc.

Hence total time taken $= 1 + (1 + \tfrac{1}{2} + \tfrac{1}{4} + \cdots)$
$$= 1 + 2$$
$$= 3\,\mathrm{s}$$

So the ball comes to rest (after an infinite number of bounces) after 3 s and travelling a total distance $8\tfrac{1}{3}\,\mathrm{m}$.

Example 2

A ball of mass 2 kg is moving with velocity $3\,\mathrm{m\,s^{-1}}$ and collides directly with another of mass 1 kg which is moving in the opposite direction with velocity $5\,\mathrm{m\,s^{-1}}$. If $e = \frac{3}{4}$, find their velocities after impact.

Let the final velocities be v_1, v_2 respectively. Then as there is no *external* force acting on the system, momentum is conserved:

$$2v_1 + v_2 = 2 \times 3 + 1 \times -5$$
$$\Rightarrow \quad 2v_1 + v_2 = 1$$

Also, by Newton's law of impact,

$$v_2 - v_1 = \tfrac{3}{4} \times (5 + 3)$$
$$\Rightarrow \quad v_2 - v_1 = 6$$

So
$$\left. \begin{array}{l} 2v_1 + v_2 = 1 \\ -v_1 + v_2 = 6 \end{array} \right\} \quad \Rightarrow \quad v_1 = -\tfrac{5}{3}, \quad v_2 = \tfrac{13}{3}$$
and

Hence the final velocities of the two balls are $-1\tfrac{2}{3}\,\text{m s}^{-1}$ and $4\tfrac{1}{3}\,\text{m s}^{-1}$.

Exercise 14.3

1 A ball of mass 4 kg moving at $3\,\text{m s}^{-1}$ catches up with another ball of mass 2 kg moving at $1\,\text{m s}^{-1}$ in the same straight line. If $e = \tfrac{3}{4}$, find their velocities after impact.

2 A ball of mass 5 kg moving at $4\,\text{m s}^{-1}$ catches up with another ball of mass 2 kg moving at $3\,\text{m s}^{-1}$ in the same straight line. If the second ball has velocity $4\,\text{m s}^{-1}$ after impact find the coefficient of restitution and the final velocity of the first ball.

3 A ball of mass 4 kg moving with speed $2\,\text{m s}^{-1}$ meets a ball of mass 2 kg moving in the opposite direction with speed $3\,\text{m s}^{-1}$. If $e = \tfrac{1}{2}$, find the final velocities of the two balls.

4 A ball of mass 3 kg moving at $5\,\text{m s}^{-1}$ meets another ball of mass 6 kg moving in the opposite direction at $3\,\text{m s}^{-1}$. If the impact brings the second ball to rest, find the final velocity of the first ball and the coefficient of restitution.

5 A smooth sphere A of mass 1 kg moves with speed $5\,\text{m s}^{-1}$ on a smooth horizontal plane directly towards a vertical wall. Before it hits the wall it is struck from behind by a smooth sphere B with the same radius as A but of mass 2 kg moving towards the wall along the same line as A with speed $10\,\text{m s}^{-1}$. Given that the coefficient of restitution between the spheres is $\tfrac{1}{2}$, find their new velocities.

The sphere A later strikes the wall and rebounds, the coefficient of restitution being $\tfrac{1}{4}$. It then collides again with B which has been following it towards the wall. Show that, after this collision, both A and B are again moving towards the wall, the speed of A being three times that of B. (MEI)

6 Three identical imperfectly elastic smooth spheres A, B, C are at rest in a

straight line, but not in contact, on a smooth horizontal plane in the order named. A is given a velocity towards B. Prove that there will be at least three collisions during the subsequent motion.

If the coefficient of restitution between each pair of spheres is $\frac{1}{2}$, determine whether there is a fourth collision. (L)

7 The centres of three equal spheres A, B, C, at rest on a smooth horizontal table, lie on a straight line. The coefficient of restitution between A and B is e and between B and C is e'. The sphere A is projected to strike B directly and B then strikes C. Show that there are no further collisions if $e' < (3e - 1)/(1 + e)$. (W)

8 A 'Supaball' when dropped on a hard floor displays a coefficient of restitution, e, of very nearly unity, and e may be assumed to be independent of the velocity of approach. When the ball is dropped vertically from a height of 2 m on to a hard floor it comes finally to rest 26.0 s after being released. What is e? What is the total distance that the ball has travelled? ($g = 9.8\,\mathrm{m\,s^{-2}}$.) (CS)

14.4 Further statics: moments, couples, and equilibrium

So far we have limited our study to that of a single particle. But we are often concerned with the behaviour of collections of particles, both at rest and in motion. These might range from the particles of a highly compressible gas injected into a turbine to those of water, nearly incompressible, in waves at sea. But the simplest such collection, with which we shall be principally concerned, is one in which the distances between every pair of particles remains constant, and this we know as a *rigid body*. First, however, we must develop the idea of *moment of a force*.

Moments of forces

When we considered a particular force, of known magnitude and direction, acting on a *single* particle, there was never any doubt about its line of action: it necessarily had to pass through the particle. But if we now consider a collection of particles, say a chair, it is a common experience that a particular force can have a very different effect if its line of action is shifted: the reaction of a man on a chair might in one position produce stability, whilst if the same force acts at a different position the chair might topple. In the case of a force acting on a single particle it was unnecessary to consider the *turning effect* of the force, because it had none. But when we consider a collection of particles, we must clearly study the turning effects of the individual forces involved.

To take a simple example, let us suppose that a beam of length 2 m is freely hinged at one of its ends O so that the beam is in a horizontal position, and let us further suppose that a force of magnitude 1000 N acts vertically upwards at its other end.

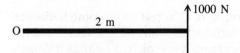

Now it is clear that the force of 1000 N exerts a turning effect about O, and we know that this effect can be altered
a) by increasing, or decreasing, the magnitude of the force;
b) by altering its direction;
c) by altering the position of its line of action, perhaps by moving it nearer to O.

The reader will already be familiar with a measure of such a turning effect called *the moment about O of the given force*, which is influenced by such changes. This was usually defined as the product of the magnitude of the force and its perpendicular distance from O:

moment about O of $\mathbf{F} = pF$

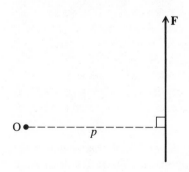

Let us now consider what will happen if we radically alter the direction of the force. For instance suppose that instead of acting vertically, the force remains perpendicular to the rod but acts in a horizontal direction:

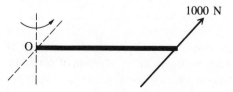

It is immediately apparent that such a change alters the turning effect from being in a vertical plane to being in a horizontal plane, as shown by the arrows in the two figures.

We therefore see that a turning effect is properly defined when we know not only its magnitude and point of application but also the plane in which it is acting. A bowler, for instance, may give equal turning effects to three successive balls, one with top spin, the next as an off-break and the last as a leg-break: their distinguishing features are the planes in which they are made to spin. Now each of these planes is perfectly defined by the direction of its normal so we see that for the turning effect to be fully defined we need to know both a magnitude and also a *direction*, that of the normal to the plane in which it acts. It is, therefore, no surprise that the moment of a force should emerge not as a scalar quantity, but as a *vector*.

Moment as a vector product

A force and its line of action are completely defined by:
a) a vector **F** representing the force; and
b) the position vector **r** of *any* point on its line of action.

Let us now consider the vector product **r** × **F**.

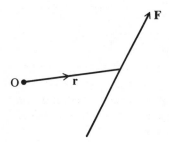

First, we notice that the vector **r** × **F** does *not* depend on the particular point which was chosen on its line of action. For suppose that, instead of the point P (with position vector **r**), we had chosen another point P′ (with position vector **r′**):

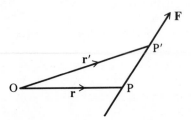

Now **PP′** is in the same direction as **F**.

So $\qquad$ **PP′** $= \lambda$**F**, where λ is a scalar

$\Rightarrow \qquad$ **r′** $-$ **r** $= \lambda$**F**

$\Rightarrow$ (**r′** $-$ **r**) $\times$ **F** $= \lambda$**F** $\times$ **F** $= 0$

$\Rightarrow \qquad$ **r′** $\times$ **F** $=$ **r** $\times$ **F**

Hence the vector product **r** $\times$ **F** is independent of the particular point chosen on **F** and depends only on **F** and its line of action.

Furthermore, if the angle between **r** and **F** is θ, we see that the magnitude of **r** $\times$ **F** is

$$rF \sin \theta = (r \sin \theta)F = pF$$

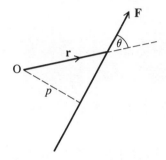

Lastly, the direction of **r** $\times$ **F** is perpendicular to **r** and to **F** and so is perpendicular to the plane containing the line of action and O.

In summary, it is evident that **r** $\times$ **F** depends only upon the vector **F** and its line of action, that its magnitude is the product of **F** and its perpendicular distance from O, and that its direction corresponds to a turning effect in the plane containing **F** and O. We therefore now *re-define* the moment about O of the force **F** acting through the point **r** as the vector **r** $\times$ **F**:

moment about O of the force **F** $=$ **r** $\times$ **F**

In particular, we notice that if **F** goes through O, its moment about O is

0 $\times$ **F** $= 0$

Also   **r** $\times$ ($-$**F**) $= -($**r** $\times$ **F**)

so that the moments about O of two equal and opposite forces are themselves equal and opposite.

More generally, an extremely valuable result follows directly from the fact that vector products are distributive with respect to addition. For suppose that **P, Q** are two forces which meet in a point and that their resultant is **R**. Let the position vector of their point of intersection be **r**.

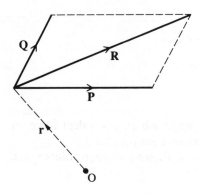

Then $\mathbf{r} \times \mathbf{P} + \mathbf{r} \times \mathbf{Q} = \mathbf{r} \times (\mathbf{P} + \mathbf{Q}) = \mathbf{r} \times \mathbf{R}$

So   the sum of the moments about any point of two concurrent forces is equal to the moment of their resultant

Equivalent systems of forces

It sometimes happens that two apparently different systems of forces have:
a) their vector sums equal;
b) their moments equal about every chosen point.

In such a case we say that the two systems are *equivalent*. In particular, the resultant of two concurrent forces has the same moment about any point as the sum of the moments of its constituent forces. So the resultant of two concurrent forces is equivalent to the forces themselves.

It can now easily be shown that two sets of forces are equivalent if:
a) their vector sums are equal, and
b) their moments are equal about any *one* particular point.
For let us take this point as origin and let the two sets be

$$\mathbf{F}_i \text{ at } \mathbf{r}_i \quad (i = 1, 2, \ldots, n)$$

and $\mathbf{F}'_i \text{ at } \mathbf{r}'_i \quad (i = 1, 2, \ldots, n')$.

Then $\sum \mathbf{F}_i = \sum \mathbf{F}'_i$ and $\sum \mathbf{r}_i \times \mathbf{F}_i = \sum \mathbf{r}'_i \times \mathbf{F}'_i$

Also the moment of the first set of forces about any other point $\mathbf{c}$

$$= \sum (\mathbf{r}_i - \mathbf{c}) \times \mathbf{F}_i$$
$$= \sum \mathbf{r}_i \times \mathbf{F}_i - \mathbf{c} \times \sum \mathbf{F}_i$$
$$= \sum \mathbf{r}'_i \times \mathbf{F}'_i - \mathbf{c} \times \sum \mathbf{F}'_i$$

$$= \sum (\mathbf{r}'_i - \mathbf{c}) \times \mathbf{F}'_i$$

= moment of the second set of forces about $\mathbf{c}$

So the two sets of forces must be equivalent.

Parallel forces

The question now arises whether we can find such an equivalent force, or resultant, when the two forces are not concurrent but parallel.

Let us consider two parallel forces $k_1\mathbf{F}$ and $k_2\mathbf{F}$ acting through points $\mathbf{r}_1$ and $\mathbf{r}_2$.

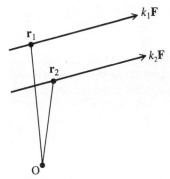

Now if a single force is equivalent to the original pair, it must be

$$k_1\mathbf{F} + k_2\mathbf{F} = (k_1 + k_2)\mathbf{F}$$

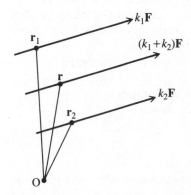

We now take a point $\mathbf{r}$ on the line of action of this force. Since it is equivalent to the original forces, we know that

$$\mathbf{r} \times (k_1 + k_2)\mathbf{F} = \mathbf{r}_1 \times k_1\mathbf{F} + \mathbf{r}_2 \times k_2\mathbf{F}$$

$$\Leftrightarrow \quad (k_1 + k_2)\mathbf{r} \times \mathbf{F} = (k_1\mathbf{r}_1 + k_2\mathbf{r}_2) \times \mathbf{F}$$

In particular, we see that this equation is true if

$$(k_1 + k_2)\mathbf{r} = k_1\mathbf{r}_1 + k_2\mathbf{r}_2$$

$$\Leftrightarrow \qquad \mathbf{r} = \frac{k_1\mathbf{r}_1 + k_2\mathbf{r}_2}{k_1 + k_2}$$

which is the point dividing the line joining $\mathbf{r}_1$ and $\mathbf{r}_2$ in the ratio $k_2 : k_1$.

As O could have been taken at any point whatsoever, it is clear that the original forces $k_1\mathbf{F}$ and $k_2\mathbf{F}$ are equivalent to a force $(k_1 + k_2)\mathbf{F}$ acting along a line which separates their lines of action in the ratio $k_2 : k_1$. So this force is again known as their resultant.

Centre of gravity

We now suppose that a system of particles consists of masses $m_1, m_2, m_3, \ldots$ at points $\mathbf{r}_1, \mathbf{r}_2, \mathbf{r}_3, \ldots$, and that the system is not so large that the forces of gravity will cease to be parallel. Then these masses have weights $m_1\mathbf{g}$, $m_2\mathbf{g}$, $m_3\mathbf{g}, \ldots$ acting at points $\mathbf{r}_1, \mathbf{r}_2, \mathbf{r}_3 \ldots$.

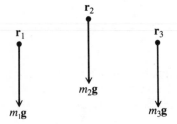

Hence, by a simple extension of the above result, we see that they are equivalent to a weight of $(m_1 + m_2 + m_3 + \cdots)\mathbf{g}$ acting at the point

$$\frac{m_1\mathbf{r}_1 + m_2\mathbf{r}_2 + m_3\mathbf{r}_3 + \cdots}{m_1 + m_2 + m_3 + \cdots}$$

which we have already met as the centre of mass.

Hence the weights of the separate particles are equivalent to their total weight acting through the centre of mass, which is therefore also called the *centre of gravity* of the system.

Couples

There is, however, one exceptional case of two parallel forces, when $k_1 = -k_2$ and the forces are equal and opposite with different lines of action. Such a system is called a *couple*.

We now consider a couple consisting of two forces, $\mathbf{F}$ at $\mathbf{r}_1$ and $-\mathbf{F}$ at $\mathbf{r}_2$.

If a point P is taken with position vector $\mathbf{p}$, the moment of the couple about P is

$$(\mathbf{r}_1 - \mathbf{p}) \times \mathbf{F} + (\mathbf{r}_2 - \mathbf{p}) \times -\mathbf{F}$$
$$= \mathbf{r}_1 \times \mathbf{F} - \mathbf{p} \times \mathbf{F} - \mathbf{r}_2 \times \mathbf{F} + \mathbf{p} \times \mathbf{F}$$
$$= (\mathbf{r}_1 - \mathbf{r}_2) \times \mathbf{F}$$

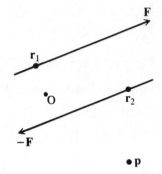

As this is independent of the position of P, it is simply called the *moment of the couple*. Note that the moment of a single force about a point on its line of action is zero, so a couple cannot possibly be equivalent to a single force and is therefore said to be *irreducible*.

Example 1

Coplanar forces $\mathbf{F}_1, \mathbf{F}_2, \mathbf{F}_3$ act through points whose position vectors are $\mathbf{r}_1$, $\mathbf{r}_2, \mathbf{r}_3$ respectively and

$$\mathbf{F}_1 = 4\mathbf{i} + 3\mathbf{j}, \qquad \mathbf{r}_1 = 3\mathbf{i} + 2\mathbf{j}$$
$$\mathbf{F}_2 = 5\mathbf{i} - 8\mathbf{j}, \qquad \mathbf{r}_2 = -\mathbf{i} + \mathbf{j}$$
$$\mathbf{F}_3 = -2\mathbf{i} + 6\mathbf{j}, \quad \mathbf{r}_3 = 2\mathbf{i} - \mathbf{j}$$

a) Show that these forces reduce to a single force and give its magnitude and direction.

b) Find a vector equation or the Cartesian equation of the line of action of the resultant.

c) Find the single force through the point $\mathbf{r} = \mathbf{i} + \mathbf{j}$ and the couple which are together equivalent to this system.

a) If these forces reduce to a single force, it must be

$$\mathbf{F}_1 + \mathbf{F}_2 + \mathbf{F}_3 = 7\mathbf{i} + \mathbf{j}$$

and so have magnitude $\sqrt{50}$ and be in the direction making $\tan^{-1} \frac{1}{7} \approx 8{\cdot}1°$ with Ox.

b) If the line of action of its resultant passes through $\mathbf{r} = x\mathbf{i} + y\mathbf{j}$,

then $\mathbf{r} \times (\mathbf{F}_1 + \mathbf{F}_2 + \mathbf{F}_3) = \mathbf{r}_1 \times \mathbf{F}_1 + \mathbf{r}_2 \times \mathbf{F}_2 + \mathbf{r}_3 \times \mathbf{F}_3$

$$\Rightarrow \qquad (x\mathbf{i} + y\mathbf{j}) \times (7\mathbf{i} + \mathbf{j}) = (3\mathbf{i} + 2\mathbf{j}) \times (4\mathbf{i} + 3\mathbf{j})$$
$$+ (-\mathbf{i} + \mathbf{j}) \times (5\mathbf{i} - 8\mathbf{j})$$
$$+ (2\mathbf{i} - \mathbf{j}) \times (-2\mathbf{i} + 6\mathbf{j})$$

$$\Rightarrow \qquad (x - 7y)\mathbf{k} = (1 + 3 + 10)\mathbf{k} = 14\mathbf{k}$$

$$\Rightarrow \qquad x - 7y = 14$$

c) Let the required force be $\mathbf{F}$ and the required couple be $\mathbf{G}$.

Again, $\mathbf{F} = \mathbf{F}_1 + \mathbf{F}_2 + \mathbf{F}_3 = 7\mathbf{i} + \mathbf{j}$

Furthermore,

$$\mathbf{r} \times (\mathbf{F}_1 + \mathbf{F}_2 + \mathbf{F}_3) + \mathbf{G} = \mathbf{r}_1 \times \mathbf{F}_1 + \mathbf{r}_2 \times \mathbf{F}_2 + \mathbf{r}_3 \times \mathbf{F}_3$$

$$\Rightarrow \quad (\mathbf{i} + \mathbf{j}) \times (7\mathbf{i} + \mathbf{j}) + \mathbf{G} = 14\mathbf{k}$$

$$\Rightarrow \qquad\qquad -6\mathbf{k} + \mathbf{G} = 14\mathbf{k}$$

$$\Rightarrow \qquad\qquad\qquad \mathbf{G} = 20\mathbf{k}$$

So required force is $7\mathbf{i} + \mathbf{j}$ and required couple is $20\mathbf{k}$.

Exercise 14.4a

1   Find the moments:

about the point	of the force	acting at
a) 0	$\mathbf{i} + \mathbf{j} - \mathbf{k}$	$\mathbf{i} - \mathbf{k}$
b) 0	$\mathbf{i} + \mathbf{j} - 2\mathbf{k}$	$\mathbf{i} + 2\mathbf{j} + \mathbf{k}$
c) $\mathbf{i} + \mathbf{j}$	$\mathbf{i} + 2\mathbf{j} - \mathbf{k}$	$\mathbf{i} + \mathbf{j} - \mathbf{k}$
d) $\mathbf{i} + 2\mathbf{j} - 3\mathbf{k}$	$3\mathbf{i} + 2\mathbf{j} - 3\mathbf{k}$	$3\mathbf{j} + \mathbf{k}$
e) $\mathbf{i} - \mathbf{j}$	$-2\mathbf{i} + 3\mathbf{j} + \mathbf{k}$	$\mathbf{j} + 2\mathbf{k}$
f) $2\mathbf{i} + \mathbf{j} - \mathbf{k}$	$-\mathbf{i} + 4\mathbf{j} - \mathbf{k}$	$-\mathbf{i} + \mathbf{j} + 2\mathbf{k}$

2   Determine in each of the following cases whether the two sets of forces are equivalent:

a) $\mathbf{i}$ acting at $\mathbf{j}$ and $\mathbf{j}$ acting at $\mathbf{i}$, $\mathbf{i} + \mathbf{j}$ acting at $2\mathbf{j}$

b) $-\mathbf{i}$ acting at $\mathbf{j}$ and $\mathbf{j}$ acting at $\mathbf{i}$, $-\mathbf{i} + \mathbf{j}$ acting at $2\mathbf{j}$

c) $2\mathbf{j}$ acting at O and $\mathbf{j}$ acting at $3\mathbf{i}$, $3\mathbf{j}$ acting at $\mathbf{i}$

d) $4\mathbf{i}$ acting at O and $-\mathbf{i}$ acting at $-3\mathbf{j}$, $3\mathbf{i}$ acting at $-4\mathbf{j}$

e) the couple formed by $-\mathbf{i}$ at $\mathbf{j}$ and $\mathbf{i}$ at $2\mathbf{j}$,   and that formed by $-\mathbf{j}$ at $2\mathbf{i}$ and $\mathbf{j}$ at $\mathbf{i}$

f) the couple formed by $-\mathbf{i}$ at $\mathbf{j}$ and $\mathbf{i}$ at $-\mathbf{j}$,   and that formed by $-\mathbf{j}$ at $\mathbf{i}$ and $\mathbf{j}$ at $\mathbf{k}$

3 Show that each of the following systems of coplanar forces is equivalent either to a single force or a single couple (which may be zero). In each case find the equivalent force (including its line of action) or the equivalent couple:

a) $\mathbf{i} + \mathbf{j}$ at $\mathbf{i} - \mathbf{j}$, $3\mathbf{i} - \mathbf{j}$ at $2\mathbf{i}$, $2\mathbf{i} + 3\mathbf{j}$ at $2\mathbf{j}$

b) $\mathbf{i} + \mathbf{j}$ at $\mathbf{i} - \mathbf{j}$, $-3\mathbf{i} + \mathbf{j}$ at $\mathbf{i} + \mathbf{j}$, $2\mathbf{i} - 2\mathbf{j}$ at $-\mathbf{i} + 2\mathbf{j}$

c) $2\mathbf{j}$ at $\mathbf{i} - \mathbf{j}$, $-\mathbf{i}$ at $\mathbf{j}$, $\mathbf{i}$ at $3\mathbf{j}$, $-2\mathbf{j}$ at O

d) $2\mathbf{i} + 3\mathbf{j}$ at $\mathbf{i}$, $-3\mathbf{i} + 4\mathbf{j}$ at $\mathbf{j}$, $4\mathbf{i} + 5\mathbf{j}$ at $\mathbf{i} + \mathbf{j}$, $-\mathbf{i} + 3\mathbf{j}$ at $-\mathbf{i} - 2\mathbf{j}$

4 ABCD is a square of side a. The anti-clockwise moment of a set of forces in the plane of the square is $10aP$ about A, $-20aP$ about B and $10aP$ about C. Taking the x-axis along AB and the y-axis along AD, determine the resultant of the set of forces and the equation of its line of action.

Determine the couple and the single force at the centre of the square which would be equivalent to this resultant.

5 a) Show that three forces represented in magnitude and direction by $k\mathbf{AB}$, $k\mathbf{BC}$, $k\mathbf{CA}$ and acting along the sides AB, BC, CA respectively of a triangle are equivalent to a couple of magnitude $2k$ times the area of the triangle.

b) Use the result from **a)** to show that four forces represented in magnitude and direction by $k\mathbf{AB}$, $k\mathbf{BC}$, $k\mathbf{CD}$, $k\mathbf{DA}$ and acting along the sides AB, BC, CD, DA respectively of a plane quadrilateral are equivalent to a couple of magnitude $2k$ times the area of the quadrilateral.   (MEI)

Statics of a system: equilibrium

Any system of particles (which may, or may not, comprise a rigid body) is said to be *in equilibrium* if, under the forces acting, it can remain at rest.

Let us consider a system in equilibrium which consists of just three particles of masses m_1, m_2, m_3 whose position vectors are $\mathbf{r}_1, \mathbf{r}_2, \mathbf{r}_3$.

The forces acting on these particles can be of two kinds:

a) *external forces*, which we call $\mathbf{F}_1, \mathbf{F}_2, \mathbf{F}_3$, acting on the particles m_1, m_2, m_3 respectively; and

b) *internal forces*, between the pairs of particles. We let the force exerted upon m_1 by m_2 be $\mathbf{F}_{12}$, and the force exerted upon m_2 by m_1 be $\mathbf{F}_{21}$, etc. So the complete set of forces can be represented:

Now, by Newton's third law, every pair of internal forces must be equal and opposite.

So $\mathbf{F}_{12} + \mathbf{F}_{21} = \mathbf{0}$

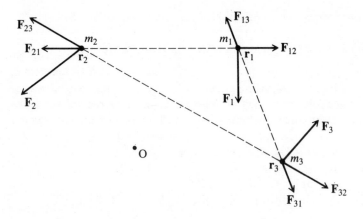

Moreover, since these forces have the same line of action, their moments about O must be equal and opposite, and

$$\mathbf{r}_1 \times \mathbf{F}_{12} + \mathbf{r}_2 \times \mathbf{F}_{21} = \mathbf{0}$$

If we now look at the equilibrium of the particles separately, we see that

$$\mathbf{F}_1 \qquad\quad + \mathbf{F}_{13} + \mathbf{F}_{12} = \mathbf{0}$$
$$\mathbf{F}_2 + \mathbf{F}_{23} \qquad\quad + \mathbf{F}_{21} = \mathbf{0}$$
$$\mathbf{F}_3 + \mathbf{F}_{32} + \mathbf{F}_{31} \qquad\quad = \mathbf{0}$$

We now add these equations and remember that $\mathbf{F}_{23} + \mathbf{F}_{32} = \mathbf{0}$, etc., so that

$$\mathbf{F}_1 + \mathbf{F}_2 + \mathbf{F}_3 = \mathbf{0}$$

Hence if a system is in equilibrium the sum of the external forces must be zero, just as when they act upon a single particle.

Moreover, we can also take the vector products of each of the above expressions with $\mathbf{r}_1, \mathbf{r}_2, \mathbf{r}_3$ respectively, and obtain

$$\mathbf{r}_1 \times \mathbf{F}_1 \qquad\qquad + \mathbf{r}_1 \times \mathbf{F}_{13} + \mathbf{r}_1 \times \mathbf{F}_{12} = \mathbf{0}$$
$$\mathbf{r}_2 \times \mathbf{F}_2 + \mathbf{r}_2 \times \mathbf{F}_{23} \qquad\qquad + \mathbf{r}_2 \times \mathbf{F}_{21} = \mathbf{0}$$
$$\mathbf{r}_3 \times \mathbf{F}_3 + \mathbf{r}_3 \times \mathbf{F}_{32} + \mathbf{r}_3 \times \mathbf{F}_{31} \qquad\qquad = \mathbf{0}$$

We now add these three equations and remember that

$$\mathbf{r}_1 \times \mathbf{F}_{12} + \mathbf{r}_2 \times \mathbf{F}_{21} = \mathbf{0}$$
$$\mathbf{r}_1 \times \mathbf{F}_{13} + \mathbf{r}_3 \times \mathbf{F}_{31} = \mathbf{0}$$
$$\mathbf{r}_2 \times \mathbf{F}_{23} + \mathbf{r}_3 \times \mathbf{F}_{32} = \mathbf{0}$$

Hence $\mathbf{r}_1 \times \mathbf{F}_1 + \mathbf{r}_2 \times \mathbf{F}_2 + \mathbf{r}_3 \times \mathbf{F}_3 = \mathbf{0}$

So the sum of the moments about O of the external forces must be zero.

Now from the start of this section, O has been any fixed point which it is wished to use as an origin. It therefore follows that the sum of the moments of the external forces about *any* point must be zero.

For simplicity, we limited ourselves to the consideration of just three particles. But the whole of the above analysis can immediately be extended to cover any number, finite or infinite. Summarising, we can see that:

> For any system of particles to be in equilibrium:
> **a)** the sum of the external forces must be zero; and
> **b)** the sum of the moments of the external forces about any fixed point must also be zero.

Example 2

A uniform beam of length 10 m and weight 2000 N has a man of weight 600 N standing on one end. If the beam is supported at two points 2 m from each end, find the reactions at these supports.

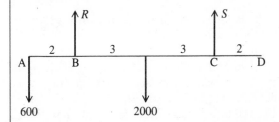

Letting the two reactions (in newtons) be R, S and taking moments about C,

$$6R = 8 \times 600 + 3 \times 2000 = 10\,800$$
$$\Rightarrow \quad R = 1800$$

Also, taking moments about B,

$$2 \times 600 + 6S = 3 \times 2000$$
$$\Rightarrow \quad 6S = 4800$$
$$\Rightarrow \quad S = 800$$

So the two reactions are 1800 N and 800 N (and we note, as we would expect, that $R + S = 600 + 2000$).

Example 3

A lightweight ladder of length 8 m is inclined at 70° to the horizontal, resting upon a rough horizontal path and a smooth vertical wall. If a woman climbs this ladder and it begins to slip when she is three-quarters of the way up, find the coefficient of friction at its lower end.

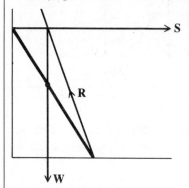

If we neglect the weight of the ladder itself, there are just three external forces acting on the system of the woman and ladder:
a) **W**, the weight of the woman;
b) **S**, the normal reaction at the smooth wall;
c) **R**, the reaction at the ground which (on account of a frictional component) can act at an inclination to the normal.

Now when the system is in equilibrium these three forces must have zero moment about every point. But **S** and **W** clearly have zero moment about their point of intersection, so **R** must also pass through this point.

Hence as the woman gradually climbs the ladder, the reaction **R** makes an increasing angle with the vertical; until, finally, when the woman is three-quarters of the way up, this angle becomes λ, the angle of friction, and the ladder starts to slip.

If we now look more closely at this limiting position, we see that (in the notation of the figure)

$$h = 8 \sin 70°$$

and

$$x = 6 \cos 70°$$

$$\Rightarrow \quad \tan \lambda = \frac{x}{h} = \frac{6 \cos 70°}{8 \sin 70°}$$

$$= \tfrac{3}{4} \cot 70°$$

$$= 0.27$$

So $\mu = \tan \lambda = 0.27$

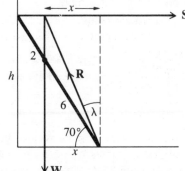

Alternatively, we can let the components of **R** be F and N, and then proceed as follows:

Resolving vertically, $N = W$

Taking moments about the point P,

$$F \times 8 \sin 70° = W \times 6 \cos 70°$$

$\Rightarrow \qquad\qquad F = \dfrac{3W}{4} \cot 70°$

$\Rightarrow \qquad\qquad \dfrac{F}{N} = \tfrac{3}{4} \cot 70° = 0.27$

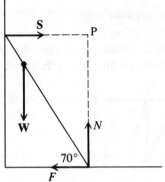

Hence, if slipping takes place at this point, $\mu = 0.27$.

Exercise 14.4b

1   A beam AB, of length 22 m and mass 150 kg has its centre of gravity at C, 11.5 m from A. It is supported in a horizontal position by two trestles at D and E, where AD = 7.5 m and AE = 15 m. Find the pressure on each trestle. (Taking $g \approx 10\,\mathrm{m\,s^{-2}}$.)

   Find the mass of the heaviest man who can sit at either end of the beam without tilting it.   (oc)

2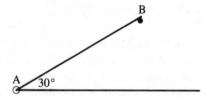

   A uniform rod of length 1 m and mass 50 g is freely hinged at A and has its other end resting on a smooth peg B. The rod in this position makes an angle of 30° with the horizontal.

   Calculate:

   a) the reaction at B;

   b) the horizontal component of the reaction at A.

   (Take $g \approx 9.81\,\mathrm{m\,s^{-2}}$.)   (c)

3   A uniform beam ABCD, where AB = a, BC = $3a$ and CD = a, is of weight $4W$ and rests horizontally in equilibrium on supports at B and C. Loads of weight xW, yW are hung from A, D respectively. Calculate, in

terms of x, y and W, the load borne by each support. If the sum of the loads borne by both supports cannot exceed $20W$, indicate by shading on the plane of the coordinate axes Oxy the region within which the point (x, y) must lie. Deduce that the greatest value of y is 14.   (OC)

4   A rod AB, hinged at A, is 4 m long and of mass 5 kg, with its centre of mass 3 m from A. A rope is attached to B, passes over a smooth pulley 3 m above A and supports a mass M hanging freely. A mass m hangs from B and keeps AB in a horizontal position.
   a) When $m = 3$, find M; and
   b) when $M = 20$, find m.

5   A non-uniform rod AB, of length a and weight W, rests making an angle β with the horizontal with its lower end B on a rough horizontal table. The centre of mass of the rod is at G where AG $= \frac{1}{3}a$. Equilibrium is maintained by a horizontal string attached to A. Show that the coefficient of friction between the rod and the table cannot be less than $\frac{2}{3}\cot\beta$. Find the tension in the string.   (OC)

6   A uniform beam AB rests in limiting equilibrium with A against a rough vertical wall and with B on rough horizontal ground. The vertical plane through the beam is perpendicular to the wall. The coefficient of friction at A is $\frac{1}{3}$ and the coefficient of friction at B is $\frac{1}{2}$. Calculate, correct to the nearest degree, the inclination of the beam to the horizontal.   (OC)

7   A uniform ladder is leaning against a vertical wall on horizontal ground. The coefficients of friction at the wall and at the ground are equal, and the ladder is on the point of slipping. Show that the total reaction at the wall must be at right angles to the total reaction at the ground.
       Draw a figure with the ladder inclined at 60° to the ground; by means of a circle with the ladder as diameter, mark on the figure the point where the two reactions intersect.   (OC)

8   A circular cylinder of weight W is held with its axis horizontal on a rough plane inclined at 30° to the horizontal by a cord wrapped round it, one end of the cord being fixed to the cylinder, the other leading away tangentially, at right angles to the axis of the cylinder. Determine the tension of the cord and the least possible coefficient of friction between the cylinder and the plane
   a) if the cord leads away horizontally;
   b) if it leads away vertically upwards;
   c) if it leads away in such a direction as to make the tension a minimum.

9   A uniform hemispherical shell rests with its curved surface against a rough vertical wall and a rough horizontal floor. If the equilibrium is limiting at both points of contact when the plane of the base of the shell makes an angle θ with the horizontal, prove that

$$\sin\theta = \frac{2\mu(1+\mu)}{1+\mu^2}$$

where μ is the coefficient of friction at both points of contact.

Prove that such a position is impossible if $\mu > \sqrt{2} - 1$.

(The centre of mass of the shell bisects the radius of symmetry.)   (oc)

10 The mass per unit length of a rod AB of length l varies linearly from m at the end A to $2m$ at the end B. Prove by means of the integral calculus that the total mass of the rod is $3\,ml/2$ and find the position of the centre of gravity.

An inextensible string of length a is attached to a point P of the rod at a distance $a\,(<l)$ from the end A. The other end of the string is fixed to the point C on a rough vertical wall and the end A of the rod rests against this wall so that A is below C and AC $= a$. Find the tension in the string and the frictional force at the point A, and show that this vanishes if $l = 18a/5$.   (MEI)

11 A uniform rod AB is suspended from a fixed point O of a rough vertical wall by a light inextensible string of length l which is fastened to B. The end A of the rod rests against the wall vertically below O with A below the level of B. If the rod is in equilibrium (not necessarily limiting) with OA $= d$ and angle AOB $= \theta$, find the tangent of the angle of inclination to the horizontal of the resultant reaction on the rod at A.

If the angle of friction at A is λ and equilibrium is limiting, with A about to slide downwards, show that

$$\cos(\theta - \lambda) = \frac{2d}{l}\cos\lambda$$

12 O, A, B, C are the four corners of a square lamina OABC of side a. Forces act in the plane of the lamina as follows: $6P$ along $\mathbf{AO}$, P along $\mathbf{OC}$, $2P$ along $\mathbf{BC}$ and $7P$ along $\mathbf{BA}$.

a) Find the magnitude of the resultant of this system of forces, and the equation of its line of action referred to OA, OC as x- and y-axes.

b) It is desired to maintain equilibrium by applying a force and a couple at one corner of the square. Find which corner should be selected so that the moment of the couple is as small as possible, and state the moment of this least couple.

c) Find the couple which must be applied to the lamina so that the resultant of the four original forces and the couple acts in a line passing through the centre of the square.   (MEI)

* 14.5 Further dynamics: moments of inertia and rotation

Dynamics of a system of particles

Suppose that a system of masses is in motion and that a typical mass m_i at point $\mathbf{r}_i$ is under the action of an external force $\mathbf{F}_i$ and an internal force $\mathbf{I}_i$.

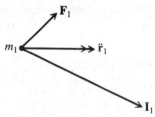

Then, by Newton's second law,

$$\mathbf{F}_i + \mathbf{I}_i = m_i\ddot{\mathbf{r}}_i$$

Adding these equations over all the particles of the system,

$$\sum \mathbf{F}_i + \sum \mathbf{I}_i = \sum m_i\ddot{\mathbf{r}}_i \tag{1}$$

But, applying Newton's third law, $\sum \mathbf{I}_i = \mathbf{0}$, and if the total mass of the system is $M\,(= \Sigma\, m_i)$ and the centre of mass is at $\bar{\mathbf{r}}$, then

$$M\bar{\mathbf{r}} = \sum m_i\mathbf{r}_i$$

$$\Rightarrow \quad M\ddot{\bar{\mathbf{r}}} = \sum m_i\ddot{\mathbf{r}}_i$$

Hence equation (1) becomes

$$\sum \mathbf{F}_i = M\ddot{\bar{\mathbf{r}}},$$

so that the centre of mass moves as though all the mass and all the external forces were concentrated upon it.

When, for example, a diving champion makes a spectacular dive from a high board, her movement may be a complicated mixture of twists and turns and somersaults, and her body is certainly far from rigid. Nevertheless, the only external forces are the weights of the various particles that make up her body, and we see from the above result that her centre of mass moves as though all these weights and all her mass were concentrated there: in other words, her centre of mass (which will itself be varying relative to her body as she curls up, 'jack-knifes', or stretches out) is bound to move in a perfect parabola.

Exercise 14.5a

1   A girl standing on perfectly smooth ice loses her balance. What is the path

of her centre of mass? If a friend helped to save her as she fell, what would be the path of their joint centre of mass?

2 Find the position vector of the centre of mass of particles of masses 4, 3, 2, 3 units at rest at the points

$$\mathbf{i} + \mathbf{j}, \quad 2\mathbf{i} - \mathbf{j}, \quad 2\mathbf{i} + \mathbf{j}, \quad 2\mathbf{i} + 3\mathbf{j}$$

respectively. If each mass is acted upon by a force directed toward the origin and proportional to its distance from the origin, find the direction of the initial acceleration of the centre of mass.   (L)

Rotation about a fixed axis: moments of inertia

It is beyond our present scope to investigate further the general motion of a system of particles, and we must restrict ourselves to the special case of a rigid body which is rotating about a fixed axis:

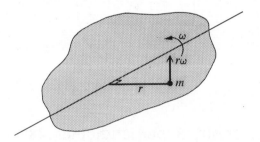

Suppose that a rigid body is rotating with angular velocity ω about a fixed axis and that a typical particle of mass m is at a perpendicular distance r from this axis.

Then the speed of m is $r\omega$ and its kinetic energy is $\frac{1}{2}mr^2\omega^2$. So the total kinetic energy of the body is $\Sigma \frac{1}{2}mr^2\omega^2 = \frac{1}{2}(\Sigma\,mr^2)\omega^2$.

Now the expression $\Sigma\,mr^2$ clearly depends only on the masses m and the way in which they are distributed in the body relative to the given axis. It is usually denoted by I, called the *moment of inertia of the body about the given axis*.

So   k.e. $= \frac{1}{2}I\omega^2$,   where $I = \Sigma\,mr^2$

Comparing the two expressions

k.e. $= \frac{1}{2}mv^2$ and   k.e. $= \frac{1}{2}I\omega^2$

it might be expected that I will, for rotational motion, play a part very similar to that taken by mass for linear motion; and just as mass can be regarded as a measure of a particle's inertia, or reluctance to accelerate when under the action of a given force, so the moment of inertia of a body about a given axis is a measure of its reluctance to accelerate rotationally when under the action of a given moment.

Our first task, however, is to calculate the moments of inertia of a number of different rigid bodies.

Circular hoop rotating about a perpendicular axis through its centre

Suppose that a circular hoop of mass M and radius a is rotating about a perpendicular axis through its centre.

If a typical particle has mass m

$$I = \sum mr^2$$

But, for every particle, $r = a$,

so $\quad I = \sum ma^2 = (\sum m)a^2$

$\Rightarrow \quad I = Ma^2$

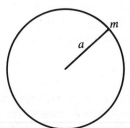

Uniform circular disc, about a perpendicular axis through its centre

Suppose again that the body has mass M and radius a, and also that its surface density is σ.

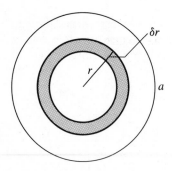

We begin by dividing the disc into a series of concentric hoops, and suppose that a typical hoop has radius r and thickness δr.

Then   area of hoop $= 2\pi r\,\delta r$

and   its mass $= 2\pi\sigma r\,\delta r$

Hence   m.i. of hoop $= (2\pi\sigma r\,\delta r)r^2$
$$= 2\pi\sigma r^3\,\delta r$$

and   m.i. of disc $= \lim\sum 2\pi\sigma r^3\,\delta r$

$$= \int_0^a 2\pi\sigma r^3\,\mathrm{d}r$$

$$= \tfrac{1}{2}\pi\sigma a^4$$

But $M = \sigma \times \pi a^2 = \pi\sigma a^2$

So   m.i. $= \tfrac{1}{2}Ma^2$

Uniform rod

Suppose that a uniform rod has mass M and length $2a$. Its moment of inertia clearly depends on the chosen axis, and we shall consider two cases, both of them perpendicular to the rod and through its end and mid-point respectively.

a) About perpendicular axis through an end

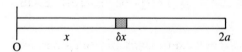

Let linear density of rod be ρ.

Then $M = 2\rho a$

Taking short elements of length δx and mass $\rho\,\delta x$, we see that

$$\text{m.i.} = \lim\sum \rho\,\delta x \times x^2$$

$$= \int_0^{2a} \rho x^2\,\mathrm{d}x$$

$$= \left[\tfrac{1}{3}\rho x^3\right]_0^{2a} = \frac{8\rho a^3}{3}$$

But $M = 2\rho a$

So   m.i. $= \dfrac{4\,Ma^2}{3}$

b) About perpendicular axis through its mid-point

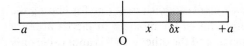

In a similar way,

$$\text{m.i.} = \lim \sum \rho \, \delta x \times x^2$$

$$= \int_{-a}^{a} \rho x^2 \, dx = \frac{2\rho a^3}{3}$$

So   m.i. $= \frac{1}{3}Ma^2$

Radius of gyration

Since $I = \sum mr^2$, we can also express it as $M\kappa^2$, where κ is a length, called the *radius of gyration* of the body about this axis.

Hence $I = M\kappa^2$

and we see that

$$\kappa^2 = \frac{\sum mr^2}{\sum m}$$

is precisely the variance of the distances r of masses m from the given axis, so that κ is their standard deviation (or 'root mean square').

Thus, for a uniform circular disc about a perpendicular axis through its centre,

$$M\kappa^2 = \frac{1}{2}Ma^2$$

$$\Rightarrow \quad \kappa = \frac{a}{\sqrt{2}} \approx 0.71a$$

and for a uniform rod of length $2a$ about a perpendicular axis through its centre,

$$M\kappa^2 = \frac{1}{3}Ma^2$$

$$\Rightarrow \quad \kappa = \frac{a}{\sqrt{3}} \approx 0.58a$$

Two general theorems

In the calculation of moments of inertia it is sometimes convenient to make use of a general theorem, and here we shall establish two: one for finding the moment of inertia about an axis in terms of the moment of inertia about a parallel axis through the centre of mass, and the other which is applicable only to a lamina.

Parallel axis theorem

Suppose that the moment of inertia of a body about a certain axis is I_O and about a parallel axis through the centre of mass is I_G.

We shall suppose that a typical mass m of the body lies at a point P and that a section taken through P perpendicular to the axes (which in our figure are supposed to be perpendicular to the page) meets them in points O and G.

Taking an origin at O, we let

$$\mathbf{OG} = \mathbf{c}, \quad \mathbf{GP} = \mathbf{r}, \quad \mathbf{OP} = \mathbf{R}$$

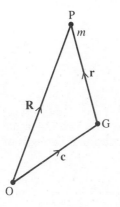

Then
$$I_O = \sum mR^2 = \sum m\mathbf{R}^2$$
$$= \sum m(\mathbf{r} + \mathbf{c})^2$$
$$= \sum m\mathbf{r}^2 + 2\sum m\mathbf{r}.\mathbf{c} + \sum m\mathbf{c}^2$$
$$= \sum m\mathbf{r}^2 + 2(\sum m\mathbf{r}).\mathbf{c} + (\sum m)\mathbf{c}^2$$

We now let the total mass of the body, $\sum m = M$. Also, since the centre of mass lies on the axis through G,

$$\sum m\mathbf{r} = \mathbf{0}$$

So
$$I_O = \sum m\mathbf{r}^2 + 2\mathbf{c}.\mathbf{0} + (\sum m)\mathbf{c}^2$$

$$\Rightarrow \quad I_O = I_G + Mc^2$$

So, for example, we could have found the moment of inertia of a uniform rod about its end from that about its centre, simply by writing

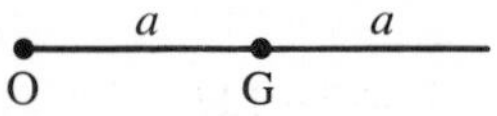

$$\begin{aligned} I_O &= I_G + Ma^2 \\ &= \tfrac{1}{3}Ma^2 + Ma^2 \\ &= \tfrac{4}{3}Ma^2 \end{aligned}$$

And the moment of inertia of a uniform hoop of mass M and radius a about a perpendicular axis through a point of its circumference

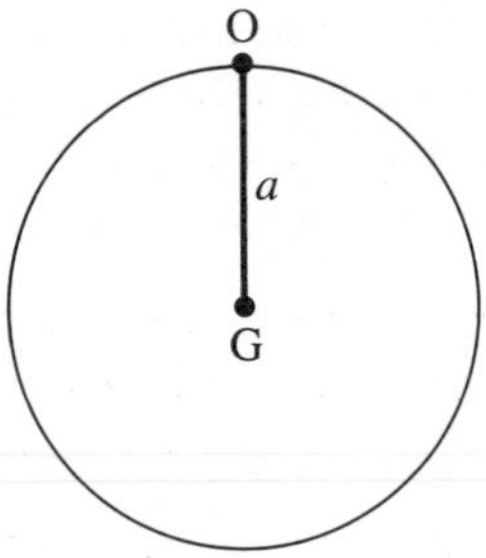

$$\begin{aligned} &= I_G + Ma^2 \\ &= Ma^2 + Ma^2 = 2Ma^2 \end{aligned}$$

'Perpendicular axes' theorem for a lamina

Suppose that perpendicular axes Ox, Oy, Oz are taken with Ox, Oy in the plane of a lamina and Oz perpendicular to its plane. Denoting moments of inertia about these axes by I_x, I_y, I_z,

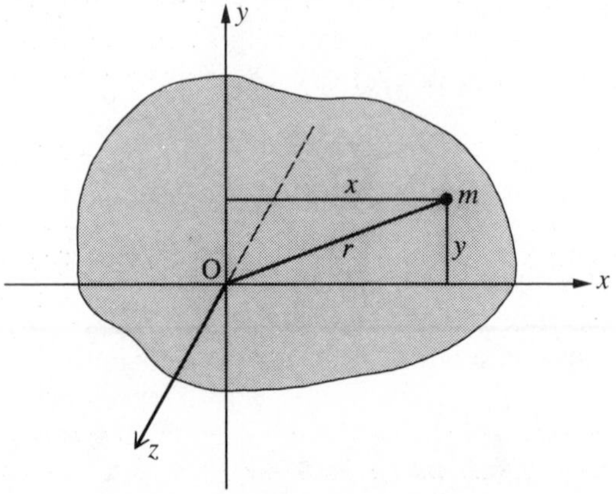

$$\begin{aligned} I_O &= \sum mr^2 \\ &= \sum m(y^2 + x^2) \\ &= \sum my^2 + \sum mx^2 \\ &= I_x + I_y \end{aligned}$$

So $\quad I_z = I_x + I_y$

Hence, for example, we can immediately find the moment of inertia I of a uniform disc about a diameter simply by observing that, since $I_x = I_y$,

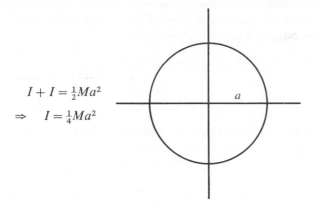

$$I + I = \tfrac{1}{2}Ma^2$$

$$\Rightarrow \quad I = \tfrac{1}{4}Ma^2$$

Similarly, we know that a uniform rectangular lamina of mass M and sides $2a$, $2b$ can be dissected into strips parallel to its sides, so that its moments of inertia about parallel axes through its centre are

$$\tfrac{1}{3}Ma^2 \qquad \text{and} \qquad \tfrac{1}{3}Mb^2$$

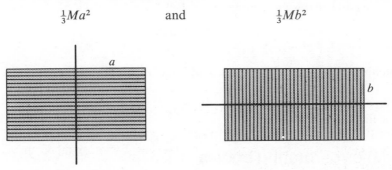

Hence, using the perpendicular axes theorem, we see that its moment of inertia about a perpendicular axis through its centre is

$$\tfrac{1}{3}Ma^2 + \tfrac{1}{3}Mb^2 = \tfrac{1}{3}M(a^2 + b^2)$$

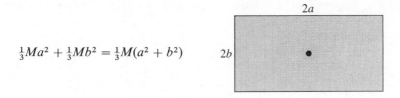

Exercise 14.5b

Calculate the moments of inertia of:

1 A uniform rectangular lamina of mass M and sides $2a$, $2b$ about a perpendicular axis through a corner.

2 A uniform circular hoop of mass M and radius a about:
   a) a diameter;
   b) a tangent;
   c) a perpendicular axis through a point on its circumference.

3 A uniform circular lamina of mass M and radius a about:
   a) a tangent;
   b) a perpendicular axis through a point on its circumference.

4 A uniform sphere of mass M and radius a about:
   a) a diameter;
   b) a tangent.

5 A uniform spherical shell of mass M and radius a about:
   a) a diameter;
   b) a parallel line through a point of the shell.

   (*Hint*: $\sum m(y^2 + z^2) = \sum m(z^2 + x^2) = \sum m(x^2 + y^2)$
   and $x^2 + y^2 + z^2 = a^2$.)

6 A uniform solid cone of mass M, height h and base-radius a, about:
   a) its axis of symmetry;
   b) a perpendicular axis through its vertex;
   c) a diameter of its base.

Dynamics of rotating bodies

Suppose that a body is rotating with angular velocity ω about a fixed axis, that the moment of the external forces about this axis has magnitude L and the moment of inertia of the body about this axis is I.

   Further, suppose that on a mass m at a distance r from the axis there are acting transversely:
a) an external force P;
b) an internal force Q.

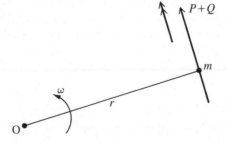

Then $P + Q = mr\dot{\omega}$

$\Rightarrow$ $r(P + Q) = mr^2\dot{\omega}$

$\Rightarrow$ $\sum rP + \sum rQ = (\sum mr^2)\dot{\omega} = I\dot{\omega}$

But since the forces Q are internal, it follows from Newton's third law that $\sum rQ = 0$.

Hence $\sum rP = I\dot{\omega}$

But $\sum rP = L$

So $\boxed{L = I\dot{\omega}}$

Now for a small rotation $\delta\theta$ the work done by P is $Pr\,\delta\theta$. So in a finite displacement, the work done by P is $\int Pr\,d\theta$; and the sum of these amounts of work for all the external forces is

$$\sum \int Pr\,d\theta = \int (\sum Pr)\,d\theta = \int L\,d\theta$$

But $L = I\dot{\omega} = I\dfrac{d\theta}{dt}\dfrac{d\omega}{d\theta} = I\omega\dfrac{d\omega}{d\theta}$

So $\int L\,d\theta = \int I\omega\dfrac{d\omega}{d\theta}\,d\theta = \int I\omega\,d\omega$

$\Rightarrow$ $\boxed{\int L\,d\theta = [\tfrac{1}{2}I\omega^2]}$

and again we see that the total work done by the external forces is equal to the gain in kinetic energy.

Example 1

A uniform flywheel of mass 200 kg and radius 0.5 m is initially at rest and is made to rotate about its fixed axis by a rope which is wrapped round the flywheel for three complete turns and then pulled with a steady force of 200 N. Find:

a) the angular acceleration of the flywheel;

b) its final angular velocity.

a) As the flywheel is uniform,

   m.i. $= \tfrac{1}{2} \times 200 \times (0.5)^2 = 25\,\text{kg}\,\text{m}^2$

But the moment about its axis of the tension in the rope

$$= 200\,\text{N} \times 0.5\,\text{m} = 100\,\text{Nm}$$

So the angular acceleration $\dot{\omega}$ is given by

$$100 = 25\dot{\omega} \quad \Rightarrow \quad \dot{\omega} = 4$$

Hence the angular acceleration is $4\,\text{rad s}^{-2}$.

b) As this angular acceleration is constant throughout three complete revolutions ($= 6\pi\,\text{rad}$), the final angular velocity ω is such that

$$\omega^2 = 0^2 + 2 \times 4 \times 6\pi = 48\pi$$

$$\Rightarrow \quad \omega = 12.3$$

Hence the final angular velocity is $12.3\,\text{rad s}^{-1}$.

Alternatively, the total work done by the tension

$$= 100\,\text{Nm} \times 6\pi\,\text{rad} = 600\pi\,\text{J}$$

and final k.e. $\qquad = \tfrac{1}{2} \times 25 \times \omega^2\,\text{J}$

So $\quad \tfrac{1}{2} \times 25\omega^2 = 600\pi$

$$\Rightarrow \qquad \omega^2 = 48\pi$$

$$\Rightarrow \qquad \omega = 12.3$$

Example 2

A flywheel whose moment of inertia is $1.01\,\text{kg m}^2$ is rotating at approximately $1200\,\text{rev min}^{-1}$, the value oscillating between 1.0 per cent above and 1.0 per cent below this figure. Calculate in joules the difference between the maximum and minimum values of the kinetic energy. If the change from minimum to maximum occurs in one-hundredth of a second calculate the average power input in kilowatts during this period.

Power to the flywheel is cut off at $1200\,\text{rev min}^{-1}$ and the wheel is brought to rest by a constant couple of $80\,\text{N m}$. Calculate how many revolutions the wheel makes in coming to rest.

Maximum angular velocity $= 1212\,\text{rev min}^{-1}$

$$= \frac{1212 \times 2\pi}{60}\,\text{rad s}^{-1}$$

$$= 40.4\pi\,\text{rad s}^{-1}$$

Minimum angular velocity $= 1188\,\text{rev min}^{-1}$

$$= 39.6\pi\,\text{rad s}^{-1}$$

So difference of kinetic energy $= \frac{1}{2} \times 1.01 \times \{(40.4\pi)^2 - (39.6\pi)^2\}$
$$= \frac{1}{2} \times 1.01 \times 80\pi \times 0.8\pi$$
$$= 0.505 \times 64\pi^2 \approx 319 \, \text{J}$$

Hence average power input in $0.01 \, \text{s} = \dfrac{319}{0.01} = 31\,900 \, \text{W} = 31.9 \, \text{kW}$

When power is cut,   k.e. $= \frac{1}{2} \times 1.01 \times (40\pi)^2 = 808\pi^2 \, \text{J}$

Now if the angle through which the flywheel turns before coming to rest is α, the work done by the braking couple is $-80\alpha \, \text{N m}$.

But   work done = gain in k.e.

So $-80\alpha = -808\pi^2$

$\Rightarrow$ $\alpha = 10.1 \, \pi^2 \, \text{rad} \approx 15.9 \, \text{rev}$

Exercise 14.5c

1   Calculate the angular momentum and the kinetic energy of:
   a) a uniform hoop of mass 8 kg and radius $\frac{1}{2}$ m which is spinning in its own plane about its centre with angular velocity 6 rad s^{-1};
   b) a uniform disc of mass 6 kg and radius 20 cm which is spinning in its own plane about its centre with angular velocity 4 rad s^{-1};
   c) the same disc rotating at the same speed, but about a diameter;
   d) a uniform bar of mass 3 kg and length 4 m rotating about one of its ends with angular velocity 6 rad s^{-1};
   e) a hollow cylinder of mass 100 kg and radius 40 cm, rotating about its axis at 100 revolutions/min;
   f) the earth, assuming it to be a uniform sphere of mass 6×10^{24} kg and radius 6.4×10^6 m.

2   A uniform flywheel has mass m and radius a, and is rotating freely about its axis with angular velocity ω. If a constant braking force F is then applied tangentially to its surface, find:
   a) its angular deceleration;
   b) the time it takes to stop;
   c) the angle through which it turns.

3   A bicycle is held with one of its wheels rotating freely with angular velocity 200 rev/min. The wheel has mass 2 kg and radius 25 cm, and when the brakes are lightly applied it is brought to rest in three revolutions. Assuming that the mass of the wheel and the constant braking force are both concentrated at its circumference, find the magnitude of this force. For how long would it need to act; and if $\mu = 0.1$, what is the total contact force between the brakes and the rim?

4   A gyroscope consists of a uniform disc of mass 2 kg and diameter 10 cm mounted on a light axle of diameter 4 mm. A string is wrapped round the axle 20 times and pulled sharply with a constant force. If this pull takes 2 s to unwrap the string, find:
a) the angular acceleration of the disc;
b) its final angular velocity;
c) the tension in the string.

5   A rope 2 m long is wound round the axle of a flywheel and pulled with a constant force of 400 N. When the string is unwound the flywheel is rotating at 200 rev/min. Find its moment of inertia.

6   A uniform rod of length $2a$ and mass M is freely pivoted at one end and is initially in a vertical position above its pivot. If it then topples over, find its loss of potential energy by the time it is **a)** horizontal, **b)** vertically below the pivot; and so find its angular velocity at these two positions. Finally, by considering the moments of its weight, find the corresponding angular accelerations.

7   A uniform cylindrical flywheel of mass 1 000 kg and radius 0.4 m is being accelerated by an electric motor which is working at a constant rate of 10 kW. Ignoring all resistances, find the angular velocity of the flywheel after one minute.

8   A heavy uniform rod of mass m and length l is suspended from one end and oscillates freely about its position of equilibrium. Find:
a) the moment of its weight about the point of suspension when the rod makes an angle θ with the downward vertical;
b) the equation of motion when the rod is in this position;
c) an approximate equation of motion when θ is small;
d) the period of small oscillations.

9   A light thread is wrapped round a uniform cylinder of mass m and radius a which is free to rotate about a horizontal axis. A mass M is tied to the end of this string and allowed to fall from rest in a vertical line. After it has fallen a distance x, find:
a) its speed v;
b) its acceleration a.

10   A uniform circular disc, of mass m and radius a, is free to rotate about a frictionless horizontal axis through its centre and perpendicular to its plane. One end of a light string is attached to a point on the circumference of the disc and part of the string is wound on the circumference. The other end of the string carries a particle A of mass $2m$ hanging freely. The system is released from rest and a restoring couple of moment $mga\theta$ acts on the disc when θ is the angular displacement of the disc. Write down the equation of motion of the disc about its centre and the equation of motion of A. Hence show that the tension in the vertical part of the string is $\frac{2}{5}(1 + 2\theta)mg$. Show also that

$$5a\frac{d^2\theta}{dt^2} = 4g - 2g\theta$$

Deduce that the motion of A is simple harmonic and find the period of this motion.

(You may assume that part of the string always remains wound on the disc during the motion.)   (OC)

14.6 Dimensions

It will have been noticed that this introduction to mechanics has been based entirely upon the three fundamental concepts, or *dimensions*, of mass, length and time. If we now use the symbol [ ] to indicate the dimensions of a quantity, we can write

[mass] $= \mathbf{M}$ [length] $= \mathbf{L}$ [time] $= \mathbf{T}$

Furthermore, we can proceed to find the dimensions of other quantities in terms of $\mathbf{M}, \mathbf{L}, \mathbf{T}$, such as

[area] $=$ [length $\times$ length] $= \mathbf{L}^2$

[volume] $=$ [area $\times$ length] $= \mathbf{L}^3$

$[\text{density}] = \left[\dfrac{\text{mass}}{\text{volume}}\right] = \dfrac{\mathbf{M}}{\mathbf{L}^3} = \mathbf{ML}^{-3}$

$[\text{velocity}] = \left[\dfrac{\text{length}}{\text{time}}\right] = \dfrac{\mathbf{L}}{\mathbf{T}} = \mathbf{LT}^{-1}$

$[\text{acceleration}] = \left[\dfrac{\text{velocity}}{\text{time}}\right] = \dfrac{\mathbf{LT}^{-1}}{\mathbf{T}} = \mathbf{LT}^{-2}$

[force] $=$ [mass $\times$ acceleration] $= \mathbf{MLT}^{-2}$

[moment of a force] $=$ [force $\times$ length] $= \mathbf{ML}^2\mathbf{T}^{-2}$

[impulse] $=$ [force $\times$ time] $= \mathbf{MLT}^{-1}$

[momentum] $=$ [mass $\times$ velocity] $= \mathbf{MLT}^{-1}$

[work] $=$ [force $\times$ distance] $= \mathbf{ML}^2\mathbf{T}^{-2}$

[kinetic energy] $=$ [mass $\times$ (velocity)^2] $= \mathbf{ML}^2\mathbf{T}^{-2}$

$[\text{power}] = \left[\dfrac{\text{work}}{\text{time}}\right] = \mathbf{ML}^2\mathbf{T}^{-3}$

It will be seen that the dimensions of impulse are identical (or *consistent*) with those of momentum, and that the dimensions of work are consistent with those of energy. This notion of consistency can be used to check, and sometimes to derive, equations in mechanics.

Example 1

A piano wire has mass m, length l, and tension F, and a student knows that its period of vibration is either

$$2\pi\sqrt{\left(\frac{ml}{F}\right)} \quad \text{or} \quad 2\pi\sqrt{\left(\frac{F}{ml}\right)}$$

Which is more likely to be correct?

Since $[m] = \mathbf{M}$, $[l] = \mathbf{L}$, $[F] = \mathbf{MLT}^{-2}$, it follows that the dimensions of these two possibilities are

$$\sqrt{\left(\frac{\mathbf{ML}}{\mathbf{MLT}^{-2}}\right)} = \sqrt{\mathbf{T}^2} = \mathbf{T} \quad \text{and} \quad \sqrt{\left(\frac{\mathbf{MLT}^{-2}}{\mathbf{ML}}\right)} = \sqrt{\mathbf{T}^{-2}} = \mathbf{T}^{-1}$$

Hence it is clear that the second possibility for a period of dimension $\mathbf{T}$ would not be consistent, so that the likelier (and in fact correct) formula is $2\pi\sqrt{(ml/F)}$.

Example 2

A simple pendulum consists of a bob of mass m attached to a string of length l and the pendulum is swinging at a point where the acceleration due to gravity is g. Investigate, by means of dimensions, the way in which the period T depends upon m, l, and g.

Firstly, we *suppose* that the period T depends jointly upon m, l, and g in such a way that

$$T \propto m^\alpha l^\beta g^\gamma, \quad \text{where } \alpha, \beta, \gamma \text{ are constants}$$

But $\quad [m^\alpha l^\beta g^\gamma] = \mathbf{M}^\alpha \mathbf{L}^\beta (\mathbf{LT}^{-2})^\gamma$
$$= \mathbf{M}^\alpha \mathbf{L}^{\beta+\gamma} \mathbf{T}^{-2\gamma}$$
and $\qquad [T] = \mathbf{T}$

So for dimensional consistency,

$$\mathbf{M}^\alpha \mathbf{L}^{\beta+\gamma} \mathbf{T}^{-2\gamma} = \mathbf{T}$$
$$\Rightarrow \qquad \alpha = 0$$
$$-2\gamma = 1 \quad \Rightarrow \quad \gamma = -\tfrac{1}{2}$$
$$\beta + \gamma = 0 \quad \Rightarrow \quad \beta = \tfrac{1}{2}$$

Hence $T \propto m^0 l^{1/2} g^{-1/2}$

$\Rightarrow \qquad T \propto \sqrt{\left(\frac{l}{g}\right)}$

It will be recognised that although this does not constitute a *proof*, it is nevertheless a highly convenient approach to such an investigation, which is confirmed by the result of p. 158 that $T = 2\pi\sqrt{(l/g)}$.

Exercise 14.6

1   State the dimensions, in terms of **M**, **L**, **T**, of the following quantities:
   a) pressure (force per unit area);
   b) line density;
   c) surface density;
   d) frequency;
   e) angular velocity;
   f) angular acceleration;
   g) angle;
   h) moment of inertia;
   i) rate of loss of mass (e.g. of a rocket);
   j) constant of gravitation (see Miscellaneous problems, chapter 9, no. 11);
   k) stiffness of a spring (see section 14.1);
   l) surface tension (energy per unit area).

2   Use the method of dimensions to predict how:
   a) the tension T in a string depends upon the mass m of a particle which is being whirled round on its end in a circle of radius r and with speed v;
   b) the height h reached by a stone will depend upon its mass m, the energy E with which it is projected vertically and the acceleration g due to gravity;
   c) the speed v of sound in a gas will depend upon its pressure p, its density ρ and the acceleration g due to gravity;
   d) the frequency f of oscillation of a light spring depends upon its stiffness k, the mass m which it is supporting and the acceleration g due to gravity;
   e) the velocity v of waves in a deep liquid depends upon its density ρ, their wavelength λ and the acceleration g due to gravity.

3   Use the method of dimensions to find an expression for the thrust of moving air on a fixed obstacle, assuming it to be the form $T = kv^\alpha \rho^\beta A^\gamma$ where v is the air velocity, ρ its density, A the frontal area which the obstacle presents to it, and k is a non-dimensional constant. Explain briefly the nature of a dynamical argument which leads us when using absolute units of force to expect the experimental value of k always to be less than unity.

Under certain conditions the compressibility c of the air can affect the thrust appreciably, c being defined as fractional change in volume produced per unit change in pressure. Find the dimensions of c and show that we can now obtain a dimensionally correct formula for T by replacing k by any polynomial in a quantity z defined as $v\sqrt{(c\rho)}$.   (MEI)

Miscellaneous problems

1   A steady uniform stream of air of density ρ and speed u strikes at right angles a plane surface of area A and proceeds, after the contact, with its effective speed still in the original direction and equal to ku (where $k > 0$). Derive an expression for the thrust of the air on the area, stating carefully the principles on which your derivation is based.

A householder has a garage with rectangular doors each 1.20 m wide by 2.25 m high, turning about vertical hinges at their edges. On a windy day she attempts to keep one door open in such a position that the wind is blowing at right angles to the door. To do this she puts a brick on the ground at the outer corner of the door, hoping that the friction of the brick on the ground will be sufficient for the task. Given the data listed below, calculate the greatest wind-speed for which the brick will be effective: state any additional assumptions you find it necessary to make in the course of your work.

Mass of brick $= 3.0\,\text{kg}$; coefficient of friction $= 0.4$; density of air $= 1.10\,\text{kg}\,\text{m}^{-3}$; $k = 0.6$; $g = 10\,\text{m}\,\text{s}^{-2}$.   (MEI)

2   A shell of mass M is at rest in space, when it bursts into two fragments, the energy released being E. Show that the relative speed of the fragments after separation cannot be less than $2\sqrt{(2E/M)}$.

Explain how your conclusion is affected if the shell is moving initially with speed U.   (CS)

3   A space craft of mass 10^4 kg is going round the moon in a circular orbit at a height of 10^6 m. Find the time of the orbit, taking g on the moon as $1.60\,\text{m}\,\text{s}^{-2}$ and the diameter of the moon as 4.14×10^6 m.

To escape from the moon the kinetic energy of the body has to be doubled. What is the magnitude and direction of the least impulse which can be applied to the body to enable it to escape?

If this impulse is spread over a period of one minute, what is the apparent weight of a 100 kg man inside the space craft during this time?   (MEI)

4   A particle of mass m is attached to the middle point of a light elastic string

of natural length a and modulus mg. The ends of the string are attached to two fixed points A and B, A being at a distance $2a$ vertically above B. Prove that the particle can rest in equilibrium at a depth $5a/4$ below A and find the period of oscillation if it is displaced slightly in the vertical direction.

5   A mountaineer falls over a cliff. He is attached to a rope which, providentially, stretches so that he just touches the ground at the foot of the cliff. Find the height of the cliff and the time taken for the mountaineer to reach the ground (in terms of his mass, the length of the unstretched rope and its elastic modulus).   (CS)

6   Three particles A, B and C each of mass m lie at rest in a straight line on a smooth horizontal table, joined by equal taut strings AB, BC. A horizontal impulse P is given to B in a direction perpendicular to the line of the strings. Describe the nature of the motion up to the instant when A and C meet, and calculate the loss of energy of the system if these masses do not separate after colliding.   (MEI)

7   A pump working effectively at P kW delivers water through a nozzle of area A cm^2 at a speed of v m s^{-1}, raising the water h m in the process. Obtain a formula for P in terms of A, v, h and g; assume that 1 m^3 of water has a mass of 10^3 kg.

The jet is directed horizontally at right angles to a vertical blade of a water wheel which is initially at rest. The wheel has a moment of inertia I about its axis which is horizontal and the point of contact of the water is c m from the axle; the water falls vertically after striking the blade.

What is the initial angular acceleration of the wheel? At what rate is energy being given to the wheel initially?   (MEI)

8   A tank containing a liquid of density ρ has a hole in the bottom. The speed u at which the liquid escapes (averaged over the cross-section of the hole) is given by $u = \lambda \rho V g$, where V is the volume of liquid in the tank and λ is a constant. Find the dimensions of λ in terms of mass, length and time.

Given that the hole has cross-sectional area A, write down a differential equation for V as a function of t, in terms of the various constants. (SMP)

9   A flywheel, whose moment of inertia about an axis perpendicular to its plane and through its centre of mass is I, is initially at rest. The axis is fixed and the flywheel rotates about it under the influence of a constant couple of magnitude C and a resisting couple of magnitude $k\omega^2$, where k is a positive constant and ω is the angular speed. Show that ω can never exceed Ω, where $\Omega = \sqrt{\dfrac{C}{k}}$. Find the time taken for the flywheel to attain

the angular speed ω and show that during this time it rotates through an angle

$$\frac{I\Omega^2}{2C}\ln\left\{\frac{\Omega^2}{\Omega^2 - \omega^2}\right\}$$ (JMB)

10   The velocity of propagation of waves of wavelength l on the surface of a liquid, is proportional to

$$\sqrt{\frac{\sigma}{\rho l}}$$

where σ is the surface tension and ρ the density. Deduce the physical dimensions of σ.

Lord Rayleigh showed that the period, T, of vibration of a small liquid drop, when given a slight distortion, depended only on σ, ρ and the radius r of its spherical equilibrium shape. Assuming that

$$T = C\sigma^\alpha \rho^\beta r^\gamma$$

where C, α, β and γ are dimensionless constants, find α, β and γ. (CS)

15 Probability distributions and further statistics

15.1 Probability distributions and generators

Take two coins, spin them as a pair ten times and record your results: two heads (HH), two tails (TT), or one of each (HT). When the authors performed this experiment they obtained the sequence

HT, HT, TT, HT, HH, HH, HT, HT, HH, HT

which can conveniently be classified according to the frequency f_r with which the number of heads obtained is x_r:

number of heads (x_r)	frequency (f_r)
$x_1 = 0$	$f_1 = 1$
$x_2 = 1$	$f_2 = 6$
$x_3 = 2$	$f_3 = 3$
	$n = 10$

In chapter 7 we called this a frequency distribution, and saw that its mean m is given by:

$$m = \frac{1}{n}\sum f_r x_r$$

$$= \tfrac{1}{10}(1 \times 0 + 6 \times 1 + 3 \times 2) = 1.2$$

and its variance s^2 by

$$s^2 = \frac{1}{n}\sum f_r (x_r - m)^2$$

$$= \frac{1}{n}\sum f_r x_r^2 - m^2$$

$$= \tfrac{1}{10}(1 \times 0^2 + 6 \times 1^2 + 3 \times 2^2) - 1.2^2$$

$$= 1.8 - 1.44$$

$$= 0.36$$

If, instead of recording frequencies, we were to tabulate the *relative* frequencies f_r/n we obtain:

number of heads (x_r)	relative frequency (f_r/n)
$x_1 = 0$	$\frac{1}{10}$
$x_2 = 1$	$\frac{6}{10}$
$x_3 = 2$	$\frac{3}{10}$

Furthermore, we can see that

$$m = \sum \frac{f_r}{n} x_r$$

and $$s^2 = \sum \frac{f_r}{n} (x_r - m)^2$$

$$= \sum \frac{f_r}{n} x_r^2 - m^2$$

So far we have made no assumptions about the coins and have not remarked upon any expectations which we might have had. Our task has been simply to record a set of events and to describe them as conveniently as possible. If, however, we now make assumptions about the coins and the way in which they are spun, we can use the probability theory of chapter 7 to predict our expectations. Suppose, for instance, we assume that the coins are both completely unbiased, and they are spun with absolute fairness; then the probabilities p_r of obtaining x_r heads are clearly:

number of heads (x_r)	probability (p_r)
$x_1 = 0$	$p_1 = \frac{1}{4}$
$x_2 = 1$	$p_2 = \frac{1}{2}$
$x_3 = 2$	$p_3 = \frac{1}{4}$

This set of probabilities is referred to as a *probability distribution*, and the set of assumptions (which must be expressed in precise probability terms) from which it arises is called the *probability model*.

If different assumptions were made, we would have a different probability model, and so a different probability distribution. For instance if one of the coins was double-headed, and the other was fair and unbiased, then the probability distribution would be as follows:

x_r	p_r
$x_1 = 0$	$p_1 = 0$
$x_2 = 1$	$p_2 = \frac{1}{2}$
$x_3 = 2$	$p_3 = \frac{1}{2}$

One of the major tasks of statistics is to compare an observed frequency distribution with the probability distribution which arises from a particular probability model, so equivalent measures to the mean and variance of a frequency distribution are required for a probability distribution. To obtain these measures, relative frequency is replaced by probability, and to maintain a clear distinction between the two distributions, the probability measures are denoted by the Greek letters μ, σ^2.

Now $m = \sum \dfrac{f_r}{n} x_r$ and $s^2 = \sum \dfrac{f_r}{n}(x_r - m)^2$

so $\mu = \sum p_r x_r$ and $\sigma^2 = \sum p_r(x_r - \mu)^2$

As in statistics, so with probability distributions, μ is called the *mean*, or *expected value* of x, σ^2 its *variance* and σ its *standard deviation*.

So for the probability distribution with two unbiased coins,

$\mu = \tfrac{1}{4} \times 0 + \tfrac{1}{2} \times 1 + \tfrac{1}{4} \times 2 = 1$

$\sigma^2 = \tfrac{1}{4} \times 1^2 + \tfrac{1}{2} \times 0^2 + \tfrac{1}{4} \times 1^2 = 0.5$

(whereas in our trial, $m = 1.2$ and $s^2 = 0.36$).

(It should be pointed out at this stage that in the calculation of μ and σ, just as for m and s, the values of x_r are not restricted to integers: see, for instance, exercise 15.1a, no. **8**.)

In summary:

statistics		probability	
relative frequency $\dfrac{f_r}{n}$		probability p_r	
mean $\quad m = \sum\limits_r \dfrac{f_r}{n} x_r$		mean (expected value) $\quad \mu = \sum\limits_r p_r x_r$	
variance $\quad s^2 = \begin{cases} \sum\limits_r \dfrac{f_r}{n}(x_r - m)^2 \\[2mm] \sum\limits_r \dfrac{f_r}{n} x_r^2 - m^2 \end{cases}$		variance $\quad \sigma^2 = \begin{cases} \sum\limits_r p_r(x_r - \mu)^2 \\[2mm] \sum\limits_r p_r x_r^2 - \mu^2 \end{cases}$	

Exercise 15.1a

1   Calculate the expected value and variance of the number of heads showing when three unbiased coins are tossed together. Conduct a series of 20 such trials and calculate the mean and variance for the frequency distribution that you obtain. Compare your theoretical and experimental results.

2   Calculate:
    a) the variance of the score obtained on throwing a single die;
    b) the variance of the total score obtained on throwing two dice;
    c) the variance of the average score on two dice.
    How are your answers to **a)** and **b)** and to **b)** and **c)** related? (Assume that the dice are 'fair'.)

3   A computer is made to produce randomly the numbers 0, 1, 2, ..., 9. What is the expected value and variance of the numbers so produced?

4   On the basis of past evidence, it is estimated that the probabilities of a certain type of plant having 4, 5, 6, 7, 8, 9 leaves are 0.13, 0.21, 0.38, 0.16, 0.09, 0.03 respectively. Find, correct to 2 decimal places, the expected value and variance of the number of leaves on such a plant.

5   Two unbiased dice are rolled and the greater score (or either if they are the same) is recorded. State the set of possible scores and the probabilities associated with these scores. Find the expected value (i.e. theoretical mean) of the recorded scores.   (SMP)

6   A boy spins a coin until he obtains a head, but impatiently gives up if he does not succeed in five attempts. What are the expected value and variance of the number of times he spins the coin?

7   What are the mean and variance of the number of different factors (other than 1 and the number itself) of an integer chosen at random in the range 1 to 30?

8   A box contains 100 tokens which differ in mass, but are otherwise identical. 20 of the tokens have a mass of 4.8 g each, 35 a mass of 5.2 g, 25 a mass of 5.7 g, 15 a mass of 6.5 g, and the remaining 5 have a mass of 8.0 g each. Calculate (to 2 decimal places) the expected value and variance of the mass of a token chosen at random from the box.

9   An examination question consists of two parts, A and B, and the probability of a pupil getting part A correct is $\frac{2}{3}$. If he gets A correct, the probability of getting B correct is $\frac{3}{4}$; otherwise it is $\frac{1}{6}$. There are three marks for a correct solution to part A, two marks for part B, and a bonus mark if both parts are correct. Calculate the expected value and variance of the pupil's total mark for the question.

10 In a game where the gambler rolls two dice, on a £1 stake the casino pays out £10 for a double six, and £3 for a score of seven (the stake money being returned as well), and for any other score the gambler loses his money. What is the casino's expected profit (to the nearest penny) on a £100 stake?

11 There are three greengrocers in town. At each shop there is (independently) a probability of 0.8 that there will be melons in stock. I go to

each shop in turn until I find one that has them; if I am unlucky at all three, then I give up and go home. Complete the table:

number of shops I will go to	1	2	3
probability	0.8		

Hence find the expected number of shops I will go to.   (SMP)

12 A bag contains 6 blue discs and 5 red discs. Three discs are randomly selected without replacement. Find the probability that the three selected discs
  a) are all of the same colour;
  b) consist of two blue discs and one red disc.
Find the mean and variance of the number of blue discs that will be selected. If instead the three discs are selected with replacement, write down the new mean and variance of the number of blue discs selected.
                                                              (JMB)

13 Two small piles of cards contain respectively the 1, 2, 3, 4, 5, 6 of diamonds and the 7, 8, 9, 10 of diamonds. Verify that these sets of numbers have expectations $3\frac{1}{2}$ and $8\frac{1}{2}$ respectively, and that the variances are $2\frac{11}{12}$ and $1\frac{1}{4}$ respectively.
  One card is drawn at random from each of the piles and the product of the numbers so formed is calculated. Find the expected value and variance of this product.
  Find also the probability that the product from any one draw will exceed its expectation.   (C)

14 A motorist drives into town and has the choice of two car parks. It takes him 20 minutes to drive from his home to car park A, which is never full, and then it takes him 15 minutes to walk to the office. If he decides to try to get to car park B it takes him 25 minutes to get there from his home. The probability that car park B is not full is p, and in this case he can walk to the office in 5 minutes. If car park B is full he drives back to car park A, which takes 5 minutes. If he always drives first to car park B find the probability distribution of the time T, in minutes, taken to get to work. Find the expected value of T, showing that its value is less than 35 if p exceeds $\frac{2}{3}$. Find also the variance of T.
  If however, the motorist tosses a coin in order to decide whether to go straight to car park A or to try car park B first, find, in this case, the probability distribution of T, and the expected value of T.   (C)

Probability generators

Suppose that we want to find the mean and variance of the geometric probability distribution (see section 7.4) of the number of throws required of a 'fair' die until a six is obtained.

The probability distribution is given by:

number of throws	probability
$x_1 = 1$	$p_1 = \frac{1}{6}$
$x_2 = 2$	$p_2 = \frac{1}{6} \times \frac{5}{6} = \frac{5}{36}$
$x_3 = 3$	$p_3 = \frac{1}{6} \times \left(\frac{5}{6}\right)^2 = \frac{25}{216}$
$\vdots$	$\vdots$
$x_r = r$	$p_r = \frac{1}{6} \times \left(\frac{5}{6}\right)^{r-1}$

Hence $\mu = \frac{1}{6} \times 1 + \frac{5}{36} \times 2 + \frac{25}{216} \times 3 + \cdots$

and $\sigma^2 = (\frac{1}{6} \times 1^2 + \frac{5}{36} \times 2^2 + \frac{25}{216} \times 3^2 + \cdots) - \mu^2$

Direct calculation of such infinite sums will be lengthy, to say the least, and even where the number of outcomes is not infinite, calculation of μ and σ^2 can be very tedious. However, when the variables x_r are integers, calculations can be eased considerably by using the *probability generator* (or as it is sometimes known, the *probability generating function*).

Let the outcomes which have associated random variables 0, 1, 2, ... have probabilities denoted by $p_0, p_1, p_2, \ldots$. Then the probability generator G(t) is defined by

$$\text{G}(t) \equiv \Sigma p_r t^r \equiv p_0 + p_1 t + p_2 t^2 + \cdots + p_r t^r + \cdots$$

In the probability generator, t is a dummy variable: it has no significance in itself, and any other letter would do just as well. We merely use the powers of t to pick out equivalent probabilities—the coefficient of t^n in G(t) giving the probability of obtaining the value n for the random variable. It should also be noted that G(t) may be a finite or infinite polynomial in t.

It is possible also to define a probability generator when the random variables are negative or non-integral (see exercise 15.1b, no. **15**). In such a case, of course, the generator will not be a polynomial, but nevertheless the properties of G(t) which we will shortly derive will still hold. In practice, however, such probability generators are not often used.

Clearly the probability generator provides us with a means of displaying a complete probability distribution in a single expression. For instance, the distribution for the number of heads obtained on throwing two unbiased coins is summarised by the generator,

$$\text{G}(t) = \frac{1}{4} + \frac{1}{2}t + \frac{1}{4}t^2$$

It is not immediately obvious how a probability generator can be used to evaluate μ and σ^2, but we shall see shortly that they are calculated from the first and second derivatives of G(t). Meanwhile, to aid differentiation, the generator will need to be written in as simple a form as possible. So let us simplify the generator of our geometric distribution:

$$G(t) = \tfrac{1}{6}t + \tfrac{1}{6} \times \tfrac{5}{6}t^2 + \tfrac{1}{6} \times (\tfrac{5}{6})^2 t^3 + \cdots + \tfrac{1}{6} \times (\tfrac{5}{6})^{r-1}t^r + \cdots$$

$$= \tfrac{1}{6}t[1 + \tfrac{5}{6}t + (\tfrac{5}{6}t)^2 + \cdots + (\tfrac{5}{6}t)^{r-1} + \cdots]$$

Now the contents of the square brackets form a geometric progression whose sum is

$$\lim_{n \to \infty} \left[\frac{1 - (\tfrac{5}{6}t)^n}{1 - \tfrac{5}{6}t} \right]$$

and since t has no significance in itself, we can choose it such that $|\tfrac{5}{6}t| < 1$, so that $(\tfrac{5}{6}t)^n \to 0$ as $n \to \infty$.

Hence $\quad G(t) = \tfrac{1}{6}t \times \dfrac{1}{1 - \tfrac{5}{6}t} = \dfrac{t}{6 - 5t}$

Derivation of μ, σ^2 from the probability generator

When the random variables x_1, $x_2, \ldots$ are 0, 1, 2, $\ldots$, the formulae for expected value and variance take the form:

$$\mu = \sum r p_r \qquad \sigma^2 = \sum r^2 p_r - \mu^2$$

Now $\quad G(t) = \sum p_r t^r$

$\Rightarrow \quad G'(t) = \sum r p_r t^{r-1} \qquad \Rightarrow \quad G'(1) = \sum r p_r$

$\Rightarrow \quad G''(t) = \sum r(r-1) p_r t^{r-2} \quad \Rightarrow \quad G''(1) = \sum r(r-1) p_r$

$\qquad\qquad\qquad\qquad\qquad \Rightarrow \quad G''(1) + G'(1) = \sum r^2 p_r$

Hence $\quad \mu = \sum r p_r \qquad = G'(1)$

and $\quad \sigma^2 = \sum r^2 p_r - \mu^2 = G''(1) + G'(1) - [G'(1)]^2$

Summarising:
$$\mu = G'(1)$$
$$\sigma^2 = G''(1) + G'(1) - [G'(1)]^2$$

Returning now to our example,

$$G(t) = \frac{t}{6 - 5t}$$

so $\quad G'(t) = \dfrac{6 - 5t + 5t}{(6 - 5t)^2}$

$$= \frac{6}{(6 - 5t)^2} \quad \Rightarrow \quad G'(1) = 6$$

and　　$G''(t) = \dfrac{60}{(6 - 5t)^3}$　$\Rightarrow$　$G''(1) = 60$

Therefore　$\mu = 6$　and　$\sigma^2 = 60 + 6 - 6^2$
$$= 30$$

Hence the expected number of throws required to obtain a six is 6, and the standard deviation of the number of throws is $\sqrt{30}$.

Binomial distribution

We saw in section 7.4 that if the chance of a particular event occurring in a single trial is p, then in n such independent trials, the probabilities of 0, 1, 2, 3, ..., n occurrences are given by the terms of the binomial expansion

$(q + p)^n$,　where $q = 1 - p$

These probabilities, therefore, are also the coefficients of successive powers of t in the expansion of

$$G(t) = (q + pt)^n$$

which is therefore the probability generator of the binomial distribution.

Now　　$G(t) = (q + pt)^n$

so　　$G'(t) = np(q + pt)^{n-1}$

$\Rightarrow$　$\mu = G'(1) = np(q + p)^{n-1} = np$,　since $q + p = 1$

Furthermore　$G''(t) = n(n - 1)p^2(q + pt)^{n-2}$

$\Rightarrow$　　　　$G''(1) = n(n - 1)p^2$

so that　　　　$\sigma^2 = G''(1) + G'(1) - [G'(1)]^2$

$$= n^2p^2 - np^2 + np - n^2p^2$$

$$= np(1 - p)$$

$$= npq$$

Hence　　$\mu = np$　　$\sigma = \sqrt{(npq)}$

The result that, for a binomial distribution,

$$G(t) = (q + pt)^n = (q + pt)(q + pt)\ldots(q + pt)$$

where $q + pt$ is the generator of a single trial, illustrates the fact that for a probability distribution arising from a series of *independent* (and not necessarily identical) trials, the overall probability generator is the product of the generators of each of the separate trials.

> ### *Example*
>
> In a certain large city, it is known that $\frac{1}{3}$ of the voters support the radical party. In a sample of 12 voters, what is the expected value and standard deviation of the number of radicals?
>
> We have a binomial probability situation with $n = 12$, $p = \frac{1}{3}$ and $q = \frac{2}{3}$.
>
> So $\mu = np = 12 \times \frac{1}{3} = 4$
>
> and $\sigma = \sqrt{(npq)} = \sqrt{(12 \times \frac{1}{3} \times \frac{2}{3})} = \sqrt{\frac{8}{3}} \approx 1.63$

Exercise 15.1b

1 Calculate the probability generator for the number of tosses of an unbiased coin needed to obtain the first head, and hence evaluate the mean and variance of the number of tosses.

2 Calculate the expected value and standard deviation of the number of heads obtained when an unbiased coin is tossed **a)** 4, **b)** 36, **c)** 100 times.

3 In the next general election suppose that 36% of the electors intend to vote Liberal. Find, in terms of n, the mean and standard deviation of the percentage of those who intend to vote Liberal in samples of size n.

(SMP)

4 Six players take it in turns to cut a pack of cards (excluding jokers), and each time the cards are replaced before the next player cuts. Calculate the expected value and standard deviation of the total numbers of spades turned up when each player has cut.

5 The probability of a pupil arriving at school late on any given day is $\frac{1}{10}$. What is the probability of his being punctual for a whole week (i.e. 5 school days)? Calculate the mean and variance of the number of days he will be late in a school term consisting of 14 weeks (i.e. 70 days). Also calculate the expected number of completely punctual weeks in the term.

6 Seedlings are planted in 10 rows of six each. The probability of a seedling dying before it flowers is $\frac{1}{8}$. Calculate the mean and variance of the number of rows in which all the seedlings flower.

7 Eight per cent of eggs sold by a certain grocer are brown.
   a) Calculate the mean and variance of the number of brown eggs obtained in a carton of 12.

b) By finding the probability of a carton containing no brown eggs, and using a suitable probability generator, find how many cartons someone can expect to buy before they find their first brown egg.

8 Prove that for a binomial probability distribution arising from n trials, the maximum possible value of the variance is $n/4$.

9 An electronics firm packs the resistors that it produces in boxes of 800. On average one component in 100 is faulty. Calculate the expected number of defective resistors in a box, and the variance of this number. How could you quickly obtain a good approximation to the variance in this case? What would the percentage error (of the actual value) be if this approximation were used?

10 A box contains a large number of screws. The screws are very similar in appearance, but are in fact of 3 different types, A, B and C, which are present in equal numbers. For a given job only screws of type A are suitable. If 4 screws are chosen at random, find the probability that:
a) exactly two are suitable;
b) at least two are suitable.
If 20 screws are chosen at random, find the expected value and variance of the number of suitable screws.   (c)

11 In a machine game of chance, when a lever is pulled, one of the numbers 1, 2, 3 appears in a window. The lever is pulled 5 times and the total score is recorded. If the probabilities associated with the numbers 1, 2, 3 are $\frac{1}{6}, \frac{1}{3}, \frac{1}{2}$ respectively, write down an expression for the probability generator $G(t)$ for the possible total scores. Evaluate $G(-1)$ and, hence or otherwise, find the probability that the total score is even.   (SMP)

12 Three balls a, b, c are placed at random in three boxes A, B, C (with one ball to each box). The variable x is defined by

$$\begin{cases} x = 1 & \text{if ball } a \text{ is in box A,} \\ x = 0 & \text{otherwise} \end{cases}$$

Find the probability generator for x.
   The variable y is defined in the table

	box			value
	A	B	C	of y
	a	b	c	1
	c	a	b	0
contents	b	c	a	0
	a	c	b	0
	b	a	c	1
	c	b	a	1

Deduce the probability generator of y.

The variable s is defined by $s = x + y + 1$. Given that x, y are independent, show that s has generator

$$\tfrac{1}{6}(t^3 + 3t^2 + 2t)$$

and deduce the mean and variance of s.   (SMP)

13 A ticket is drawn at random from a set of ten, numbered 1, 2, 3, ..., 10. Write down the probability generator for the number on the ticket as a series, and sum it.

Four tickets are now drawn, one at a time with replacement, from the set. Write down the probability generator for the total on the four tickets. What is the probability that the total of the numbers on the four tickets is 11?   (SMP)

14 By using the fact that the probability generator for the number of throws of a die required to obtain a six is $t/(6 - 5t)$, or otherwise, obtain the probability generator for the number of throws required to obtain two sixes (not necessarily consecutively). Hence calculate the expected value and variance of the number of throws required to obtain two sixes.

15 A multiple-choice examination paper consists of 25 questions, each with 5 possible answers, only one of which is correct. A student gains 4 marks for a correct answer, and loses one mark for an incorrect answer.

a) An entirely ignorant student has to guess the answer for each question. Find the probability generator for the number of marks he obtains on the paper, and hence calculate his expected total mark, and the standard deviation of that mark. (*Hint:* first find the probability generator for the marks on a single question.)

b) A rather more intelligent student has twice as much chance of choosing the correct answers as he has of choosing any one of the wrong answers; find the expected value and standard deviation of his total score on the paper.

16 A petrol company gives away one medal with every purchase of petrol. Each medal is equally likely to be any one of a set of n medals. At a stage when a customer already has s different medals, show that the expected number of purchases to get the first medal different from those already held is $n/(n - s)$. Deduce that the expected number of purchases for a complete set is

$$n\left(1 + \frac{1}{2} + \frac{1}{3} + \cdots + \frac{1}{n}\right)$$ (C)

15.2 Continuous probability distributions

Statement 1: The probability of throwing a 5 with a fair die is $\frac{1}{6}$.

Statement 2: The probability of a voter, chosen at random in London, being a Liberal is $\frac{1}{6}$.

Statement 3: The probability of the new machine turning out a component of length 241 mm is $\frac{1}{6}$.

Superficially the three statements above seem very similar. They all make predictions about what may occur if we conduct certain trials, and we presume that there is some foundation, theoretical or experimental, for each. But let us examine the statements separately, and in a little more detail.

Statement 1 is presumably based on theoretical considerations (themselves supported by past experience) and uses the notion of 'equal likelihood'. It tells us that if we conduct a large number of trials, about $\frac{1}{6}$ of the outcomes will be 5s; and that, in general, the larger the number of trials, the closer we get to this fraction.

Statement 2 can only be based on statistical evidence, and tells us that in a large, randomly chosen group of voters, we could expect about $\frac{1}{6}$ to be Liberals.

Both of these statements are concerned with *discrete* (i.e., definite and distinct) outcomes: when a die is thrown there are only six distinct results possible, and similarly a voter has political views which can be expressed in certain distinct choices (and we would regard 'don't know' as a category of choice).

Statement 3 however, is concerned with length, a *continuous* quantity, and presumably tells us that if, for instance, we look at 150 components produced by the machine, about 25 will be 241 mm in length. But if we take these 150 components, and measure each by a micrometer to 0.001 mm, then we would certainly not expect 25 of them to be exactly 241 mm in length; so the statement in its present form seems to be meaningless. Perhaps it is more likely to mean that if measurements are taken *to the nearest millimetre*, then the probability of a length of 241 mm is $\frac{1}{6}$?

So although we can never be absolutely certain about the *exact* length of any object, and must say that

P(length = 241 mm) = 0

nevertheless our original statement can be interpreted as:

P(240.5 ≤ length ≤ 241.5) = $\frac{1}{6}$

Indeed for any continuous quantity, we cannot talk (as we did for discrete quantities) about the probability of it attaining a specific value. We can only meaningfully talk about the probability of length, time, etc., lying *within a certain range of values*.

Cumulative probability

Even though, for a continuous variable like length, we cannot use P(length = 241 mm), it would clearly be valuable if we could associate some form of probability with any given length. This is done by using the idea of cumulative probability.

The *cumulative probability* of length 241 is defined to be the probability that the length is less than (or equal to) 241, and is written

$$F(241) = P(\text{length} \leqslant 241)$$

More generally, the *cumulative probability function* (or *distribution function*) for any variable X is defined by

$$F(x) = P(X \leqslant x)$$

so that $0 \leqslant F(x) \leqslant 1$ for all x

(It should be noted that we can also meaningfully talk about the cumulative probability of a discrete variable; and for instance if we throw a single die, $F(5) = \frac{5}{6}$. But clearly this notion is less useful, and is rarely used.)

Example 1

In a race, the times of the competitors are measured by stop-watch, to the nearest $\frac{1}{10}$ of a second. What is the cumulative probability function for error in measured time?

Let the error be denoted by x, which is uniformly distributed over the interval -0.05 to 0.05.

Therefore $F(-0.05) = 0$ and $F(x) = 0$ for any $x \leqslant -0.05$

$$F(0.05) = 1 \quad \text{and} \quad F(x) = 1 \quad \text{for any } x \geqslant 0.05$$

and $F(x)$ increases steadily in value from 0 to 1 over the interval $-0.05 \leqslant x \leqslant 0.05$.

So $F(x) = 10(x + 0.05)$,
   if $-0.05 \leqslant x \leqslant 0.05$

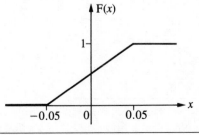

Example 2

Particles are emitted uniformly from a radioactive source in the corner of a room. A screen (of theoretically infinite length) is positioned one metre away from the source, so that all particles must strike the screen. What is the cumulative probability function of the distance of the point of impact from the bottom of the screen?

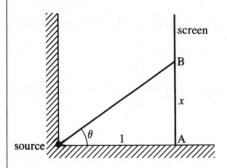

Since the particles are emitted uniformly over the complete arc of $\frac{1}{2}\pi$ radians,

$F(x) = P(\text{particle striking AB})$

$\qquad = \text{fraction of total angle } \frac{1}{2}\pi \text{ subtended by AB}$

$\qquad = \dfrac{\theta}{\frac{1}{2}\pi} = \dfrac{2\theta}{\pi}$

But since $\quad \tan\theta = \dfrac{x}{1} = x$

then $\qquad\qquad \theta = \tan^{-1}x$

Hence $\quad F(x) = \dfrac{2}{\pi}\tan^{-1}x$, in the interval $x \geqslant 0$

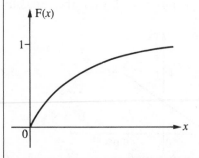

Probability density function

The cumulative probability function, however, does have the disadvantage of often being awkward to use. From its graph, for instance, we cannot easily picture the overall probability distribution, and we normally use instead the *probability density function* (p.d.f.), derived as follows:

By definition of $F(x)$, we know that

$$P(a \leqslant X \leqslant b) = F(b) - F(a)$$

Hence $P(x \leqslant X \leqslant x + \delta x) = F(x + \delta x) - F(x)$

$$= \frac{F(x + \delta x) - F(x)}{\delta x} \delta x$$

Now as $\delta x \to 0$, we know that $\dfrac{F(x + \delta x) - F(x)}{\delta x} \to F'(x)$

Hence $P(x \leqslant X \leqslant x + \delta x) \approx F'(x) \delta x$

Clearly the function $F'(x)$ is very useful, since it enables us to find the probability of x lying in any small interval; and since the probability is equal to this function multiplied by the width of the interval, the function is referred to as the probability density function, and is rewritten $f(x)$.

Since $f(x) = F'(x)$,   it follows that

$$F(x) = \int_L^x f(x) \, dx, \quad \text{where } L \text{ is the least possible value of } x$$

Furthermore, the probability that x lies in the interval $[a, b]$ is clearly seen to be

$$\int_a^b f(x) \, dx$$

and is therefore represented by the area shaded on the diagram.

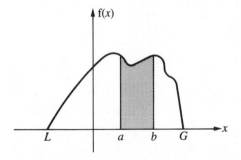

It follows, since $F(x)$ is an increasing function and $f(x)$ is its derivative, that

$f(x) \geqslant 0$; and, since the total probability must be equal to 1, that

$$\int_{L}^{G} f(x)\,dx = 1$$

where L and G are the least and greatest possible values of x.

Returning to example 1,

$$F(x) = 10(x + 0.05) \quad \text{for } -0.05 \leqslant x \leqslant 0.05$$

so $f(x) = F'(x)$

$$= \begin{cases} 10 & \text{for } -0.05 \leqslant x \leqslant 0.05 \\ 0 & \text{elsewhere} \end{cases}$$

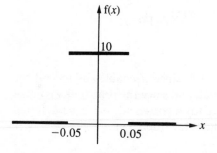

This continuous probability distribution is therefore analogous to the discrete rectangular distribution discussed in chapter 7.

Similarly, in example 2,

$$F(x) = \frac{2}{\pi}\tan^{-1}x \qquad \text{for } x \geqslant 0$$

So $f(x) = F'(x)$

$$= \begin{cases} \dfrac{2}{\pi(1 + x^2)} & \text{for } x \geqslant 0 \\ 0 & \text{for } x < 0 \end{cases}$$

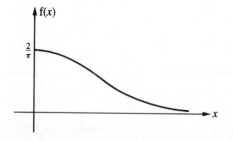

Example 3

The departure time of pupils from a school can be modelled as in the diagram below, where t is the time, in minutes, after the final bell. Find the equation of the probability density function, and use it to find the probability of

a) a pupil leaving less than 2 minutes after the bell;
b) a pupil leaving between 4 and 7 minutes after the bell.

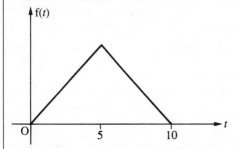

The total area beneath the graph must equal 1, so if the maximum value of the probability density function (when $t = 5$) is h, then

$$\tfrac{1}{2} \times 10 \times h = 1$$

$$\Rightarrow \qquad h = \tfrac{1}{5}$$

So
$$f(t) = \begin{cases} \frac{1}{25}t & \text{for } 0 \leqslant t \leqslant 5 \\ \frac{1}{25}(10 - t) & \text{for } 5 \leqslant t \leqslant 10 \\ 0 & \text{for } t < 0 \text{ or } t > 10 \end{cases}$$

Therefore

a)
$$P(t < 2) = \int_0^2 \frac{t}{25}\, dt$$

$$= \left[\frac{t^2}{50} \right]_0^2$$

$$= 0.08$$

b)
$$P(4 < t < 7) = \int_4^5 \frac{t}{25}\, dt + \int_5^7 \frac{10 - t}{25}\, dt$$

$$= \left[\frac{t^2}{50} \right]_4^5 + \left[\frac{1}{25}\left(10t - \frac{t^2}{2} \right) \right]_5^7$$

$$= 0.18 + 0.32$$

$$= 0.5$$

Exercise 15.2a

1 A p.d.f. is given by

$$f(x) = \begin{cases} \dfrac{x+1}{4} & \text{for } 0 \leqslant x \leqslant a \\ 0 & \text{for } x < 0 \text{ and } x > a \end{cases}$$

By using the property of the p.d.f. that $\int f(x)\,dx = 1$, find the value of a. Obtain an expression for the cumulative probability function. Calculate:
a) $P(x < \frac{1}{2})$ **b)** $P(x > 1)$

2 Verify that

$$f(x) = \begin{cases} \dfrac{3}{x^4} & \text{for } x \geqslant 1 \\ 0 & \text{for } x < 1 \end{cases} \quad \text{is a p.d.f.}$$

Find the value of X such that $P(x < X) = 0.5$. Calculate:
a) $P(x > 10)$ **b)** $P(2 < x < 4)$

3 A p.d.f. is given by

$$f(x) = \begin{cases} \frac{1}{6}x & \text{for } 0 \leqslant x \leqslant 3 \\ \frac{1}{2}(4 - x) & \text{for } 3 \leqslant x \leqslant 4 \\ 0 & \text{for } x < 0 \text{ and } x > 4 \end{cases}$$

Sketch the graph of $f(x)$. Calculate:
a) the probability that x occurs in the interval $[1, 2]$;
b) the probability that $x > 2$.
Obtain the cumulative probability function, and hence, or otherwise, find the median of the distribution.

4 The p.d.f. of a distribution is given by $f(x) = k(1 - x^2)$ for $-1 \leqslant x \leqslant 1$, and $f(x) = 0$ elsewhere.
   Find the value of k, and hence calculate:
a) $P(-\frac{1}{2} \leqslant x \leqslant \frac{1}{2})$ **b)** $P(x > \frac{1}{3})$
Obtain an expression for the cumulative probability function of the distribution.

5 The p.d.f of a distribution is given by $f(x) = a \sin \pi x$ for $0 \leqslant x \leqslant 1$, and $f(x) = 0$ elsewhere.
   Find the value of a, and obtain an expression for the cumulative probability function. Calculate:
a) $P(x < \frac{1}{3})$ **b)** $P(\frac{1}{2} < x < \frac{2}{3})$

6 The probability that a transistor in a radio lasts less than t hours is $1 - e^{-t/2\,000}$. Find the p.d.f. for the lifetime of a transistor.
a) What is the probability that a transistor lasts more than 4000 hours?

b) What is the probability that a transistor ceases to function after 2000 hours of use but before 3000 hours?

c) If a radio contains 8 transistors, what is the probability that none of them fails before 1000 hours of use?

7   If a point is chosen at random on a circular disc, of radius 4 cm, find the cumulative probability function and the probability density function of its distance (r) from the centre of the disc. What is the probability that a point lies between 2 and 3 cm from the centre of the disc?

8   The random variable X has probability density function f given by

$$f(x) = \frac{c}{(x + 1)^2 + 1} \quad (-\infty < x < \infty)$$

where c is a constant. By considering the derivative of arctan $(x + 1)$, or otherwise, show that $c = \pi^{-1}$.

Write down the median value of X, and determine the value X_0 of X such that there is a probability 0.1 that $X > X_0$.

Calculate the probabilities that **a)** $X > 0$, and **b)** $X > -2$.

Three independent measurements of X are taken. What is the probability that at least one of them is greater than 0? (MEI)

Parameters of a continuous distribution

In a discrete probability distribution, the expected value and variance were defined by

$$\mu = \sum_r x_r p_r \quad \text{and} \quad \sigma^2 = \sum_r (x_r - \mu)^2 p_r$$
$$= \sum_r x_r^2 p_r - \mu^2$$

In a continuous distribution, the probability corresponding to p_r is given by $f(x)\,\delta x$.

So $\mu = \lim \sum x f(x)\,\delta x \quad \text{and} \quad \sigma^2 = \lim \sum (x - \mu)^2 f(x)\,\delta x$
$$= \lim \sum x^2 f(x)\,\delta x - \mu^2$$

$$\Rightarrow \quad \mu = \int_L^G x f(x)\,dx \quad \text{and} \quad \sigma^2 = \int_L^G (x - \mu)^2 f(x)\,dx$$
$$= \int_L^G x^2 f(x)\,dx - \mu^2$$

Example 4

The p.d.f. of the lifetime (t), in minutes, of atoms of a radioactive element is given by $f(t) = 3e^{-3t}$. Find the expected lifetime of an atom, its variance, and the half-life of the element (the time it takes for half of the material to decay).

The expected lifetime, μ, is given by $\quad \mu = \int_0^\infty 3t e^{-3t} dt$

Integrating by parts,

$$\mu = \left[-t e^{-3t} \right]_0^\infty + \int_0^\infty e^{-3t} dt$$

$$= 0 + \left[-\frac{e^{-3t}}{3} \right]_0^\infty$$

$$= \tfrac{1}{3} \min$$

and $\quad \sigma^2 = \int_0^\infty 3t^2 e^{-3t} dt - \tfrac{1}{9}$

$$= \left[-t^2 e^{-3t} \right]_0^\infty + \int_0^\infty 2t e^{-3t} dt - \tfrac{1}{9}$$

$$= 0 + \tfrac{2}{3} \times \tfrac{1}{3} - \tfrac{1}{9}$$

$$= \tfrac{1}{9} \min^2$$

Now if T is the half-life, then

$$\int_0^T 3 e^{-3t} dt = \tfrac{1}{2}$$

$$\Rightarrow \quad \left[-e^{-3t} \right]_0^T = \tfrac{1}{2}$$

$$\Rightarrow \quad 1 - e^{-3T} = \tfrac{1}{2}$$

$$\Rightarrow \quad e^{-3T} = \tfrac{1}{2}$$

$$\Rightarrow \quad e^{3T} = 2$$

$$\Rightarrow \quad 3T = \ln 2$$

$$\Rightarrow \quad T = \tfrac{1}{3} \ln 2 \approx 0.23 \min$$

Exercise 15.2b

1–5   Calculate the mean and variance of each of the probability distributions defined in exercise 15.2a nos. **1–5**.

6   A random variable x has cumulative probability function

$$F(x) = \begin{cases} 0 & (x \leqslant a) \\ \dfrac{x - a}{b - a} & (a < x < b) \\ 1 & (x \geqslant b) \end{cases}$$

Find the p.d.f. $f(x)$, and sketch the graph of $f(x)$. Obtain the mean and variance of x.   (C)

7   The probability density function, $f(x)$, of a random variable x, is given by

$$\begin{aligned} f(x) &= \alpha x(4 - x) \quad &\text{if } 0 < x < 4 \\ &= 0 \quad &\text{otherwise} \end{aligned}$$

Find the value of α, and hence find the mean and variance of x. What is the probability that x lies between 0 and 1?   (AEB)

8   A random variable x has the cumulative probability function

$$F(x) = \begin{cases} 0 & (x < 0) \\ kx^3 & (0 \leqslant x \leqslant 2) \\ 1 & (x > 2) \end{cases}$$

where k is a constant. Find the mean, median, and variance of x.   (MEI)

9   A probability distribution has the probability density function

$$f(x) = \begin{cases} 0 & (x < 2) \\ k\,e^{-\lambda x} & (x \geqslant 2, \quad k \text{ constant}) \end{cases}$$

Find, in terms of λ, the mean, median and standard deviation of the distribution.   (MEI)

10   The probability that a light bulb lasts longer than t hours is $e^{-t/\mu}$. Find the probability density function for the lifetime of a bulb. Show that the mean lifetime is μ.

If the mean lifetime is 1500 hours, how unlikely is it that the bulb will last more than 3000 hours?

If the manufacturer wants to ensure that less than one in a thousand bulbs fail before 5 hours, what is the lowest mean lifetime he can allow his bulbs to have?   (SMP)

11   A random variable x has the probability distribution

$$\begin{aligned} f(x)\,\mathrm{d}x &= C\,\mathrm{d}x \quad (-2a \leqslant x \leqslant -a, \quad a \leqslant x \leqslant 2a) \\ f(x)\,\mathrm{d}x &= 0 \quad \text{elsewhere} \end{aligned}$$

Sketch the distribution. Obtain the value of the constant C, and the standard deviation, σ, of the distribution.

Find the probabilities of obtaining values of x

a) within one standard deviation of the mean;

b) within two standard deviations of the mean.

Find the value of k such that $P(|x| < k\sigma) = 0.95$.   (MEI)

12 A large swarm of bees forms a uniformly dense spherical cluster of radius R with the queen at the centre. Show that the probability that a particular bee finds itself at a distance between r and $r + \delta r$ from the queen is $(3r^2/R^3)\,\delta r$. [You may assume that the volume of a thin spherical shell, of radius r and thickness δr, is approximately $4\pi r^2 \delta r$.]

Calculate the mean distance of the bees from the queen, and the variance of these distances.   (SMP)

13 The mass X kg of a particular substance produced per hour in a chemical process is a continuous random variable whose probability density function is given by

$$f(x) = 3x^2/32 \qquad (0 \leqslant x < 2)$$
$$f(x) = 3(6 - x)/32 \quad (2 \leqslant x \leqslant 6)$$
$$f(x) = 0 \qquad\qquad (\text{otherwise})$$

a) Find the mean mass produced per hour.

b) The substance produced is sold at £2 per kg and the total running cost of the process is £1 per hour. Find the expected profit per hour and the probability that in an hour the profit will exceed £7.   (JMB)

14 A continuous random variable, X, has probability density function $\lambda \sin x\,(0 \leqslant x \leqslant \pi)$, and zero outside this range. Find a value of the constant λ, the mean, the variance, the median, and the quartiles. What is the probability that a random observation lies within one standard deviation of the mean?   (O)

15 A mathematical model for the fraction x of the sky covered with cloud $(0 < x < 1)$ assigns to this a p.d.f.

$$f(x) = \frac{k}{\sqrt{(x - x^2)}}$$

Calculate:

a) the value of k;

b) the expected fraction covered by cloud;

c) the probability that not more than $\frac{1}{4}$ of the sky is covered.

(*Hint:* Your integrations may be made easier by using the substitution $x = \sin^2 \theta$. You may assume that this substitution is valid, even though the function to be integrated may be discontinuous at the ends of the interval of integration.)   (SMP)

16 The probability density function of a distribution is given by

$$f(x) = \frac{e^{-x}x^{\lambda-1}}{(\lambda-1)!} \quad (x \geqslant 0, \quad \lambda \text{ integer} > 0)$$

Find the expected value and variance of x. Sketch $f(x)$ when $\lambda = 2$.

(MEI)

15.3 The Normal distribution

In chapter 7 we saw how the Normal function was used as a suitable model to describe certain statistical data, drawn from a wide variety of sources. This function is of course a probability density function. It is written $\phi(x)$ and it will be shown that it has the equation

$$\phi(x) = \frac{1}{\sqrt{(2\pi)}\sigma} e^{-(x-\mu)^2/2\sigma^2}$$

where μ and σ are the mean and the standard deviation of the distribution.

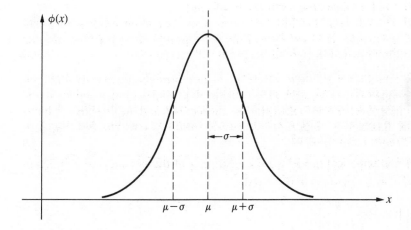

In the case of the standardised Normal function (with $\mu = 0$ and $\sigma = 1$), the equation is

$$\phi(t) = \frac{1}{\sqrt{(2\pi)}} e^{-t^2/2}$$

This probability function is not integrable, which is why we have to use tables of values of the corresponding cumulative probability function, $\Phi(t)$ (see over):

t	$\Phi(t)$	t	$\Phi(t)$	t	$\Phi(t)$	t	$\Phi(t)$
0·0	0·5000	1·0	0·8413	2·0	0·9772	3·0	0·99865
0·1	0·5398	1·1	0·8643	2·1	0·9821	3·1	0·99903
0·2	0·5793	1·2	0·8849	2·2	0·9861	3·2	0·99931
0·3	0·6179	1·3	0·9032	2·3	0·9893	3·3	0·99952
0·4	0·6554	1·4	0·9192	2·4	0·9918	3·4	0·99966
0·5	0·6915	1·5	0·9332	2·5	0·9938	3·5	0·99977
0·6	0·7257	1·6	0·9452	2·6	0·9953	3·6	0·99984
0·7	0·7580	1·7	0·9554	2·7	0·9965	3·7	0·99989
0·8	0·7881	1·8	0·9641	2·8	0·9974	3·8	0·99993
0·9	0·8159	1·9	0·9713	2·9	0·9981	3·9	0·99995
1·0	0·8413	2·0	0·9772	3·0	0·9986	4·0	0·99997

We now see how the Normal function arises purely from considerations of probability, as the limiting case (for large values of n) of the binomial distribution.

The limiting binomial distribution

The probability diagrams below illustrate the binomial distributions with $p = \frac{1}{2}$ and $\frac{1}{5}$ for values of $n = 4$, 10 and 20.

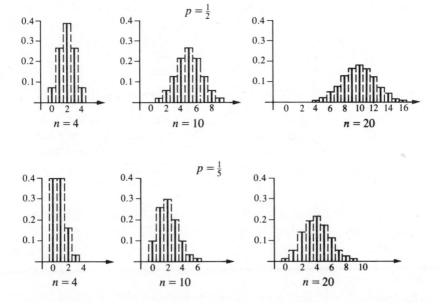

In the first case, with $p = \frac{1}{2}$, the distinctive Normal bell-shape is immediately apparent, even for relatively small values of n. And for $n = 20$ the outline of the probability diagram approximates very closely to the Normal curve. With a symmetrical binomial distribution perhaps this is not so surprising, but with an apparently asymmetrical distribution, such as we have with $p = \frac{1}{5}$, the result is much more striking. Indeed, for small values of n (e.g. $n = 4$), the probability diagram bears no resemblance at all to the symmetrical Normal shape; but when $n = 10$ we see that the diagram is much less skew, and when $n = 20$ we once again have a diagram whose outline is very similar to the characteristic Normal shape. It is not difficult to envisage that as n becomes larger and larger, and the horizontal and vertical scales are suitably adjusted (maintaining a total area of 1), the diagram tends closer and closer to the Normal curve. The two diagrams with $n = 20$ are reproduced below, with Normal curves superimposed.

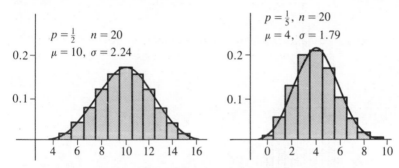

We know, from section 15.1, that for a binomial distribution

$$\mu = np \quad \text{and} \quad \sigma = \sqrt{(npq)}$$

which therefore give us the values of μ and σ indicated on the diagram.

* Derivation of the Normal probability function

We can now derive the equation, stated earlier in this section, of the Normal probability function, by regarding it as the limit of the binomial distribution.

We have seen that as $n \rightarrow \infty$ the profile of the binomial distribution tends:
a) to move to the right (since $\mu = np$);
b) to spread (since $\sigma = \sqrt{(npq)}$); and
c) to become flatter.

To counteract these effects we will standardise the variable r (the number of successes), letting

$$t = \frac{r - \mu}{\sigma} = \frac{r - np}{\sqrt{(npq)}}$$

This transformation diminishes all widths by a factor $1/\sqrt{(npq)}$; so if the area beneath the profile is to remain 1, each of the ordinates must be stretched by a corresponding factor $\sqrt{(npq)}$ which will also counteract the flattening.

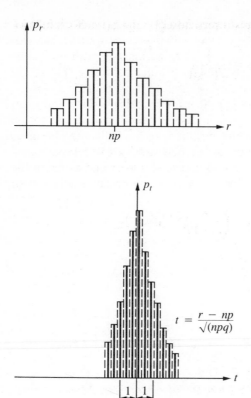

By this means we can bring each distribution into a standard position with $\mu = 0$ and $\sigma = 1$, and so can more easily investigate the alterations of its shape (rather than of its position and spread) as $n \to \infty$.

If we now consider the two neighbouring ordinates p_r and p_{r+1} of the original distribution, we see that each is stretched by a factor $\sqrt{(npq)}$, whilst the distance between them is reduced from 1 to $1/\sqrt{(npq)}$. Furthermore, we suppose that as these ordinates draw closer together, the distribution tends to one with probability density function $\phi(t)$:

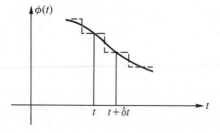

If the ordinates p_r and p_{r+1} are transformed into $\phi(t)$ and $\phi(t + \delta t)$, it follows that

$$\phi(t) = \sqrt{(npq)}p_r \quad \text{and} \quad \delta t = \frac{1}{\sqrt{(npq)}}$$

$$\phi(t + \delta t) = \sqrt{(npq)}p_{r+1}$$

Hence $\quad \delta\phi = \phi(t + \delta t) - \phi(t)$

$$\approx \sqrt{(npq)}(p_{r+1} - p_r)$$

$$\Rightarrow \qquad \frac{\delta\phi}{\phi} \approx \frac{p_{r+1} - p_r}{p_r} = \frac{p_{r+1}}{p_r} - 1$$

But $\quad p_r = \binom{n}{r}p^r q^{n-r} \quad \text{and} \quad p_{r+1} = \binom{n}{r+1}p^{r+1}q^{n-r-1}$

So $\quad \dfrac{\delta\phi}{\phi} \approx \dfrac{\binom{n}{r+1}p^{r+1}q^{n-r-1}}{\binom{n}{r}p^r q^{n-r}} - 1$

$$\approx \frac{(n-r)p}{(r+1)q} - 1$$

$$\approx \frac{np - rp - rq - q}{(r+1)q} = \frac{np - r - q}{(r+1)q}$$

But $\quad r = np + \sqrt{(npq)}t \quad \text{and} \quad \delta t = \frac{1}{\sqrt{(npq)}}$

So $\quad \dfrac{1}{\phi}\dfrac{\delta\phi}{\delta t} \approx \dfrac{\{-\sqrt{(npq)}t - q\}\sqrt{(npq)}}{\{np + \sqrt{(npq)}t + 1\}q}$

$$\approx \frac{-npt - \sqrt{(npq)}}{np + \sqrt{(npq)}t + 1}$$

$$\approx \frac{-t - \sqrt{(q/np)}}{1 + \sqrt{(q/np)}t + 1/np}$$

Letting $n \to \infty$ (and so $\delta t \to 0$),

$$\frac{1}{\phi}\frac{d\phi}{dt} = -t$$

$$\Rightarrow \quad \ln\phi(t) = -\tfrac{1}{2}t^2 + \ln c$$

$$\Rightarrow \qquad \phi(t) = c\,e^{-\frac{1}{2}t^2}$$

Hence $\quad \displaystyle\int_{-\infty}^{\infty} \phi(t)\,dt = c\int_{-\infty}^{\infty} e^{-\frac{1}{2}t^2}\,dt$

But $\int_{-\infty}^{\infty} \phi(t)\,dt = 1$, since $\phi(t)$ is a p.d.f., and we have also seen (Miscellaneous problems, chapter 10) that $\int_{-\infty}^{\infty} e^{-x^2}\,dx = \sqrt{\pi}$, which implies that $\int_{-\infty}^{\infty} e^{-\frac{1}{2}t^2}\,dt = \sqrt{(2\pi)}$.

Hence $c = \dfrac{1}{\sqrt{(2\pi)}}$

and the limiting curve is

$$\phi(t) = \frac{1}{\sqrt{(2\pi)}}\,e^{-\frac{1}{2}t^2}$$

Finally, we can reverse the process of standardisation by the transformation

$$t = \frac{x - \mu}{\sigma}$$

and thereby obtain the probability density function of the more general Normal distribution as

$$\phi(x) = \frac{1}{\sqrt{(2\pi)}\sigma}\,e^{-(x-\mu)^2/2\sigma^2}$$

We are now able to put the limiting property of the Normal distribution to practical use, as illustrated in the following example.

Example

Forty per cent of university mathematics students are female. Assuming that these young ladies are scattered randomly throughout the country, what is the probability that a mathematics department of 100 students contains at least 50 women?

The probability model that describes this situation is a binomial one, with $n = 100, p = 0.4, q = 0.6$.
  The required probability is therefore given by:

$$P \text{ (at least 50 women)} = p_{50} + p_{51} + \cdots + p_{100}$$

$$= \binom{100}{50}(0.4)^{50}(0.6)^{50}$$

$$+ \binom{100}{51}(0.4)^{51}(0.6)^{49} + \cdots + \binom{100}{100}(0.4)^{100}$$

The direct evaluation of such a probability would be extremely tedious. But to ease the calculation, we can now use the Normal approximation to this binomial distribution, with

$$\mu = np = 100 \times 0.4 = 40$$

and $\sigma = \sqrt{npq} = \sqrt{(100 \times 0.4 \times 0.6)} = 4.9$

Now if we denote the number of female students by f, then we require

$$P(f \geqslant 50) = 1 - P(f \leqslant 49)$$

Using the Normal distribution therefore it would seem reasonable to say that

$$P(f \leqslant 49) = \Phi\left(\frac{49 - 40}{4.9}\right)$$

but this is not quite true and the discrepancy is illustrated in the diagram.

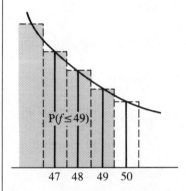
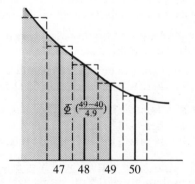

The shaded areas represent, on the left $P(f \leqslant 49)$ and on the right $\Phi[(49 - 40)/(4.9)]$. The difference between them is fairly evident, and arises from the fact that we are using a continuous curve to approximate to a discrete probability distribution. What we need to do is to shift the boundary of the shaded area on our right-hand diagram half a unit to the right. We are then considering $\Phi[(49.5 - 40)/(4.9)]$, which gives a better approximation to $P(f \leqslant 49)$. Such a modification is known as a *continuity correction*. (For very large values of n the correction is insignificant, and so may be disregarded.)

Hence $P(f \leqslant 49) = \Phi\left(\dfrac{9.5}{4.9}\right)$

$$= \Phi(1.94)$$

$$= 0.974$$

> $\Rightarrow$ $P(f \geqslant 50) = 1 - 0.974 = 0.026$
>
> So the probability of a department of 100 containing at least 50 women is 0.026.

For a general binomial distribution, where n is large, and the probability of 'success' $= p$,

$$P \text{ (at most } N \text{ successes)} = \Phi\left(\frac{N + \frac{1}{2} - np}{\sqrt{npq}}\right)$$

Exercise 15.3a

1   If an unbiased coin is tossed 100 times, what is the probability that
   a) there will be more than 60 heads?
   b) there will be at least 45 and at most 55 heads?
   c) there will be fewer than 43 heads?

2   If an unbiased coin is tossed 1000 times, what is the probability that
   a) there will be more than 600 heads?
   b) there will be at least 450 and at most 550 heads?
   c) there will be fewer than 520 heads?

3   If a 'fair' die is thrown 300 times, what is the probability that
   a) there will be more than 60 sixes?
   b) there will be fewer than 45 sixes?

4   In an examination which consists of 100 questions, a student has a probability of 0.6 of getting each question correct. The student fails the examination if he obtains a mark less than 55, and obtains a distinction for a mark of 68 or more. Calculate:
   a) the probability that he fails the examination;
   b) the probability that he obtains a distinction.

5   If it is known that 30% of people are short-sighted, what is the probability that more than 33% of a sample, chosen at random, will be short-sighted if:
   a) the sample consists of 100 people?
   b) the sample consists of 1000 people?
   c) the sample consists of 10000 people?

6   It is known that 72% of TV viewers watch a particularly popular programme. What is the probability that in a sample of 500 viewers, chosen at random,
   a) more than 350 watch the programme?

b) more than 375 watch the programme?
c) fewer than 340 watch the programme?

7 A playing card is drawn at random from a full pack, containing no jokers, its suit is recorded and it is then returned to the pack; the pack is then shuffled and the procedure repeated. In 60 such draws calculate, using a Normal approximation, the probability that:
a) no more than 12 hearts appear;
b) more than 20 spades are obtained.

8 A confectionery firm produces three types of toffees, liquorice, nut and plain, and mixes them together in the ratio 1:2:5 before packing them into boxes. If there are 80 toffees to a box, what percentage of boxes will contain:
a) more than 25 nut toffees?
b) fewer than 58 plain toffees?
c) more nut and liquorice than plain toffees?

9 A roulette wheel contains the numbers 0, 1, 2, ..., 36. When the wheel is spun and a ball introduced it is equally likely to come to rest on any of these numbers when the wheel stops. What is the probability that the outcome is divisible by 4 or 5 (or both)? If the wheel is spun 100 independent times what is the probability of getting more than 45 outcomes which are divisible by 4 or 5 as above? (AEB, 1982)

10 In a certain manufacturing process a 10% rate of defectives is regarded as just tolerable. It is decided to accept a day's batch if in a random sample of size n taken from it the proportion of defectives does not exceed 12%. Calculate the value of n if there is a 0.05 probability of rejecting a batch which is in fact producing 10% of defectives.

The percentage of defectives rises to 14%. The procedure described above is used with a sample of size 600; calculate, to 2 decimal places, the probability that as a result of the test the batch is accepted. (MEI)

11 A grain merchant has a batch of seed which he knows is either of strain A or of strain B. On average 60% of strain A and 90% of strain B will germinate. He plans to plant n seeds, selected at random from the batch, and to regard the batch as strain A if y or fewer germinate, and as strain B if more than y germinate.

If the seed is strain A, the merchant wishes to plant just sufficient seed to be 99% certain that his test does not mislead him into regarding it as strain B. Show that n and y must satisfy

$$y + \tfrac{1}{2} \approx 0.6n + 1.14\sqrt{n}$$

[You may assume that the Normal approximation to binomial probability is valid for this value of n. Take 0.99 as $\Phi(2.33)$.]

The merchant also wishes to be 95% certain that, should the seed be of strain B, his test does not mislead him into regarding it as strain A. Show that

$$y + \tfrac{1}{2} \approx 0.9n - 0.49\sqrt{n}$$

and hence find suitable values for n and y.   (SMP)

12   A test consists of multiple-choice questions. In each question there are five choices, one of which is correct. A candidate has probability p of knowing the correct answer to any question. If he does not know the answer, he makes a random guess from among the five choices given.
a) For a question selected at random, find the probability of the candidate selecting the correct answer.
b) If he selects the correct answer, find the conditional probability that he has guessed.
c) Find his expected score if there are 40 questions in the test, and one mark is given for a correct answer, zero for an incorrect answer.
d) Use a Normal approximation to estimate the probability of a candidate with $p = \tfrac{1}{2}$ scoring more than 27 marks in the test.
e) In order to make the expected score of an entirely ignorant candidate zero, the examiner decides to alter the marking scheme to give n marks for a correct solution and -1 mark for an incorrect solution. Establish what value he should choose for n.   (SMP)

The central limit theorem

A computer may be used to produce randomly the binary digits 0 and 1. Then if x denotes the value of the digit so obtained,

$$P(x = 0) \quad = \quad P(x = 1) \quad = \quad \tfrac{1}{2}$$

and we have a simple example of a rectangular distribution.

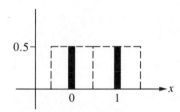

For this distribution,

$$\mu_x = \tfrac{1}{2} \times 0 + \tfrac{1}{2} \times 1 = 0.5$$

$$\text{and} \quad \sigma_x^2 = \tfrac{1}{2}(1 - \tfrac{1}{2})^2 + \tfrac{1}{2}(0 - \tfrac{1}{2})^2 = 0.25$$

$$\Rightarrow \quad \sigma_x = 0.5$$

Now suppose that instead of looking at the digits individually, we group them together in sets of, say, 10. And for each group we will consider a new variable, y, the sum of the digits.

y can therefore take values from 0 to 10, and since each digit can have only one of two values, y has a binomial probability distribution with $n = 10$, $p = \frac{1}{2}, q = \frac{1}{2}$.

Hence $\mu_y = np = 5$

and $\sigma_y = \sqrt{(npq)} = \sqrt{2.5}$

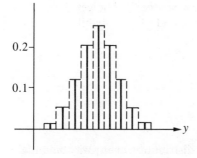

y represents the sum of 10 digits. We now define a third variable, z, to be the average of each group of digits, so that $z = \frac{1}{10}y$.

Since z is a constant fraction of y, its probability distribution will have the same shape as that of y, shown in the diagram, but the horizontal scale will be reduced by a factor 10.

So $\mu_z = \frac{1}{10}\mu_y = 0.5$

and $\sigma_z = \frac{1}{10}\sigma_y = \frac{1}{10}\sqrt{2.5} = \dfrac{0.5}{\sqrt{10}}$

Hence $\mu_z = \mu_x$ and $\sigma_z = \dfrac{\sigma_x}{\sqrt{10}}$

Therefore if we consider the averages of groups of 10 digits, the expected value is the same as that of the digits taken individually, but the standard deviation of the averages (or 'sample means') is a fraction $1/\sqrt{10}$ (i.e. about $\frac{1}{3}$) of the standard deviation of the individual digits.

Moreover, despite the fact that the original distribution was rectangular, the distribution of sample means appears to be approaching the Normal distribution. This represents a special case of one of the most important and remarkable theorems in statistics, known as *the central limit theorem*, that if samples of size n are drawn at random from *any* background population then, as $n \to \infty$, the distribution of the sample means tends to the Normal distribution. (Although we generally refer to the sample *means* in the central limit theorem, the theorem clearly also holds for the sample *sums*.)

Having already discovered the limiting property of the binomial distribution it is perhaps hardly surprising that the theorem should hold for our example of the binary digits. But suppose we consider the sample means of sets of 10 random digits in the range 0–9; the frequency diagram below illustrates the sample means, suitably grouped, of 50 samples taken from a table of random digits.

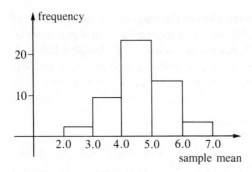

The sample means in this case do not have a distribution in any way binomial; nevertheless, as predicted by the central limit theorem, the Normal shape of the distribution is again apparent.

So, in yet another way, is emphasised the importance of the Normal distribution.

Exercise 15.3b

1   Using a table of random digits (0–9), obtain the sample means of 100 sets of 10 digits. Display these sample means on a frequency diagram similar to that shown in the text.

2   What would have been the values of the expected value and standard deviation of the sample mean of the binary digits, described in the text, if the sample had been   a) of size 25,   b) of size 100?

3   A computer is now programmed to produce randomly the digits 0, 2. Calculate the values of μ and σ for this simple rectangular distribution, and hence obtain the expected value and standard deviation of the mean of a sample of size 10.

4   The digits 1, 2, 3, 4 are produced randomly. By applying the results obtained for binary digits, calculate the expected value and standard deviation of the mean of a sample of 100 such numbers. Hence, using the central limit theorem, estimate the probability that the mean of such a sample exceeds 2.7.

5

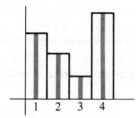

Simulate the asymmetrical distribution in the diagram by using a table of random digits, and recording 0, 1, or 2 as a score of '1', 3 or 4 as a score of '2', 5 as a score of '3', and 6, 7, 8, or 9 as a score of '4'. Display 100 such scores on a frequency diagram. Then obtain the sample means of 100 sets of 10 scores, and display these sample means on a suitable frequency diagram.

6 Explain the main idea of the central limit theorem.

Many random samples, each of a thousand pebbles, are taken from a very large source and the total weight of each sample is recorded. These total weights are found to have a mean of 157 kg with a standard deviation of 1.2 kg. The mean weight of the pebbles in each sample is calculated. Calculate the mean and standard deviation of these means.

Obtain an estimate for the standard deviation of the whole set of pebbles.

What proportion of the samples would be expected to have a total weight in excess of 160 kg?

Find the weight which would be exceeded by the total sample weight of 98% of the samples. (AEB, 1981)

15.4 The Poisson distribution

In section 15.3 we discovered that, for large values of n, the Normal distribution gave us a good approximation to the binomial distribution, and is in fact the limit of the binomial distribution as $n \to \infty$.

Consider, however, the binomial probabilities with $n = 1000$ and $p = 0.001$ and their Normal approximations.

probability	binomial	Normal
p_0	0.3677	0.2417
p_1	0.3681	0.3831
p_2	0.1840	0.2417
p_3	0.0613	0.0605
p_4	0.0153	0.0059
p_5	0.0030	0.0002
p_6	0.0005	0.0000

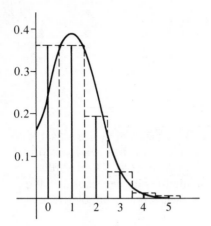

Here, despite the fact that n is very large, the Normal distribution seems to give us a poor approximation. This apparent contradiction of our previous result arises from the fact that in section 15.3 we assumed that as n became very large, so did the mean $\mu(= np)$, with the effect that the binomial distribution became approximately symmetrical about this mean. For the distribution illustrated above, however, this is not so; for, even though n is large,

$$\mu = np = 1000 \times 0.001 = 1$$

a relatively small value, and the distribution cannot be symmetrical about μ.

In fact, the limiting Normal property of the binomial distribution holds only if

$$\mu = np \to \infty \quad \text{as } n \to \infty$$

(an assumption which was made in the last section). In many cases, of course, this is so, but if np is finite as $n \to \infty$, then the limit of the binomial is quite different, as we shall now show.

We know that for the general binomial distribution,

$$p_r = \binom{n}{r} p^r (1 - p)^{n-r}$$

$$= \frac{n(n - 1) \ldots (n - r + 1)}{r!} p^r (1 - p)^{n-r}$$

$$= \frac{1\left(1 - \dfrac{1}{n}\right) \ldots \left(1 - \dfrac{r - 1}{n}\right) n^r}{r!} p^r (1 - p)^{n-r}$$

$$= \left(1 - \frac{1}{n}\right) \ldots \left(1 - \frac{r - 1}{n}\right) \frac{(np)^r}{r!} (1 - p)^{n-r}$$

$$= \left(1 - \frac{1}{n}\right)\dots\left(1 - \frac{r-1}{n}\right)\frac{\mu^r}{r!}\left(1 - \frac{\mu}{n}\right)^{n-r}$$

Now since r is finite, as $n \to \infty$

$$\left(1 - \frac{1}{n}\right)\dots\left(1 - \frac{r-1}{n}\right) \to 1$$

and $\left(1 - \frac{\mu}{n}\right)^{n-r} = 1 - (n-r)\frac{\mu}{n} + \frac{(n-r)(n-r-1)}{2!}\frac{\mu^2}{n^2} - \cdots$

$$= 1 - \left(1 - \frac{r}{n}\right)\mu + \left(1 - \frac{r}{n}\right)\left(1 - \frac{r-1}{n}\right)\frac{\mu^2}{2!} - \cdots$$

$$\to 1 - \mu + \frac{\mu^2}{2!} - \frac{\mu^3}{3!} + \cdots$$

$$= e^{-\mu}$$

Therefore, in this limiting case,

$$p_r = \frac{\mu^r}{r!}e^{-\mu}$$

and the distribution so defined is known as the *Poisson probability distribution*.

The distribution can also be obtained by considering its probability generator.

The probability generator of the general binomial distribution is given by

$$G(t) = (q + pt)^n$$

$$= (1 + p(t-1))^n \quad \text{since } p + q = 1$$

$$= \left(1 + \frac{\mu(t-1)}{n}\right)^n \quad \text{since } \mu = np$$

Now we know (Miscellaneous problems, chapter 10) that if x is finite,

$$\left(1 + \frac{x}{n}\right)^n \to e^x \quad \text{as } n \to \infty$$

So $G(t) \to e^{\mu(t-1)} \quad \text{as } n \to \infty$

Now $\quad e^{\mu(t-1)} = e^{\mu t} \times e^{-\mu}$

$$= \left(1 + \mu t + \frac{\mu^2 t^2}{2!} + \cdots + \frac{\mu^r t^r}{r!} + \cdots\right) e^{-\mu}$$

$$= e^{-\mu} + \mu e^{-\mu} t + \cdots + \frac{\mu^r}{r!} e^{-\mu} t^r + \cdots$$

so the coefficient of t^r gives $p_r = \dfrac{\mu^r}{r!} e^{-\mu}$.

Returning to the original binomial distribution, if we now use the Poisson distribution with $\mu = 1$, we obtain probabilities which are very good approximations to the binomial, and these equivalent binomial and Poisson probabilities are set out in the table below.

probability	binomial	Poisson
p_0	0.3677	0.3679
p_1	0.3681	0.3679
p_2	0.1840	0.1839
p_3	0.0613	0.0613
p_4	0.0153	0.0153
p_5	0.0030	0.0031
p_6	0.0005	0.0005

So we now have, under different circumstances, another limit to the binomial distribution, but with the important difference that this limiting distribution, like the binomial itself, but unlike the Normal distribution, is discrete.

The probability diagrams for the Poisson distribution with $\mu = 0.5$, 2 and 5 respectively, are illustrated in the diagrams below.

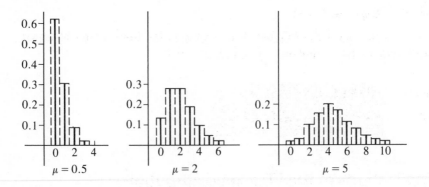

It can be seen from these diagrams that, as μ becomes larger, the Poisson distribution becomes more symmetrical in shape and, like the binomial, approximates to the Normal distribution.

Calculation of Poisson probabilities

In the actual computation of probabilities, the Poisson distribution has one advantage over the Normal and binomial distributions; it involves the use of only one parameter, μ, as opposed to two in each of the other cases (μ, σ for the Normal and n, p for the binomial). And though the formula for p_r may seem at first glance rather daunting, in practice the calculation of probabilities can be carried out very conveniently:

Suppose $\mu = 3$

then $\quad p_0 = e^{-3} \qquad\qquad = 0.0498$

$\qquad p_1 = 3e^{-3} \ = 3p_0 = 0.1494$

$$p_2 = \frac{3^2 e^{-3}}{2!} = \tfrac{3}{2}p_1 = 0.2241$$

$$p_3 = \frac{3^3 e^{-3}}{3!} = \tfrac{3}{3}p_2 = 0.2241$$

$$p_4 = \frac{3^4 e^{-3}}{4!} = \tfrac{3}{4}p_3 = 0.1681$$

and $\quad p_{r+1} = \dfrac{3}{r+1}p_r \qquad\qquad \left(\text{so in general}\quad p_{r+1} = \dfrac{\mu}{r+1}p_r\right)$

Mean and variance of Poisson distribution

For the binomial distribution,

$$\mu = np$$

and $\sigma^2 = npq = np(1 - p)$

Now we can regard the Poisson distribution as the limit of the binomial distribution as $n \to \infty$ and $np = \mu$ (so that $p \to 0$).

Hence $\qquad$ mean $= \mu$

and $\qquad$ variance $= \lim\limits_{n \to \infty} np(1 - p) = \mu$

Applications of the Poisson distribution

We have derived the Poisson distribution as a limiting case of the binomial distribution. But it arises more fundamentally in its own right, whenever there is a continuous variable—say time or distance—and discrete events occur *randomly* and *independently*. In such a case, the number of events in a given interval has a Poisson distribution. (By imagining the continuous variable split into a very large number of equal small intervals, readers can verify for themselves that these conditions are consistent with those of a limiting binomial distribution.) Examples of such situations, in which the Poisson distribution is applicable, are provided by flaws occurring in lengths of steel piping, people arriving in a shop queue in a certain time interval, and radioactive particles being emitted from a source in some time interval.

Consider the first of these examples:

Example 1

It is found experimentally that there is an average of one flaw in every 2 metres of a certain type of steel piping. If the piping is cut into 1-metre lengths, what is the probability of 0, 1, 2 flaws in such a length? Find the equivalent probabilities if the pipe is cut into 4-metre lengths.

a) For a 1-metre length of pipe, the expected number of flaws, $\mu = \frac{1}{2}$.

Hence $p_0 = e^{-\frac{1}{2}} = 0.6065$

Then using the relationship we have recently obtained between successive Poisson probabilities,

$$p_1 = \tfrac{1}{2}p_0 = 0.3032$$

and $p_2 = \dfrac{\frac{1}{2}p_1}{2} = 0.0758$

b) For 4-metre lengths of pipe, $\mu = 2$.

Hence $p_0 = e^{-2} = 0.1353$

$\Rightarrow$ $p_1 = 2p_0 = 0.2706$

$\Rightarrow$ $p_2 = \dfrac{2p_1}{2} = 0.2706$

It was quite clear that example 1 should be tackled by means of the Poisson distribution. By comparison, in the following example we can use either a binomial or a Poisson probability model, and the equivalent results for the two models are compared. In example 3 we fit a Poisson model to experimental data.

Example 2

A firm of wholesale fruit distributors has found that on average one apple in 50 is bruised on arrival from the growers. If the apples arrive in cartons of

100, calculate the probabilities of a carton having 0, 1, 2, 3, more than 3 bruised apples.

Using a binomial model, $n = 100$ and $p = \frac{1}{50}$.

Using a Poisson model, $\mu = 100 \times \frac{1}{50} = 2$.

The calculations for the two models are carried out simultaneously:

binomial		Poisson

$$p_0 = \left(\frac{49}{50}\right)^{100} \qquad = 0.1316 \qquad p_0 = e^{-2} = 0.1353$$

$$p_1 = \binom{100}{1}\left(\frac{49}{50}\right)^{99}\frac{1}{50} \qquad = 0.2706 \qquad p_1 = 2p_0 = 0.2706$$

$$p_2 = \binom{100}{2}\left(\frac{49}{50}\right)^{98}\left(\frac{1}{50}\right)^{2} \qquad = 0.2727 \qquad p_2 = p_1 \;\;= 0.2706$$

$$p_3 = \binom{100}{3}\left(\frac{49}{50}\right)^{97}\left(\frac{1}{50}\right)^{3} \qquad = 0.1818 \qquad p_3 = \tfrac{2}{3}p_2 = 0.1804$$

$$p_{>3} \qquad\qquad\qquad\qquad = 0.1433 \qquad p_{>3} \qquad = 0.1431$$

The results clearly are very similar but, since its calculations are so much easier, we would usually adopt the Poisson model.

Example 3

The following data (without which any account of Poisson probability would be incomplete!) gives the number of deaths by horse-kick of Prussian cavalrymen in ten corps during the years 1875–1894.

number of deaths per corps per year	0	1	2	3	4
frequency of occurrence	109	65	22	3	1

We will fit a Poisson probability model to the data, and calculate the equivalent theoretical frequencies.

The mean number of deaths per year is given by

$$m = \frac{109 \times 0 + 65 \times 1 + 22 \times 2 + 3 \times 3 + 1 \times 4}{200}$$

$$= \frac{122}{200}$$

$$= 0.61$$

We will use this mean value as our parameter μ in the Poisson distribution.

To obtain the theoretical (or expected) frequencies, the probabilities are multiplied by the total frequency, 200, and rounded off to the nearest whole number.

number of deaths	probability		expected frequency	observed frequency
0	$e^{-0.61}$	$= 0.5434$	109	109
1	$0.61\,p_0$	$= 0.3315$	66	65
2	$\dfrac{0.61}{2}p_1$	$= 0.1011$	20	22
3	$\dfrac{0.61}{3}p_2$	$= 0.0206$	4	3
4	$\dfrac{0.61}{4}p_3$	$= 0.0031$	1	1

This close correspondence between observed and expected frequencies supports the assumption on which the Poisson model is based: that the deaths occurred randomly.

* Alternative derivation of Poisson distribution

To complete our account of the Poisson distribution, we now derive it directly from our basic assumptions.

Let us suppose that a series of events, such as the flashes of a Geiger counter, occur independently; and yet, over a long interval, at a uniform rate, so that in an interval of unit length the expected number of flashes is λ. Then in a very short interval δt, the probability

of a single flash is $\lambda \delta t$,

of two or more flashes is 0,

and   of no flashes is $1 - \lambda \delta t$.

We now denote the probability of having r flashes in the first t units of time by $p_r(t)$. Then, if we consider a further small interval δt, it is clear that

$p(r$ flashes in time $t + \delta t)$
  $= p(r$ flashes in time t and no flashes in interval $\delta t)$
    $+ p(r - 1$ flashes in time t and 1 flash in interval $\delta t)$

$\Rightarrow \quad p_r(t + \delta t) = p_r(t)(1 - \lambda \delta t) + p_{r-1}(t)\lambda \delta t$

$$\Rightarrow \quad \frac{p_r(t + \delta t) - p_r(t)}{\delta t} + \lambda p_r(t) = \lambda p_{r-1}(t)$$

Letting $\delta t \to 0$, we obtain

$$p_r'(t) + \lambda p_r(t) = \lambda p_{r-1}(t)$$

$$\Rightarrow \quad \frac{d}{dt}[e^{\lambda t} p_r(t)] = \lambda e^{\lambda t} p_{r-1}(t)$$

$$\Rightarrow \quad \left[e^{\lambda t} p_r(t) \right]_0^t = \lambda \int_0^t e^{\lambda t} p_{r-1}(t) \, dt$$

But $p_r(0) = 0$ if $r \neq 0$

so $e^{\lambda t} p_r(t) = \lambda \int_0^t e^{\lambda t} p_{r-1}(t) \, dt$ (1)

Furthermore,

$p(0$ flashes in time $t + \delta t)$
$= p(0$ flashes in time t *and* 0 flashes in interval $\delta t)$

$$\Rightarrow \quad p_0(t + \delta t) = p_0(t)(1 - \lambda \delta t)$$

$$\Rightarrow \quad \frac{p_0(t + \delta t) - p_0(t)}{\delta t} + \lambda p_0(t) = 0$$

Letting $\delta t \to 0$,

$$p_0'(t) + \lambda p_0(t) = 0$$

$$\Rightarrow \quad \frac{d}{dt}[e^{\lambda t} p_0(t)] = 0$$

$$\Rightarrow \quad e^{\lambda t} p_0(t) = p_0(0)$$

But we certainly start with 0 flashes, so $p_0(0) = 1$.

Hence $e^{\lambda t} p_0(t) = 1$ (2)

If we now put $P_r(t) = e^{\lambda t} p_r(t)$, we can rewrite equations (1) and (2) as

$$P_r(t) = \lambda \int_0^t P_{r-1}(t) \, dt$$

and $P_0(t) = 1$

So $P_1(t) = \lambda \int_0^t dt = \lambda t \quad \Rightarrow \quad p_1(t) = \lambda t \, e^{-\lambda t}$

$$P_2(t) = \lambda \int_0^t \lambda t \, dt = \frac{\lambda^2 t^2}{2} \quad \Rightarrow \quad p_2(t) = \frac{\lambda^2 t^2}{2} e^{-\lambda t}$$

$$P_3(t) = \lambda \int_0^t \frac{\lambda^2 t^2}{2} \, dt = \frac{\lambda^3 t^3}{6} \Rightarrow \quad p_3(t) = \frac{\lambda^3 t^3}{6} e^{-\lambda t}$$

and, more generally,

$$P_r(t) = \frac{\lambda^r t^r}{r!} \qquad \Rightarrow \quad p_r(t) = \frac{\lambda^r t^r}{r!} e^{-\lambda t}$$

Finally, regarding t as fixed, we can write $p_r(t)$ as p_r; and since the expected number of flashes in this interval is λt, we write $\mu = \lambda t$ and obtain

$$p_r = \frac{\mu^r}{r!} e^{-\mu}$$

Exercise 15.4

1 It is known that 0.1% of all people react adversely to a certain type of drug. What is the probability that, out of a sample of 1000 people,
   a) none will react to the drug;
   b) just one person will react;
   c) more than two people will react?

2 A hiker on a walking holiday in a country area estimates that in walking 150 kilometres she has passed a total of 60 public houses. Assuming that these public houses are distributed fairly randomly about the countryside, what is the probability of her being able to quench her thirst
   a) within the next kilometre;
   b) within the next five kilometres?

3 During a certain period of the day, an average of four people enter a supermarket each minute. Calculate the probability that
   a) no-one enters during a particular minute;
   b) three people enter during a particular minute;
   c) five people enter during a particular minute;
   d) no-enters during a two-minute period.

4 Over a period of time, the number of serious road accidents in a city averages 1.8 per day. What is the probability of there being
   a) no accidents on a given day;
   b) more than two accidents on a given day;
   c) fewer than five accidents in a three-day period?

5 The average proportion of bad eggs in an egg packing station is one in 2000. The eggs are packed in boxes containing six eggs each.
   a) Evaluate to two significant figures the probability that a box contains one bad egg.

b) A customer complains if she obtains two or more boxes, with one bad egg each, per 100 boxes. What is the probability that she complains? (MEI)

6 A manufacturer of electrical components knows that he can expect 0.2% of his products to be faulty. What is the maximum number of components he can pack into a box, if he wishes to ensure that at least 90% of the boxes contain no defectives?

7 During the Second World War, 535 flying bombs fell on South London. The distribution of the numbers of hits in 576 areas, each of $0.25\,\text{km}^2$ is given in the table. Show that the aim was effectively random within the South London area by calculating the expected frequencies for a Poisson distribution with the same average number of hits.

number of hits	0	1	2	3	4	5	6 or more
frequency	229	211	93	35	7	1	0

(MEI)

8 The incidence of plumbing repairs in 78 council houses over a period of ten years is given in the table. Do the data support the view that there are good and bad tenants or the view that there are lucky and unlucky tenants?

number of repairs	0	1	2	3	4	5	6	7	8	9	10
number of houses	3	13	16	16	10	9	3	5	1	1	1

(Use a Poisson probability model and compare theoretical and observed frequencies.) (MEI)

9 At a stage in the mass production of lamp holders, random samples, each of 40 articles, are examined and the number of defective articles recorded. The numbers of defective articles in each of 200 samples are shown in the following frequency table:

defectives	0	1	2	3	4	5	6	7
number of samples	29	56	42	42	23	7	0	1

Find the mean number of defectives per sample and the variance. Give reasons for thinking that the distribution approximates to a Poisson distribution.

Show that on this assumption there is a probability of about 5% of a sample containing more than four defectives and a probability of less than 1% of a sample containing more than six defectives. (OC)

10 Spot blemishes occur randomly along a steel wire. The number counted in consecutive centimetre lengths had the following distribution:

number of spots	0	1	2	3	4	5	6	7
frequency	102	150	112	56	21	6	2	1

If the count had been over consecutive 2 cm lengths, estimate the frequency of occurrence of 3 spots per 2 cm. (You may assume that the frequencies are close to a Poisson distribution.)   (SMP)

11   The following are the numbers of breakdowns of a machine in 40 12-hour periods:

0 4 0 1 1 0 0 2 2 0 1 1 0 1 1 0 1 3 2 0
2 1 2 2 1 1 0 0 2 0 0 1 0 2 1 1 3 1 0 2

Calculate the mean number of breakdowns per 12-hour period and, using this value for μ in a Poisson probability model, calculate the probabilities of 0, 1, 2, 3, 4 breakdowns. Hence estimate the number of breakdown-free days in a year (taken to consist of 300 12-hour working days). Also estimate the number of days in the year when there will be more than 2 breakdowns.

12   A footballer finds that the number of goals he scores in a match has a Poisson distribution with mean $\frac{1}{4}$. What is the distribution of the number of goals he scores in n matches? How many matches must he play in order to be 95% sure of scoring at least 20 goals?   (O)

13   Using probability generators, show that if x, y have independent Poisson distributions with means μ_x, μ_y then $x + y$ has a Poisson distribution with mean $\mu_x + \mu_y$.

14   Instruments A and B are set to record radiation from separate and independent sources. The number of particles recorded by the instruments in one second have Poisson distributions with means 2 and 4 respectively. Calculate the probabilities that
a) at least one particle is recorded on one of the two instruments in a second;
b) a total of 4 or more particles is recorded on the two instruments in a second.   (OC)

15   Lorry drivers call at random at a roadside transport cafe on a north–south main road; the arrival rates are constant at 6 drivers per hour northbound and 2 drivers per hour southbound over the whole time that the cafe is open. Find the probabilities that, in a half-hour period,
a) 1 or more northbound and 1 or more southbound drivers call,
b) a total of at least 2 drivers calls.
The cafe manager has found that southbound drivers spend more than northbound ones, and that the cafe will run profitably in a half-hour period provided a total of at least 3 drivers or at least 2 southbound drivers calls. What is the probability of making a profit in a half-hour period? Find, using a suitable approximation, the probability of making a profit in at least 32 such periods out of a working day of 36 periods.   (MEI)

16 A factory produces high-quality china dishes. Flaws occur in the dishes themselves with a Poisson probability distribution, mean μ, and flaws occur independently in the glazing, again with a Poisson probability distribution with mean μ. A dish is immediately rejected if it has a flaw of either kind. What is the probability that a dish is rejected? Hence find the probability that a rejected dish only had flaws in the glazing (expressing your answer in as simple a form as possible).

17 In a simplified probability model of the service in a barber's shop it is supposed that all haircuts take exactly six minutes and that a fresh batch of customers arrives at six-minute intervals. The number of customers in a batch is described by a Poisson probability function, the mean number being 3. Any customer who cannot be served instantly goes away and has his hair cut elsewhere. The shop is open for 40 hours a week. Calculate the theoretical frequencies with which batches of 0, 1, 2, 3, 4, 5 and more than 5 customers will arrive per week.

The proprietor reckons that it costs him £25 a week to staff and maintain each chair in his shop, and he charges 25p for each haircut. Calculate his expected weekly profit if he has **a)** 3, **b)** 4, **c)** 5 chairs.

(SMP)

18 Show that when an event occurs randomly in time (i.e., the number of events occurring in a fixed interval of time has a Poisson distribution—with mean λ) then the time interval between successive occurrences has the exponential distribution:

$$f(x) = \lambda e^{-\lambda x} \quad (x > 0)$$

The mean number of calls per minute made on a certain telephone exchange is two. Find:
a) the probability that in any given one minute interval not more than one call will be made;
b) the probability that in any given half-minute interval exactly two calls will be made;
c) the probability that there will be no call in the half-minute immediately following a call.

(MEI)

15.5 Samples

In practical statistics we rarely consider individual readings or measurements in isolation; the normal procedure is to use a *sample*, or set of readings, and from these readings to obtain a *statistic*, or representative value. The most common, and generally most useful, statistic is the *mean*. Intuitively we expect it to give us a good approximation to the expected value (or mean), μ, of the background population, and we also feel that somehow the mean is a more reliable estimate of μ than is a single reading. We will now give some substance to these intuitive notions by extending our ideas of expectation to cover the expected value and variance of the mean of a sample. But before examining the mean itself, we must first establish some basic results in expectation algebra.

Expectation algebra

Suppose that X is a random variable, and takes values

$$x_1, x_2, x_3, \ldots$$

with probabilities $p_1, p_2, p_3, \ldots$.

We now use the notation $E(X)$ for the *expected value*, or *expectation* of x, so that

$$E(X) = \sum x_i p_i = \mu$$

Furthermore, we denote the variance of X by $\text{Var}(X)$ so that

$$\text{Var}(X) = \sum (x_i - \mu)^2 p_i = \sum x_i^2 p_i - \mu^2$$
$$= E[(X - \mu)^2] = E(X^2) - \mu^2$$

Hence

$$E(X) = \mu$$

and

$$\text{Var}(X) = E[(X - \mu)^2] = E(X^2) - \mu^2$$

We now let a be constant, so that aX takes values ax_1, ax_2, $\ldots$ with probabilities $p_1, p_2, \ldots$.

Hence $E(aX) = \sum ax_i p_i$

$$= a \sum x_i p_i$$

$$= a E(X)$$

Similarly $\text{Var}(aX) = \sum (ax_i - a\mu)^2 p_i$

$$= \sum a^2 (x_i - \mu)^2 p_i$$

$$= a^2 \sum (x_i - \mu)^2 p_i$$

$$= a^2 \text{Var}(X)$$

Summarising,

$$E(aX) = a E(X)$$

$$\text{Var}(aX) = a^2 \text{Var}(X)$$

The same results can be shown to hold for a continuous variable, using the probability density function, and integration instead of summation.

Example 1

If measurements are made in centimetres, the expected value and variance of length of leaf of a certain type of rhododendron are 5.5 cm and 2.4 cm² respectively. What will the expected value and variance be if measurements are made in millimetres?

Let X denote the leaf length in centimetres,

then $E(X) = 5.5$ and $Var(X) = 2.4$

The length in millimetres is given by $10X$, so the expected length in millimetres is given by $E(10X)$

and $E(10X) = 10E(X)$

$$= 55$$

Similarly $Var(10X) = 10^2 \, Var(X)$

$$= 240$$

These results of course are hardly surprising, and simply reinforce the above expectation results with common sense.

Two random variables

Let X, Y be two random variables, each of them discrete. (An equivalent analysis is also possible for continuous variables, but it involves the use of more advanced mathematics, with which the reader is probably not yet acquainted; it can be found in most specialised probability texts.)

Let X take possible values $x_1, x_2, \ldots, x_i, \ldots$
and Y take possible values $y_1, y_2, \ldots, y_j, \ldots$

(a finite or infinite number in each case).

We define the *joint probability distribution* of X and Y by $P(X = x_i, Y = y_j)$, which can be written $P(x_i, y_j)$ or more simply p_{ij}.

Then $P(x_i) = \sum_j p_{ij}$,   summing over all possible values of j

and $P(y_j) = \sum_i p_{ij}$,   summing over all possible values of i

This situation can now be summarised in the table:

$X \backslash Y$	y_1	y_2	$\cdots$	y_j	$\cdots$	
x_1	p_{11}	p_{12}		p_{1j}		$P(x_1)$
x_2	p_{21}	p_{22}		p_{2j}		$P(x_2)$
.						
.						
.						
x_i	p_{i1}	p_{i2}		p_{ij}		$P(x_i)$
.						
.						
.						
	$P(y_1)$	$P(y_2)$	$\cdots$	$P(y_j)$	$\cdots$	1

Using this notation, let us now investigate the expected value and variance of $X + Y$, the sum of the two random variables.

Considering the values of $X + Y$ which arise from all the combinations of values of X and Y,

$$E(X + Y) = \sum_i \sum_j (x_i + y_i)p_{ij}$$

$$= \sum_i \sum_j x_i p_{ij} + \sum_i \sum_j y_j p_{ij}$$

$$= \sum_i \left(x_i\left(\sum_j p_{ij}\right)\right) + \sum_j \left(y_j\left(\sum_i p_{ij}\right)\right)$$

$$= \sum_i x_i P(x_i) + \sum_j y_j P(y_j)$$

$$= E(X) + E(Y)$$

Also $E(XY) = \sum_i \sum_j p_{ij} x_i y_j$

But if X, Y are independent then $p_{ij} = P(x_i)P(y_j)$

$$\Rightarrow \quad E(XY) = \sum_i \sum_j P(x_i)P(y_j)x_i y_j$$

$$= \sum_i P(x_i)x_i \sum_j P(y_j)y_j$$

$$= E(X)\,E(Y)$$

Moreover, $\text{Var}(X + Y) = \text{E}[(X + Y)^2] - [\text{E}(X + Y)]^2$

$$= \text{E}(X^2 + 2XY + Y^2) - [\text{E}(X) + \text{E}(Y)]^2$$

$$= \{\text{E}(X^2) - [\text{E}(X)]^2\} + \{\text{E}(Y^2) - [\text{E}(Y)]^2\}$$
$$+ 2\{\text{E}(XY) - \text{E}(X)\text{E}(Y)\}$$

$$= \text{Var}(X) + \text{Var}(Y) + 2\,\text{Cov}(X, Y)$$

where $\text{Cov}(X, Y) = \text{E}(XY) - \text{E}(X)\,\text{E}(Y)$ and is known as the *covariance* of X and Y, which is a measure of their interdependence.

In particular, we have seen that if X, Y are independent, then

$$\text{E}(XY) = \text{E}(X)\text{E}(Y)$$

$\Rightarrow$ $\quad \text{Cov}(X, Y) = 0$

$\Rightarrow$ $\quad \text{Var}(X + Y) = \text{Var}(X) + \text{Var}(Y)$

Summarising: in general

$$\text{E}(X + Y) = \text{E}(X) + \text{E}(Y)$$

$$\text{Var}(X + Y) = \text{Var}(X) + \text{Var}(Y) + 2\,\text{Cov}(X, Y)$$

but if X, Y are independent $\quad \text{Var}(X + Y) = \text{Var}(X) + \text{Var}(Y)$

Similarly, $\text{E}(X - Y) = \text{E}(X) - \text{E}(Y)$

$$\text{Var}(X - Y) = \text{Var}(X) + \text{Var}(Y) - 2\,\text{Cov}(X, Y);$$

and note that for independent X, Y

$$\text{Var}(X - Y) = \text{Var}(X) + \text{Var}(Y)$$

so that the difference of two independent variables has exactly the same variance as their sum.

Example 2

A normal 'fair' 6-faced die is thrown together with a 'fair' tetrahedral (4-faced) die. What is the expected value and variance of the total score registered on the two dice? (For the tetrahedral die, the score registered is that of the face on which it lands.)

Let the score on the 6-faced die be denoted by X,

then $\text{E}(X) = \frac{1}{6}(1 + 2 + 3 + 4 + 5 + 6)$

$$= 3.5$$

and $\quad \mathrm{Var}(X) = \frac{1}{6}(1^2 + 2^2 + 3^2 + 4^2 + 5^2 + 6^2) - 3.5^2$

$$= \frac{91}{6} - 3.5^2$$

$$= 15.17 - 12.25$$

$$= 2.92$$

Now let the score on the 4-faced die be denoted by Y,

then $\quad \mathrm{E}(Y) = \frac{1}{4}(1 + 2 + 3 + 4)$

$$= 2.5$$

and $\quad \mathrm{Var}(Y) = \frac{1}{4}(1^2 + 2^2 + 3^2 + 4^2) - 2.5^2$

$$= \frac{30}{4} - 2.5^2$$

$$= 7.5 - 6.25$$

$$= 1.25$$

Now the total score is given by $X + Y$

and $\quad \mathrm{E}(X + Y) = \mathrm{E}(X) + \mathrm{E}(Y)$

$$= 3.5 + 2.5$$

$$= 6$$

and if we assume that the two scores are independent, then

$\quad \mathrm{Var}(X + Y) = \mathrm{Var}(X) + \mathrm{Var}(Y)$

$$= 2.92 + 1.25$$

$$= 4.17$$

So the expected value and variance of the total score are respectively 6 and 4.17.

Example 3

A firm which produces garden tools manufactures the blades and shafts (including handle) of its spades separately. The blade lengths (up to the point of attachment to the shaft) are approximately Normal, with mean 30 cm and standard deviation 0.1 cm, and the lengths of the shafts are also Normally distributed with mean 78 cm and standard deviation 0.5 cm. What percentage of spades, when assembled, have lengths exceeding 108.2 cm?

We will assume (without proof) that the sum of two Normally distributed variables is itself Normal. We will denote the length of the blade by X, and the length of the shaft by Y. Therefore $X + Y$ is the total length of the spade.

Hence $E(X + Y) = E(X) + E(Y) = 30 + 78 = 108$

and $\text{Var}(X + Y) = \text{Var}(X) + \text{Var}(Y) = 0.5^2 + 0.1^2 = 0.26$

So the mean total length $= 108$, and its standard deviation is $\sqrt{0.26} = 0.51$.

Now since we have assumed that $X + Y$ is Normally distributed,

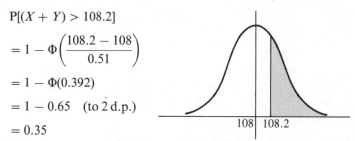

$P[(X + Y) > 108.2]$

$= 1 - \Phi\left(\dfrac{108.2 - 108}{0.51}\right)$

$= 1 - \Phi(0.392)$

$= 1 - 0.65 \quad \text{(to 2 d.p.)}$

$= 0.35$

Therefore 35% of the spades are longer than 108.2 cm.

Exercise 15.5a

1   Prove that, if a is a constant,
   a) $E(X + a) = E(X) + a$
   b) $\text{Var}(X + a) = \text{Var}(X)$

2   If X and Y are random variables, and a, b are constants, obtain expressions, in terms of $E(X)$, $E(Y)$, $\text{Var}(X)$ and $\text{Var}(Y)$ for $E(aX + bY)$ and $\text{Var}(aX + bY)$:
   a) when X, Y are independent;
   b) when X, Y are not independent.

3   Extend no. 2 to three mutually independent random variables X, Y, Z and constants a, b, c to obtain expressions for:
   a) $E(aX + bY + cZ)$
   b) $\text{Var}(aX + bY + cZ)$

4   Find the expected total score obtained when three standard, 'fair' dice are thrown together. What is the variance of this score?

5   Two standard, 'fair' dice, one blue, the other red, are thrown together, and a total score is obtained by doubling the number shown on the blue die and adding to it the number on the red die. What are the expected value and variance of this score? Compare your results with those obtained in no. 4.

6 Two standard 'fair' dice are thrown, and their scores recorded. What are the expected value, and standard deviation of the difference between the two scores?

7 Alan and Bob play a game in which Alan uses a 6-faced die and Bob a 4-faced die (as described in the text in example 2). To make the game 'fair', they decide to double all Alan's scores and treble Bob's. Would you expect this to lead to a fair game? If not, who has the advantage? Evaluate the variance of each (increased) score on the single throw of a die.

8 If X and Y are independent random single digit numbers other than zero, calculate the variance of
   a) X b) $X + Y$ c) $2X$ d) $2X - Y$ (SMP)

9 A woman travels from her London office to her home by a tube journey from station A to station B. Her walking times to A and from B add up to 5 minutes with negligible variation, the variable factors in the journey being as follows, measured in minutes:

	mean time	standard deviation
waiting for a train	8	2.6
train journey	47	1.8

Assuming that these two factors are independent and normally distributed, find the mean time and standard deviation of her whole journey.
   Estimate the probability of the *whole* journey taking
   a) less than 52 minutes;
   b) more than 65 minutes;
   c) between 57 and 62 minutes.   (MEI)

10 The discrete random variable, X, has the following probability distribution.

x	-1	0	1
$P(X = x)$	$\frac{1}{4}$	$\frac{1}{2}$	$\frac{1}{4}$

Find the mean and variance of X.
   If two independent random variables, X_1 and X_2, have the same distribution as X, find the distribution of $X_1 - X_2$, and give its mean and variance.
   What is the distribution of $X_1 + X_2$? Comment on the relationship between this distribution and that of $X_1 - X_2$.   (O)

11 The random variable X is distributed Normally with mean 2 and standard deviation 4. Find the probabilities that a) $X > 6$ and b) $|X| > 6$.

A second random variable Y, independent of X, is distributed Normally with probability density function f, where

$$f(y) = \frac{1}{(18\pi)^{\frac{1}{2}}} e^{-(y-1)^2/18}$$

Write down the value of the mean and standard deviation of Y, and hence determine the mean and standard deviation of $(X + Y)$. Find the probabilities that c) $X + Y > 6$ and d) $|X + Y| > 6$.   (MEI)

12 Three independent components are placed in an electrical circuit in such a way that, as soon as the first fails, the second (reserve) component is automatically switched in, and then when the second component itself fails, the third is automatically switched in. The circuit functions as long as one of the components works. If the lifetimes of the components are Normally distributed with mean 2000 hours and standard deviation 300 hours, what is the probability that:
a) the circuit will function for more than 6500 hours;
b) the circuit will break down within 5000 hours?

13 If $X_1, X_2, \ldots, X_n$ are independently distributed variables with means μ_1, $\mu_2, \ldots, \mu_n$ and standard deviations $\sigma_1, \sigma_2, \ldots, \sigma_n$ respectively, write down the mean and standard deviation of $Y = \lambda_1 X_1 + \lambda_2 X_2 + \cdots + \lambda_n X_n$ where $\lambda_1, \lambda_2, \ldots, \lambda_n$ are constants.

In the manufacture of certain metal tubes there are three distinct processes. The mean times for each process and the standard deviations of these times are given in the following table:

	mean time/s	standard deviation/s
process 1	60	6
process 2	10	1
process 3	40	5

During the course of its manufacture each tube is subjected to processes 1 and 3 once each, and to process 2 five times. Assuming that the times spent at each stage are distributed independently, calculate the mean and standard deviation of the total time taken in the manufacture of a tube.

Assuming Normal variation, estimate:
a) the percentage of tubes which would take less than 130 seconds to manufacture;
b) the percentage of tubes which would take more than 160 seconds to manufacture.   (MEI)

14 In a factory, cylindrical pins are made to fit into circular holes in blocks of metal. The diameter of a pin is a Normal variable with mean 9.80 mm and standard deviation 0.10 mm. The diameter of a hole is a Normal variable

with mean 10.00 mm and standard deviation 0.12 mm. A pin and hole 'fit' if the diameter of the hole exceeds that of the pin by not more than 0.30 mm. Find the probability of getting a 'fit'. (Assume difference in diameters to be Normally distributed.) (C)

15 Two independent random variables X_1 and X_2 have zero mean and the same variance, σ^2. Given that $Y_1 = a_1 X_1 + a_2 X_2$ and $Y_2 = b_1 X_1 + b_2 X_2$, where a_1, a_2, b_1, b_2 are constants, show:
a) that $\text{Var}(Y_1) = (a_1^2 + a_2^2)\sigma^2$;
b) that the condition for $\text{Cov}(Y_1, Y_2) = 0$ is $a_1 b_1 + a_2 b_2 = 0$.
Obtain an expression for $\text{Var}(Z)$ where $Z = \lambda_1 Y_1 + \lambda_2 Y_2$ and λ_1, λ_2 are constants.

Mean of n independent variables

We will consider a sample of n random variables $X_1, X_2, \ldots, X_n$, chosen independently from the same population. Then each variable individually will have the same expected value μ and variance σ^2 as the population from which it is drawn. Let m be the mean of the sample.

Then
$$E(m) = E\left(\frac{1}{n}\sum_1^n X_i\right)$$

$$= \frac{1}{n}E(\sum X_i)$$

$$= \frac{1}{n}\sum_1^n E(X_i)$$

$$= \frac{1}{n}n\mu$$

$$= \mu$$

and
$$\text{Var}(m) = \sum \text{Var}\left(\frac{X_i}{n}\right)$$

$$= \frac{1}{n^2}\sum_1^n \text{Var}(X_i)$$

$$= \frac{1}{n^2}n\sigma^2$$

$$= \frac{\sigma^2}{n}$$

So
$$E(m) = \mu \quad \text{and} \quad \text{Var}(m) = \frac{\sigma^2}{n}$$

and the standard deviation of the mean (also known as the *standard error* of the mean) is given by $\sigma/\sqrt{n}$.

Hence we are absolutely justified in using m as an estimate of μ; and since the variance of the mean is less than that of individual readings, it follows, as expected, that there is less variation in the values of m than in the individual values of X.

Small and large samples

Although we have been able to specify how the expected value and variance of any sample mean may be obtained, for a small sample there is no immediate way of stating its probability distribution; each must be analysed individually.

For large samples however, we can use the central limit theorem, which tells us that any mean of a large sample has an approximate Normal distribution, *irrespective of the population from which it is drawn.*

Samples from a Normal distribution

Since the central limit theorem tells us that the mean of a large sample is Normally distributed, it seems reasonable to assume (as can be proved) that the mean of *any sample* (large or small) drawn from a population which is itself Normally distributed, will also be Normally distributed with expected value μ and standard deviation $\sigma/\sqrt{n}$.

Example 4

A firm produces jars of jam in such a way that the net mass of jam per jar is known to be Normally distributed with mean 340 g and standard deviation 12 g. The jars are packed in cartons of 16. In what percentage of the cartons will the average mass of jam per jar be less than 336 g?

For a sample of 16 jars, the mean mass m will be Normally distributed with $\mu = 340$ and $\sigma = 12/\sqrt{16} = 3$.

Therefore $P(m < 336) = \Phi\left(\dfrac{336 - 340}{3}\right)$

$$= \Phi(-1\cdot33)$$
$$= 1 - \Phi(1.33)$$
$$= 1 - 0.91 \quad \text{(to 2 d.p.)}$$
$$= 0.09$$

So the average mass of jam per jar will be less than 336 g in 9% of cartons.

Example 5

A laboratory obtains a certain chemical solution from the manufacturers in $100\,cm^3$ containers. The manufacturers state that on average the mass of impurity in each container is 3.8 mg, and its standard deviation is 0.9 mg. For a certain experiment the laboratory wants to ensure that there is at most a 5% chance that the mass of impurity in each $100\,cm^3$ of solution it uses exceeds 4.0 mg. How many containers of solution need to be mixed together to ensure this?

Suppose that we mix n containers of chemical. We will assume that n is large, so that the mean mass of impurity in the sample so obtained is approximately Normally distributed, with

$$\mu = 3.8\,mg \quad \text{and} \quad \sigma = 0.9/\sqrt{n}\,mg$$

We require $P(m > 4.0) \leqslant 0.05$
$\Rightarrow \qquad\qquad P(m \leqslant 4.0) > 0.95$

$$\Rightarrow \qquad \Phi\left(\frac{4.0 - 3.8}{0.9/\sqrt{n}}\right) > \Phi(1.67)$$

$$\Rightarrow \qquad\qquad \frac{0.2\sqrt{n}}{0.9} > 1.67$$

$$\Rightarrow \qquad\qquad \sqrt{n} > 1.67 \times 4.5 = 7.52$$
$$\Rightarrow \qquad\qquad n > 56.7$$

So at least 57 containers of the solution need to be mixed together.

Exercise 15.5b

1   Assuming that the heights of male college students are Normally distributed with mean 173 cm and standard deviation 8 cm, what is the probability that the mean height of a sample of 40 students will
   a) exceed 175 cm;
   b) lie between 172 and 174 cm?

2   The marks obtained in a mathematics examination are approximately Normally distributed with mean 54 and standard deviation 13. What is the probability that the average mark for a group of 10 students
   a) exceeds 60;
   b) is less than 50?

3   A certain industrial process takes a worker an average of 24 minutes to complete, and the standard deviation of the completion time is 6 minutes. If the process is repeated 100 times, use the central limit theorem to find

approximately the probability that the mean process time exceeds 25 minutes.

4 Each schoolboy in a class of 36 spins an unbiased coin 100 times. What is the probability that the average number of heads obtained per boy
a) is less than 49;
b) is greater than 52?
(A continuity correction is not necessary in this case.)

5 Capsules of a certain drug have a nominal net mass of 10 g. They are packed in batches of 100 capsules. A machine fills the capsules with the drug in such a way that the masses are normally distributed with mean net mass 10.06 g and standard deviation 0.2 g. Find the probability that:
a) a batch does not contain any 'underweight' capsules;
b) the mean net mass of the capsules in a batch does not fall below the nominal net mass.

6 A coal merchant sells his coal in bags marked '50 kg'. In fact he claims that the average mass is 50 kg, with a standard deviation of 1 kg. A suspicious trade inspector has 60 of the bags weighed, and finds that their mean mass is 49.6 kg. What is the probability that such a result, or a more extreme one, could have arisen if the coal merchant's claim is true? Do you think that the inspector's suspicions were justified?

7 Electric light bulbs are produced with a mean life-time of 1600 hours and a standard deviation of 250 hours. Calculate the minimum sample size necessary to ensure that there is at most a 1% chance of the average life-time of such a sample being less than 1550 hours. (Assume that the sample will be large enough for the mean to have an approximately Normal distribution.)

8 How large a sample would you take in order to estimate the mean of a population, so that the probability is 0.95 that the sample mean will not differ from the true mean of the population by more than 0.2 of the standard deviation of the population?

Unbiased estimates

If T is some statistical estimate calculated from data, and θ a probability parameter, then if $E(T) = \theta$, T is said to be an *unbiased estimate* of θ.

It should be noted, however, that an unbiased estimate is not necessarily a particularly good estimate—its variance, for instance, may be quite large. It is simply the best we have.

It therefore follows, since $E(m) = \mu$, that the sample mean m is an unbiased estimate of μ, the mean, or expected value, of the parent population.

Since m is an unbiased estimate of μ, it would seem to follow that s^2 should be an unbiased estimate of σ^2; after all, both are referred to as variances, and as we saw in section 15.1, their forms are very similar. To pursue this further, we note that s^2 is itself a random variable, depending only on the sample from which it is calculated, and we can therefore consider its expected value.

Now $s^2 = \dfrac{1}{n}\sum(X_i - m)^2$

but in this form it is a little difficult to manage, since we cannot say anything directly about $E[(X_i - m)^2]$. However, we do know that

$\quad E[(X_i - \mu)^2] = \sigma^2$

Also, recalling that X_i has mean m and variance s^2, it follows that $X_i - \mu$ has mean $m - \mu$ and variance s^2.

So $s^2 = \dfrac{1}{n}\sum(X_i - \mu)^2 - (m - \mu)^2$

$\Rightarrow\quad E(s^2) = E\left[\dfrac{1}{n}\sum(X_i - \mu)^2\right] - E[(m - \mu)^2]$

$\qquad\qquad = \sigma^2 - \mathrm{Var}(m)$

$\qquad\qquad = \sigma^2 - \dfrac{\sigma^2}{n}$

$\qquad\qquad = \dfrac{n-1}{n}\sigma^2$

So s^2 is *not* an unbiased estimate of σ^2, and in fact s^2 will tend to give a low value for σ^2 (as can readily be seen by considering the zero value of s^2 that would arise if the sample had only one member).

But $E\left(\dfrac{n}{n-1}s^2\right) = \dfrac{n}{n-1} \times \dfrac{n-1}{n}\sigma^2 = \sigma^2$

so that an 'unbiased' estimate of σ is

$\hat{\sigma} = \sqrt{\left(\dfrac{n}{n-1}\right)}\, s$

Exercise 15.5c

1   The lengths, in millimetres, of eight full-grown fish of a certain species were found to be:

   56, 49, 68, 58, 63, 60, 55, 59

   Obtain unbiased estimates for the mean and variance of the length of the species.

2 A sample of 15 leaves was obtained from a certain plant, and the areas of these leaves, in square centimetres, are given below:

4.60, 5.85, 5.13, 3.94, 5.61, 4.08, 4.71, 4.29, 6.01, 5.46, 5.80, 6.33, 4.43, 4.90, 5.09

Obtain unbiased estimates of the mean and variance of leaf area (correct to 2 decimal places).

3 In a sample of bars of chocolate produced in a particular factory, the mean mass was found to be 101.3 g and the standard deviation 1.6 g. Obtain an unbiased estimate (correct to 2 decimal places) of the standard deviation of all bars of chocolate produced in the factory if the sample size was:
a) 5   **b)** 20   **c)** 100

4 t is an unbiased estimator for a parameter θ, and t_i, $i = 1, 2, \ldots, n$ are values of t obtained from n experiments.
a) Prove that

$$\frac{1}{n} \sum_{i=1}^{n} t_i \quad \text{is an unbiased estimate of } \theta$$

b) Find the relationship between the numbers λ_i, $(i = 1, \ldots n)$ if $\Sigma \lambda_i t_i$ is an unbiased estimate of θ.

5 X and Y are independent random variables, each having a Poisson distribution with parameter μ. Show that both $(X + Y)/2$ and $(X - Y)^2/2$ are unbiased estimators of μ. (O)

6 What are the values of the mean and variance of the number of successes in n independent repetitions of a trial in which the probability of success is p?

An ordinary shaped die is biased in such a way that the probabilities of throwing a 1 and a 6 are equal but different from $\frac{1}{6}$. Denote this common probability by θ. Two methods are suggested for estimating θ. These are:

Method 1: Toss the die 20 times; record the number X of sixes that occur; take $P_1 = (X/20)$ as the estimate of θ;

Method 2: Toss the die 10 times; record the total number Y of ones and sixes that occur; take $P_2 = (Y/20)$ as the estimate of θ.

Show that $E(P_1) = E(P_2) = \theta$. Calculate the variances of P_1 and P_2, and hence decide which of the methods is the better one for estimating θ.
(MEI)

7 The random variable X has a probability density function given by

$$f(x) = \begin{cases} \dfrac{1}{\theta} e^{-x/\theta} & (x \geqslant 0, \text{ where } \theta > 0) \\ 0 & (\text{otherwise}) \end{cases}$$

A random sample $X_1, X_2, \ldots, X_n$ is taken from a population with the above distribution. The estimator T is defined by

$$T = k \sum_{i=1}^{n} X_i^2, \quad \text{where } k \text{ is a constant.}$$

Find the value of k such that T is an unbiased estimator of θ^2.   (AEB, 1981)

8   A random variable X has expectation μ and variance σ^2. Find the variance of

$$\bar{X} = \frac{1}{n} \sum_{i=1}^{n} X_i$$

and the expected value of

$$s^2 = \frac{1}{n-1} \sum (X_i - \bar{X})^2$$

where $X_1, X_2, \ldots, X_n$ are n independent observations from the distribution of X.

A further set of m independent observations $Y_1, Y_2, \ldots, Y_m$ is taken from the same distribution and

$$t^2 = \frac{1}{m-1} \sum_{j=1}^{m} (Y_j - \bar{Y})^2$$

If $u^2 = \lambda s^2 + (1 - \lambda)t^2$, $(0 < \lambda < 1)$, show that s^2, t^2 and u^2 have the same expectation and find λ such that $V[u^2]$ is a minimum.

$$\left(\text{Take } V[s^2] = \frac{2\sigma^4}{n-1}. \right)$$

(MEI)

* 15.6 Significance and confidence

In a television advertisement, out of a sample of eight people interviewed leaving a supermarket, six were found to prefer the new washing powder, 'Squelch'. Is this convincing evidence that more people use Squelch than all other powders?

This question involves firstly practical and experimental considerations, and secondly mathematical considerations. The first thing we must know is how the eight people were chosen. Were they chosen randomly? (For the present we must be content with an intuitive notion of randomness.) Or did the interviewer particularly look out for shopping bags containing packets of Squelch? Such questions must be asked before any meaningful mathematical deductions can be made. However since we are primarily concerned with the mathematics of such situations, we will assume that these practical questions have been settled, and for the purposes of this example we will assume that the eight people were chosen randomly.

We shall not (and indeed we cannot) answer the question about preference for Squelch directly. Instead we invert the question and ask:

'Assuming that exactly 50% of people prefer Squelch to other powders, what is the probability that this result (six out of eight choices for Squelch), or a more extreme one, will occur by chance?'

(The phrase 'this result, *or a more extreme one*' though it seems to complicate the problem, does in fact provide us with a standard method for dealing with such problems and the necessity of its inclusion will become increasingly obvious.)

If this probability is large, then there is no reason to doubt the assumption, but if the probability is small then there is considerable evidence against our assumption; we become sceptical about the assumption that people are indifferent, and increasingly certain that they do indeed prefer Squelch to other powders.

The phrase 'it is not true that more people prefer Squelch to any other powder' could be interpreted mathematically in a number of ways. We will consider the most extreme case for which the statement holds, when exactly half prefer Squelch, so that the statement is only just true. (Apart from any other considerations of course, this will simplify the mathematics.) We therefore have a binomial situation, with 'success' being regarded as finding someone who prefers Squelch.

So $n = 8$, $p = \frac{1}{2}$, and the probability distribution takes the form shown in the diagram.

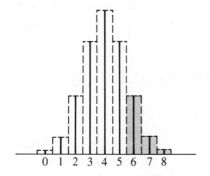

$$P(\geqslant 6 \text{ 'successes'}) = \binom{8}{6} \times \left(\frac{1}{2}\right)^8 + \binom{8}{7} \times \left(\frac{1}{2}\right)^8 + \binom{8}{8} \times \left(\frac{1}{2}\right)^8$$

$$= 28 \times \frac{1}{2^8} + 8 \times \frac{1}{2^8} + 1 \times \frac{1}{2^8}$$

$$= \frac{37}{256}$$

$$= 0.145 \approx \frac{1}{7}$$

Therefore there is a probability of $\frac{1}{7}$ that six out of a sample of eight people would choose Squelch even when there is no overall preference for the powder. So we should regard the evidence of the advertisement as not particularly convincing.

This analysis is one example of a very simple *significance test*. The probability $\frac{1}{7}$ is a measure of the significance of the experimental result on the assumption of no overall preference. More generally, significance testing requires the use of two basic concepts: *null hypothesis* and *level of significance*.

Null hypothesis

This is the basic hypothesis upon which we build a probability model. It should be a precise statement, which lends itself to mathematical formulation. In our example the null hypothesis was that exactly half of people prefer Squelch. The term 'null' arises primarily from contexts such as medical research, where new treatments or drugs are tested for effectiveness, the null hypothesis being that such a treatment has no effect; and generally speaking a null hypothesis takes the form of just such a negative statement.

Levels of significance

Using the null hypothesis, we evaluate the probability of obtaining such an extreme value as our sample statistic: the smaller this probability, the greater is the evidence against the null hypothesis. We normally use three standard levels of significance: 5%, 1%, and 0.1%, corresponding to probabilities of 0.05, 0.01, and 0.001 respectively. If, for instance, we obtain a probability of 0.03, we say that this result is significant at the 5% level (but not at the 1% level), and we can therefore reject the null hypothesis at the 5% level. Which significance level is used generally depends on the nature of the experiment or research.

Errors

There are two distinct types of error which may arise from significance testing:
a) A sample may not provide significant evidence against the null hypothesis, and so the hypothesis may be accepted when in fact it is incorrect.
b) If we reject the null hypothesis at, say, the 5% level, then in 5 out of 100 trials the hypothesis could be correct, and the extreme probability will have occurred purely by chance. The incidence of this type of error can of course be reduced by using a higher level of significance, for example the 1% level.

But further detailed analysis of such errors is beyond the scope of this text, and for the present we must just accept that such unavoidable errors may arise.

The following two examples are typical significance tests.

Example 1

Before the introduction of a new drug, the success rate in treatment of patients for a certain disease was 43%. However, of 88 patients treated with the drug, 51 recovered. On the basis of this evidence, is the drug significantly effective in treating the disease?

The null hypothesis is that the drug has no effect, i.e. the probability of a patient recovering is still 0.43.

We therefore have a binomial situation, with $n = 88, p = 0.43$.

Since n is large we can approximate by a Normal distribution with

$$\mu = 88 \times 0.43 = 37.8$$

and $\sigma = \sqrt{(88 \times 0.43 \times 0.57)} = 4.64$

Therefore   P(51 or more recoveries) $= 1 - \mathrm{P}(\leqslant 50 \text{ recoveries})$

$$= 1 - \Phi\left(\frac{50.5 - 37.8}{4.64}\right)$$

(using a continuity correction)

$$= 1 - \Phi(2.74)$$
$$= 1 - 0.997$$
$$= 0.003$$

As $0.003 < 0.01$, this result is significant at the 1% level. So the probability, on the basis of the null hypothesis, of obtaining the observed result is very remote, and we can say with considerable confidence that the drug is effective.

Example 2

The machine described in example 4 of section 15.5, filling jars with jam such that the mass of jam is Normally distributed with $\mu = 340\,\mathrm{g}$ and $\sigma = 12\,\mathrm{g}$, breaks down. After it is repaired, it is found that the average mass of jam in the first 100 jars it fills is 342 g. Has the performance of the machine significantly altered?

The null hypothesis is that the machine is unchanged, i.e., the means of samples, size 100, are Normally distributed, with mean 340 and standard deviation $12/\sqrt{100} = 1.2$.

We need to find the probability of a result of 342, or one more extreme, occurring by chance. In this case, however, a more extreme result could reasonably be one less than 338, since we are not really concerned whether the mass of jam is increased or decreased, but simply whether it has changed, and the relevant probability is indicated by the shaded areas in the diagram.

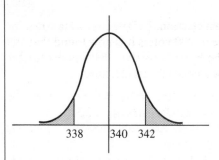

For fairly obvious reasons, therefore, this is referred to as a *two-tailed* (or two-sided) test, as opposed to the previous example which was one-tailed.

Now $\quad P(|\text{mass} - 340| \geqslant 2) = 2\left[1 - \Phi\left(\dfrac{342 - 340}{1.2}\right)\right]$

$$= 2[1 - \Phi(1.67)]$$
$$= 2[1 - 0.952]$$
$$= 0.096$$

This is not significant at the 5% level, and we can say that this evidence does not indicate any significant alteration in the machine's performance.

Exercise 15.6a

1   A coin is tossed eight times, and a head is obtained each time. Test, at the 1% level, the hypothesis that the coin is unbiased. (Use a two-tailed test.)

2   An amateur magician claims that he can read people's minds. To prove this he tells a person to think of one of the colours blue, yellow or red, and he then predicts what that colour is. In ten such mind-reading sessions he is correct seven times. Is this sufficient evidence to accept his claim?

3   Test, at the 5% level, the hypothesis that the variable x is Normally distributed with mean 80 and standard deviation 10, if we obtain:
   a) a value of x equal to 100;
   b) a value of x equal to 65;
   c) three values of x, all less than 70;
   d) three values of x, whose average is equal to 70.

4 Over a number of years an average of 40 out of 100 patients who underwent a difficult operation survived. Last year new medical techniques were introduced and 73 out of 150 patients survived the operation. Explain whether or not the maintaining of these techniques is statistically justified.   (SMP)

5 In a certain constituency, in the last election, $\frac{5}{8}$ of the electorate voted for the Nationalist Party. In a sample of 200 voters it is now found that 107 support the Nationalists. Test the hypothesis that the support for the Nationalist Party has not decreased since the last election.

6 A tyre company claims that the lifetimes of its tyres are Normally distributed with mean 34000 km and standard deviation 6000 km.
a) A motorist finds that one of their tyres lasts only 26000 km. Does this lead you to doubt the manufacturers' claims?
b) A car-hire firm using the tyres finds that the average life-time of 100 such tyres is 32200 km. Does this evidence lead you to doubt the manufacturers' claims?

7 The masses of a certain species of rabbit are known to be Normally distributed with mean 1.68 kg and standard deviation 0.24 kg. Nine rabbits are fed on specially enriched foodstuffs, and their average mass after two months is found to be 1.85 kg. Test at   a) the 5% level, b) the 1% level, the hypothesis that the foodstuff does not increase the rabbits' masses.

8 The following experiment was carried out 100 times by each of 16 pairs of children in a class. 'Each pair has an ordinary pack of cards from which one child selects a card and replaces it; the other child does likewise, and they score a point if the suit is the same for both'. The average number of points scored per pair was 27.2. Test at the 5% significance level, the hypothesis that all the cards were drawn at random.   (SMP)

9 A random sample of 200 persons from a large population is examined for eye colour. In this sample 50 persons are found to have blue eyes. Is this result consistent with the hypothesis that the proportion of blue-eyed persons in this population is 0.2? What would have been your conclusion if there had been 200 blue-eyed persons in a random sample of 800 persons?   (MEI)

10 A grower sows 200 Carlton lettuce seeds under conditions such that the average germination rate is 75%. By using a suitable approximation, find
a) the probability that more than 170 seeds germinate;
b) the probability that less than 140 seeds germinate.
The grower also sows 120 Alberni lettuce seeds under the same conditions as the Carlton seeds, and finds that 82 seeds germinate. Test whether the Alberni seeds have a germination rate less than 75%.   (C)

11 The protein intakes for a sample of six unemployed men were observed to be 86, 82, 66, 66, 94, 104 g day^{-1}. A large number of employed men had an average intake of 97.6 g day^{-1} and the variance of their intake was 225.

Do these data support the view that at the time of the study unemployed men were underfed by comparison with men in employment? (MEI)

12 You are engaged as an expert witness for the prosecution in a court case in which a gaming club is accused of running an unfair roulette wheel. The evidence is that out of 3700 trial spins, zero (on which the club wins) turned up 140 times. There are 37 possible scores on a trial spin, labelled 0 to 36, and these should have equal probability. Test whether there is evidence that the wheel is biased. (O)

13 According to a certain hypothesis the variable x is uniformly distributed in $(0, 1)$. Determine the probability that the smallest member of a sample of n observations of x has value not less than X when $0 \leqslant X \leqslant 1$.

Ten observations were taken of x; the values of the smallest and largest were 0.2 and 0.7. Calculate the probability that ten observations:
a) are each not less than 0.2;
b) are each not greater than 0.7;
c) lie within an interval of length 0.5.
State whether your results would lead you to reject the hypothesis.   (C)

14 In a sample of 1000 voters, an opinion pollster finds that 520 intend to vote for Jones and 480 for Smith in the forthcoming election. What is the probability of a result such as this, or one even more favourable for Jones, if in fact Jones has only the support of 50% of the electorate? How many people in the sample would need to support Jones to make it significant   **a)** at the 5% level,   **b)** at the 1% level, that she has more support than Smith?

15 A children's test which had been standardised some years ago to have a mean of 100 and a standard deviation of 15 was applied recently to a random group of children, and their mean score was recorded as 101.6, but no record was made of the number of children tested. What is the least number of children tested if the result shows that the population mean is unlikely still to be 100? (You should assume that the mean value will certainly not have decreased. Work at the 1% significance level.)

(SMP)

16 A sawing machine cuts timber into planks. The lengths of the planks are Normally distributed with variance 25 cm^2; the mean length is intended to be exactly 100 cm. A random sample of 25 planks was found to have mean length 101.2 cm. Test at the 5% level of significance whether the mean has been correctly set.

Suppose in fact the mean has been set to 102 cm. What is the probability of accepting the null hypothesis (that the mean is 100 cm) with your test? What would this probability be if the sample size had been 100? Comment.   (MEI)

17   To test whether a damaged coin is unbiased it is decided to toss the coin nine times and to judge it as being unbiased only if 4, 5 or 6 heads are tossed. Calculate the probability that the coin will be judged as being biased when in fact it is unbiased.

A close examination of the coin shows that if it is biased at all then a head will be more probable than a tail. This suggests testing the hypothesis that the coin is unbiased against the alternative that a head is more probable than a tail. Given that the test is to be based on the outcomes of twenty tosses of the coin and is to have a significance level of 5%, determine, either by referring to tables, or otherwise, the least number of heads that must occur in the twenty tosses for the coin to be judged as being biased in favour of a head.   (JMB)

18   Independent observations $x_1, x_2, \ldots, x_n$ are taken from the Normally distributed population $N(\mu, \sigma^2)$ where the value of σ^2 is known to be 4. The null hypothesis value of μ is zero and the only other possible value of μ is 1.

α denotes the probability that the null hypothesis will be rejected when in fact it is true; β denotes the probability that the null hypothesis will be accepted when in fact it is false.

a) Suppose $n = 25$ and that it is decided to reject the null hypothesis if the observed sample mean $\bar{x}$ exceeds 0.4. Determine α and β.

b) Suppose $n = 4$ and that it is required that $\alpha = 0.05$. Determine the constant k where observed values of $\bar{X}$ exceeding k will lead to rejection of the null hypothesis. Also find the value of β.

c) Suppose it is required that $\alpha = \beta = 0.01$. Determine the necessary sample size n and the constant k where observed values of $\bar{X}$ exceeding k will lead to rejection of the null hypothesis.   (MEI)

Confidence intervals

The links in lengths of chain produced in a factory are known to have breaking tensions which are Normally distributed with standard deviation 0.24 kN. Eight links are tested to destruction, and are found to have breaking tensions of 3.61, 3.58, 3.66, 3.65, 3.89, 3.48, 3.50, 3.59 kN. What can we say, if anything, about the average breaking tension of all links produced in the factory?

Let us denote the average breaking tension (of the background population) by μ.

Now the sample mean (which for this sample is 3.62) is distributed

Normally, with the same μ as the background population, and $\sigma = 0.24/\sqrt{8} = 0.085$.

We know that $(m - \mu)/\sigma$ has a standard Normal distribution, and from the properties of the standard Normal curve, it is 95% certain that $(m - \mu)/\sigma$ lies between -1.96 and 1.96.

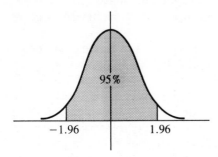

So $\quad P\left(-1.96 < \dfrac{m - \mu}{\sigma} < 1.96\right) = 0.95$

$\Rightarrow \quad P(-1.96\sigma < m - \mu < 1.96\sigma) = 0.95$

Rewriting the orderings,

$\quad P(m - 1.96\sigma < \mu < m + 1.96\sigma) = 0.95$

Therefore for the chain links, it is 95% certain that

$\quad 3.62 - 1.96 \times 0.085 < \mu < 3.62 + 1.96 \times 0.085$

$\Rightarrow \quad\quad\quad 3.62 - 0.17 < \mu < 3.62 + 0.17$

$\Rightarrow \quad\quad\quad\quad 3.45 < \mu < 3.79$

The interval 3.45 to 3.79 is referred to as the 95% confidence interval for μ, and 3.45 and 3.79 are the 95% confidence limits.

Similarly the 99% confidence limits for μ are

$\quad 3.62 \pm 2.58 \times 0.085 = 3.62 \pm 0.22 \quad$ (i.e. 3.40 and 3.84)

The 95% and 99% confidence intervals are the standard intervals used, but there is no reason why others should not be employed. Quite naturally, the more certain we need to be of the interval within which μ lies, the larger that interval becomes. So for any experimental results, we have to balance the usefulness of the result with the confidence we can put in it, in choosing a suitable interval. Also, of course, the larger the sample used, the narrower will be the confidence interval so obtained.

The general procedure for establishing confidence limits for μ (assuming a Normal distribution, with known standard deviation) is as follows:

a) Obtain a representative value—this may be a single reading or (better) the mean of a sample—and determine its standard deviation.

b) Decide on a confidence coefficient: 0.95 is for example the confidence coefficient corresponding to a 95% confidence interval. Denote this coefficient by $1 - 2\alpha$ (so for a 0.95 coefficient, $\alpha = 0.025$).

c) Define c_α by

$$\Phi(-c_\alpha) = \alpha$$

$$\Rightarrow \quad \Phi(c_\alpha) = 1 - \alpha$$

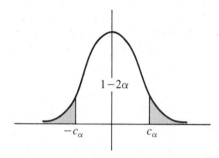

therefore the unshaded area is $1 - 2\alpha$, and the shaded areas are each α.

d) Then, in the case of the sample mean

$$P\left(m - \frac{\sigma}{\sqrt{n}}c_\alpha < \mu < m + \frac{\sigma}{\sqrt{n}}c_\alpha\right) = 1 - 2\alpha$$

and the confidence limits are $m \pm \sigma c_\alpha/\sqrt{n}$.

Similarly, for a single observation x, the confidence limits for μ are $x \pm \sigma c_\alpha$. However the confidence intervals obtained from *single* observations are usually so wide as to be of little real use.

This technique of obtaining confidence intervals is very useful. But it does have limitations, the most obvious of which is that the standard deviation of the population has to be known; in practice, however, σ is often not known, and has to be estimated from the sample. A more complicated technique has then to be employed (using what is known as the 't' distribution), and the effect of using an estimated value of σ is to widen the equivalent confidence intervals for a known value of σ. However, for *large* samples the process described above can give good approximations to the confidence limits even when the standard deviation has to be estimated.

One important situation where the method can be used, with an estimated standard deviation, is in obtaining a proportion from a large sample.

Suppose that we wish to estimate the proportion (P) of a population with a certain attribute. In a sample from the population, of size n (where n is large), the number with the attribute is found to be m, giving an estimate for the proportion of $m/n = p$.

Assuming a Normal approximation, 95% confidence limits for $\mu (= nP)$ are

$$m \pm 1.96\sigma$$

so 95% confidence limits for P are given by

$$\frac{m}{n} \pm 1.96\frac{\sigma}{n} = p \pm 1.96\frac{\sigma}{n}$$

But since m is obtained from a binomial distribution, σ can be approximated by $\sqrt{(np(1 - p))}$.

Hence 95% confidence limits for P take the form

$$p \pm 1.96 \sqrt{\left(\frac{p(1 - p)}{n}\right)}$$

Example 3

A market researcher in a certain town finds that in a random sample of 400 households, 120 own freezers. Find 95% confidence limits for the proportion of households owning freezers.

The estimated proportion, $p = 120/400 = 0.3$. So the 95% confidence limits are

$$0.3 \pm 1.96 \sqrt{\left(\frac{0.3 \times 0.7}{400}\right)} = 0.3 \pm 0.045$$

Exercise 15.6b

1 It is known that the variable x is Normally distributed with standard deviation 8. The average size of a sample of 10 values of x is found to be 63. Find **a)** 90%, **b)** 95% and **c)** 99% confidence limits for the mean value of x.

2 The melting points, in °C, of 10 samples of a certain metal were found to be 1154, 1151, 1154, 1150, 1148, 1152, 1155, 1153, 1149, 1154. Past experience indicates that these observations will be Normally distributed, with standard deviation equal to 3 °C. Find **a)** 95% and **b)** 99% confidence limits for the mean melting point of the metal.

3 It is known that the heights of 14-year old schoolboys are Normally distributed, with a standard deviation of 9.2 cm. The average height of 16

such schoolboys, chosen at random in a certain school, is found to be 159.1 cm. Find   **a)** 95% and   **b)** 99% confidence limits for the average height of 14-year old schoolboys.

4   The average length of 200 nails produced by a certain machine was found to be 7.42 cm. If the standard deviation of nail lengths is known to be 0.53 cm, find   **a)** 95% and   **b)** 99% confidence limits for the mean length of all nails produced by the machine.

5   It is known that an examination paper is marked in such a way that the standard deviation of the marks is 15.1. In a certain school, 80 candidates take the examination, and they have an average mark of 57.4. Find   **a)** 95% and   **b)** 99% confidence limits for the mean mark in the examination.

6   A factory manufacturing ammeters tests them for zero errors in their calibration. From past routine tests, it is known that the standard deviation of these errors is 0.3.

  A batch of 9 ammeters, taken from one worker's production has zero errors of $1.0, -0.1, -0.3, 1.6, 0.5, 0.4, 0.5, 0.2, -0.2$. Test whether there is evidence of bias in the ammeters produced by this worker, and establish a 95% confidence interval for the mean zero error of these ammeters.   (o)

7   A random sample of 100 capacitors, each of nominal capacitance 2 μF was taken from a very large batch. The capacitance of each capacitor in the sample was measured. The results of these measurements are summarised in the following table:

capacitance/μF (mid-interval value)	1.85	1.90	1.95	2.00	2.05	2.10	2.15
number of capacitors	2	12	23	31	20	10	2

Calculate the mean and the standard deviation of these capacitances. Calculate also the standard error of the mean, and use it to determine 95% confidence limits for the mean capacitance of capacitors in the batch. (Use a Normal approximation.)   (MEI)

8   An opinion pollster finds that in a sample of 200 voters, 115 support the Independent candidate in the next election. Find 95% confidence limits for the percentage of all voters in the constituency supporting the Independent candidate.

9   A survey in Aldervale revealed that 86 houses out of a random sample of 500 were fitted with some form of double glazing. Find 95% approximate confidence limits for the true proportion of all houses in Aldervale which have double glazing.   (AEB, 1982)

10   An insurance company has sold a certain specialised form of insurance to a large number of customers, and wishes to investigate the proportion p of them who are women. Records for a random sample of 100 customers are examined. State the distribution of the number of women in samples of this size, and give the mean and variance of the Normal distribution that may be used to approximate this. Given that there are in fact 32 women in the sample, use a 1% significance test to determine whether it is reasonable to assume that this type of insurance appeals equally to men and to women. Provide also an approximate two-sided 95% confidence interval for p.   (MEI)

11   In a random sample of 1000 people from a large population 300 stated that they used a certain detergent. Show that 95% confidence limits for the proportion of the population using this detergent are (approximately) 0.272 to 0.328. (Use a Normal approximation to the binomial distribution, and an estimated value of σ.)

   Following an advertising campaign a second sample of 800 people was taken, and of these 260 stated that they used the detergent. Is this evidence of the success of the campaign?

   Subsequently it was decided to give a free gift with each packet of the detergent, and later in a random sample of 600 people 216 stated that they used the detergent. Does this indicate the success of the free gift scheme?   (MEI)

12   The probability density function of a variable x takes the form shown in the diagram, the constant a being unknown. A single observation x_1 is made of x. Give an unbiased estimate of a, and calculate 95% confidence limits for a. (Use first principles.)

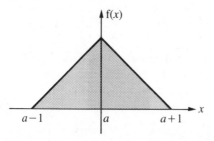

After a total of 24 observations have been made, the mean $\bar{x}$ is calculated. Use this mean to obtain an unbiased estimate of a, and calculate approximate 95% confidence limits for a.   (C)

* 15.7 Correlation and regression

Bivariate data

The marks of a class of 20 pupils in a mathematics examination and a French examination are given in the table below.

mathematics	10	13	17	19	21	22	22	22	23	24
French	10	16	11	16	17	17	18	20	22	18

mathematics	24	26	27	28	31	31	31	34	35	37
French	25	20	21	18	18	24	26	29	22	25

Since the marks are paired for each pupil, such data is referred to as bivariate data (as opposed to univariate data, where each item is treated individually; all the data so far dealt with in this chapter has been univariate, though we rarely use this term). It is generally convenient to represent bivariate data in the form of ordered pairs. Thus the data above would be $(10, 10)$, $(13, 16)$, ...

Scatter diagrams

To display bivariate data pictorially, we use a scatter diagram. This is obtained by regarding the ordered pairs of bivariate data as Cartesian coordinates, and marking the associated points on a graph:

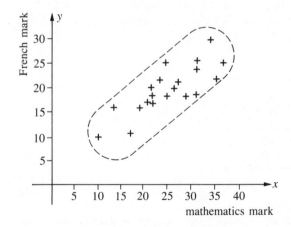

The points in the diagram all lie within a relatively narrow band, and this is emphasised by the boundary which has been superimposed around the points. This suggests some association between the x and y coordinates (i.e. between the mathematics and French marks). Such association is generally referred to as *correlation*.

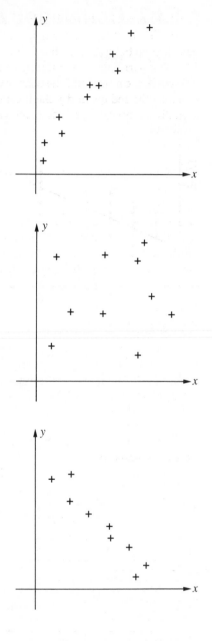

This scatter diagram illustrates bi-variate data which is closely correlated (also referred to as *positive* or *direct* correlation), i.e. small values of *x* correspond to small values of *y* and large *x* to large *y*. An example of such data might be the number of pupils and number of staff in secondary schools.

The second diagram shows data for which there is little or no correlation, for example the number of pupils and the number of trees planted in the grounds of a group of schools.

The third diagram illustrates *negative* (or inverse) correlation—the data here for example might be the numbers of pupils in schools, and the average percentages of the school's pupils per class.

Regression

A precise measure of correlation follows naturally from investigation of the related topic of regression. Whilst in the study of correlation both variables are regarded as having equal status, in regression one (say *x*, though either may be chosen) is regarded as the independent variable, and the dependence of the other variable, y, on x is investigated. To be precise, we need to find a relationship between each fixed value of x, and the expected (or mean) value of y for that particular x. Let us illustrate this by a theoretical example.

Two dice, one red and the other blue, are thrown together: let x denote the score on the red die and y the total score on the two dice. Then the possible (equally likely) values of y for each value of x are shown in the diagram:

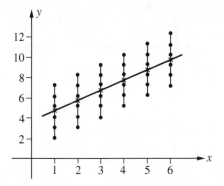

Now when $x = 1$, the expected value of y is equal to

$$\tfrac{1}{6}(2 + 3 + 4 + 5 + 6 + 7) = 4.5$$

This point is marked on the diagram by a cross, and equivalent points for the other values of x are similarly marked. The straight line upon which all these points lie is then referred to as the *regression line of y on x*, and its equation in this case is $y = x + 3.5$.

Similarly in the diagram below we have the regression line of x on y, which we see has equation $y = 2x$.

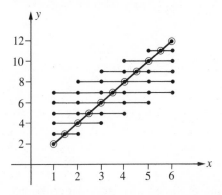

In this case y has been regarded as the independent variable, and for a particular total score, we are considering the expected value of the score on the red die.

This example, of course, is an idealised probability model. In general, the relationship between x and the equivalent expected value of y will not necessarily be linear. But whatever the relationship, linear or otherwise, we shall restrict our task (at this stage) to finding a straight line approximation. Such an approximation is generally referred to as the line of regression of y on x, though it must be emphasised that such a line, derived from a bivariate sample, will only provide an approximation to the true regression relationship in the background population from which the sample is drawn.

There is of course no single criterion for determining the 'best' line of regression for a particular set of data, but in practice the most useful one is known as the *least squares criterion*, and the line so obtained is called the *least squares line of regression of y on x*:

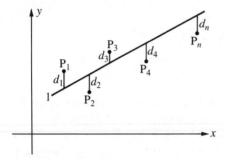

If d_i is the distance of the point P_i from the line l in the y-direction, and l is chosen in such a way that $\sum d_i^2$ is a minimum, then l is the least squares line of regression of y on x.

If instead of the line of regression of y on x, we need that of x on y, the least squares criterion is still used, but the deviations from the regression line are in this case measured in the x-direction:

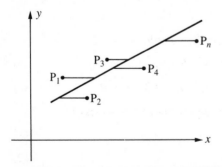

The obvious question arises: why should we choose this particular criterion? Why not the more obvious one of minimising the sum of perpendicular distances from the line? Our original definition of the regression relationship throws some light on the choice of deviation parallel to the axes, but the main justification for the criterion is on more advanced mathematical grounds. On a practical level, however, the algebra following from the least squares criterion is simpler than that arising, for example, from considerations of perpendicular distance, and the very neat and convenient results are in themselves some justification for the method.

Derivation of the line of regression of y on x

Let the points $P_1, P_2, \ldots, P_i, \ldots, P_n$ of the scatter diagram have coordinates $(x_1, y_1), (x_2, y_2), \ldots (x_i, y_i), \ldots (x_n, y_n)$. Now if we take the equation of the regression line to be $y = a + bx$, then

$$\text{deviation} \quad d_i = y_i - (a + bx_i)$$
$$= y_i - a - bx_i$$

We therefore need to find values of a, b such that

$$S = \sum_{i=1}^{n} (y_i - a - bx_i)^2 \quad \text{is a minimum}$$

We minimise S by regarding it as a function of a, b and then equating the *partial derivatives* of S with respect to a, b (written $\partial S/\partial a$, $\partial S/\partial b$) to zero.

($\partial S/\partial a$ denotes the derivative of S with respect to a, *while regarding b as a constant*. Similarly $\partial S/\partial b$ is the derivative of S with respect to b, *keeping a constant*. A thorough treatment of partial differentiation can be found in most books on advanced calculus, but a detailed knowledge of the subject is not essential for the purposes of this section.)

Now $\quad \dfrac{\partial S}{\partial a} = -2\sum (y_i - a - bx_i)$

and $\quad \dfrac{\partial S}{\partial b} = -2\sum x_i(y_i - a - bx_i)$

So if $\partial S/\partial a = 0$ and $\partial S/\partial b = 0$, we obtain the equations:

$$\sum (y_i - a - bx_i) = 0 \quad \Rightarrow \quad \sum y_i - na - b\sum x_i = 0 \tag{1}$$

$$\sum x_i(y_i - a - bx_i) = 0 \quad \Rightarrow \quad \sum x_iy_i - a\sum x_i - b\sum x_i^2 = 0 \tag{2}$$

and in their most convenient form, these equations can be written:

$$\sum y_i = na + b\sum x_i \tag{1}$$

$$\sum x_iy_i = a\sum x_i + b\sum x_i^2 \tag{2}$$

Dividing equation (1) by n, we obtain

$$\frac{1}{n}\sum y_i = a + b\frac{1}{n}\sum x_i$$

and writing the means of x and y as $\bar{x}$ and $\bar{y}$, the equation becomes

$$\bar{y} = a + b\bar{x}$$

This tells us that the regression line passes through the 'mean point', $(\bar{x}, \bar{y})$, so that the equation of the regression line can be rewritten as

$$y - \bar{y} = b(x - \bar{x})$$

Hence we only need find the value of b from equations (1) and (2). This is done by multiplying equation (1) by $\sum x_i$ and equation (2) by n, and then subtracting (1) from (2). This gives:

$$n\sum x_i y_i - \left(\sum x_i\right)\left(\sum y_i\right) = b[n\sum x_i^2 - \left(\sum x_i\right)^2]$$

and dividing through by n^2,

$$\frac{\sum x_i y_i}{n} - \left(\frac{\sum x_i}{n}\right)\left(\frac{\sum y_i}{n}\right) = b\left[\frac{\sum x_i^2}{n} - \left(\frac{\sum x_i}{n}\right)^2\right]$$

Now $\dfrac{\sum x_i^2}{n} - \left(\dfrac{\sum x_i}{n}\right)^2 = s_x^2$, the *variance* of x

and we also write

$$\frac{\sum x_i y_i}{n} - \left(\frac{\sum x_i}{n}\right)\left(\frac{\sum y_i}{n}\right) = s_{xy}, \quad \text{which is called the } \textit{covariance} \text{ of } x, y.$$

(s_{xy} is a simplified statistical version of the covariance of x, y which arose in section 15.5.)

So $\quad s_{xy} = b s_x^2 \quad \Rightarrow \quad b = \dfrac{s_{xy}}{s_x^2}$

This quantity, s_{xy}/s_x^2, known as the *coefficient of regression* of y on x, is therefore the gradient of the line of regression.

The equation of the line of regression can therefore be written in its most convenient form as:

$$y - \bar{y} = \frac{s_{xy}}{s_x^2}(x - \bar{x})$$

If, alternatively, y is taken as the independent variable, it is clear by symmetry that the line of regression of x on y is given by

$$x - \bar{x} = \frac{s_{xy}}{s_y^2}(y - \bar{y})$$

Now let us return to our original bivariate data, where the two variables were mathematics (x) and French (y) marks. We will see how the French mark depends linearly on the mathematics mark by obtaining the line of regression of y on x.

x	y	x^2	xy
10	10	100	100
13	16	169	208
17	11	289	187
19	16	361	304
21	17	441	357
22	17	484	374
22	18	484	396
22	20	484	440
23	22	529	506
24	18	576	432
24	25	576	600
26	20	676	520
27	21	729	567
28	18	784	504
31	18	961	558
31	24	961	744
31	26	961	806
34	29	1156	986
35	22	1225	770
37	25	1369	925

$$\sum x_i = \ 497 \qquad \sum y_i = \ 393 \qquad \sum x_i^2 = 13315 \qquad \sum x_i y_i = 10284$$

$$\Rightarrow \bar{x} = \ 24.85 \qquad \bar{y} = \ 19.65 \qquad \frac{\sum x_i^2}{20} = \ 665.75 \qquad \frac{\sum x_i y_i}{20} = \ 514.2$$

Hence $\quad s_x^2 = 665.75 - (24.85)^2, \qquad s_{xy} = 514.20 - 24.85 \times 19.65$

$$= 665.75 - 617.52 \qquad\qquad\quad = 514.20 - 488.30$$

$$= 48.23 \qquad\qquad\qquad\qquad\quad = 25.90$$

$$\Rightarrow \quad \frac{s_{xy}}{s_x^2} = \frac{25.90}{48.23} = 0.537$$

Therefore the line of regression of y on x is

$$y - 19.65 = 0.537(x - 24.85)$$

$$= 0.537x - 13.35$$

$$\Rightarrow \qquad y = 0.54x + 6.30, \quad \text{to 2 decimal places}$$

In a similar way, we can calculate the regression line of x on y to be

$$y = 0.84x - 1.29$$

The scatter diagram for the data is reproduced on p. 282 with the two regression lines superimposed. The intersection of the two lines is the point $(\bar{x}, \bar{y})$, since of course we have shown that this point must lie on both lines.

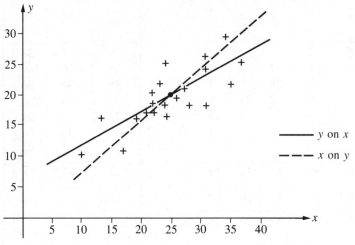

Example 1   (Grouped data)

In a sample of 150 married couples, the relationship between age of husband and wife was investigated. The results are illustrated in the table below, where the numbers in the table are frequencies (e.g. 5 couples with both husband and wife in the age-range 16–19).

age of wife (Y)	age of husband (X)						
	16–19	20–29	30–39	40–49	50–59	60–69	70–79
16–19	5	3					
20–29	1	25	7	2			
30–39		1	28	5	1	1	
40–49		1	2	24	4		
50–59				3	18		
60–69					1	11	2
70–79						1	4

Find the two lines of regression for the data.

We first of all label each interval by its average value, so that for instance the interval 20–29 (which actually includes ages up to 30), is labelled as 25. We then choose 45 as a working zero for both X and Y and scale down the values of X, Y by 10 to obtain the transformed (or coded) variables x, y. Working in terms of x, y will simplify the necessary arithmetic. The table then takes the form shown below, and we add the columns to obtain the frequencies f_x, and the rows to obtain f_y. The columns and rows signifying

xf_x, x^2f_x, yf_y and y^2f_y can then be easily completed and their totals calculated.

x \ y	−2.7	−2	−1	0	1	2	3	f_y	yf_y	y^2f_y
−2.7	5	3						8	−21.6	58.3
−2	1	25	7	2				35	−70	140
−1		1	28	5	1	1		36	−36	36
0		1	2	24	4			31	0	0
1				3	18			21	21	21
2					1	11	2	14	28	56
3						1	4	5	15	45
f_x	6	30	37	34	24	13	6	150	−63.6	356.3
xf_x	−16.2	−60	−37	0	24	26	18	−45.2		
x^2f_x	43.7	120	37	0	24	52	54	330.7		

So $\sum f_x = \sum f_y = \sum f = 150$.

and $\sum xf_x = -45.2 \Rightarrow \bar{x} = \dfrac{\sum xf_x}{\sum f} = -0.301$

$\sum x^2f_x = 330.7 \Rightarrow \dfrac{\sum x^2f_x}{\sum f} = 2.205$

$\sum yf_y = -63.6 \Rightarrow \bar{y} = \dfrac{\sum yf_y}{\sum f} = -0.424$

$\sum y^2f_y = 356.3 \Rightarrow \dfrac{\sum y^2f_y}{\sum f} = 2.375$

It remains to calculate $\sum xyf$. This is done simply by drawing out the table again, but with the entries this time representing xyf rather than just the frequency f. Then the numbers in each row are added to obtain a further column, and the total for this column gives $\sum xyf$.

x \ y	−2.7	−2	−1	0	1	2	3	
−2.7	36.4	16.2						52.6
−2	5.4	100	14	0				119.4
−1		2	28	0	−1	−2		27
0		0	0	0	0			0
1				0	18			18
2					2	44	12	58
3						6	36	42
								317

So $\sum xyf = 317$ $\Rightarrow$ $\dfrac{\sum xyf}{\sum f} = 2.113$

Hence $s_x^2 = 2.205 - (-0.301)^2 = 2.114$
$\qquad s_y^2 = 2.375 - (-0.424)^2 = 2.195$
$\qquad s_{xy} = 2.113 - (-0.301) \times (-0.424) = 1.986$

Now transforming back to the original variables X and Y,

$\bar{X} = 10\bar{x} + 45 = 41.99$
$\bar{Y} = 10\bar{y} + 45 = 40.76$
$s_X^2 = 100 s_x^2 = 211.4$
$s_Y^2 = 100 s_y^2 = 219.5$
$s_{XY} = 100 s_{xy} = 198.6$

So the regression coefficient of Y on $X = \dfrac{198.6}{211.4} = 0.939$

and the regression coefficient of X on $Y = \dfrac{198.6}{219.5} = 0.905$

Therefore the equation of the regression line of Y on X is

$Y - 40.76 = 0.939 (X - 41.99)$

$\Rightarrow \qquad Y = 0.9X + 1.3$ to 1 decimal place

and the equation of the regression line of X on Y is

$X - 41.99 = 0.905 (Y - 40.76)$
$\Rightarrow \qquad Y = 1.1X - 5.6,$ to 1 decimal place

Exercise 15.7a

1   a) Obtain the heights (x) and masses (y) of 10 people of about the same age. Represent this data on a scatter diagram, and draw by eye through the points of your diagram what you consider to be the line of 'best fit'. Calculate the equations of the lines of regression of y on x and of x on y, and draw them on the same diagram.
    b) Obtain the heights and masses of a larger number of people, and group this data in a suitable way. Then calculate the equations of the two regression lines, and compare with the results of part a).

2   An experiment was conducted to discover the effect of adding a certain compound to the diet of mice. The results of the experiment are:

number of units of compound added (x)	1	2	3	4	5	6	7	8
mean gain in mass of mice (y)/mg	8.7	10.9	9.2	10.9	11.6	11.1	12.3	13.8

Find the equation of the line of regression of y on x.

3 In the following table W g is the mass of a certain chemical substance which dissolved in water at $T°C$:

$T/°C$	10	20	30	40	50	60	70	80	90
W/g	45	46	50	56	59	63	64	67	74

Calculate the equation of the line of regression of W on T. Use this equation to obtain a tentative value for W when $T = 56$. (MEI)

4 A sample of eight families of a certain monkey species was chosen in which each family had just a single fully grown male offspring, and the heights of father and son for each family are given below.

height of father/cm (x)	67	64	70	71	65	67	68	69
height of son/cm (y)	68	62	70	73	64	69	71	69

Plot the data on a scatter diagram. Find the equation of the line of regression of y on x, and superimpose this line on the diagram.

5 The breadth (x) and length (y) of 12 leaves from a certain tree are given below:

breadth/mm	39	36	35	35	30	31	34	28	33	30	19	29
length/mm	81	80	74	89	71	73	76	75	84	81	77	82

Plot the data on a scatter diagram. Find the equations of the regression lines of a) y on x, and b) x on y, and draw these lines on the scatter diagram.

6 A certain machine processes raw material. The purities (x) of 10 batches of raw material are given below along with the yield (y) from each batch.

purity (x)	0.49	0.41	0.53	0.46	0.47	0.39	0.51	0.42	0.49	0.52
yield (y)	24	21	26	25	20	22	25	23	25	27

Plot the data on a scatter diagram. Find the equation of the line of regression of y on x, and superimpose this line on the diagram.

7 The chronological ages and the reading ages (in years and months) of 10 children from a primary school are given below:

chronological age	7.5	7.11	8.1	8.2	8.8	9.0	9.4	9.10	10.2	10.6
reading age	7.8	7.10	8.0	7.9	8.9	9.1	8.8	9.10	12.1	10.7

Find the equation of the line of regression of reading age (y) on chronological age (x), if x and y are measured in months.

8 To test the effect of a new drug, twelve patients were examined before the drug was administered and given an initial score (I) depending on the severity of various symptoms. After taking the drug they were examined again and given a final score (F). A decrease of score represented an improvement. The scores for the twelve patients are given in the table.

patient	1	2	3	4	5	6	7	8	9	10	11	12
score initial (I)	61	23	8	14	42	34	32	31	41	25	20	50
final (F)	49	12	3	4	28	27	20	20	34	15	16	40

Calculate the equation of the line of regression of F on I.

On the average, what improvement would you expect for a patient whose initial score was 30?   (MEI)

9 The table gives the results of 4 estima-
tions of a variable y at each of four values
of a variable x. Calculate the equation of
the regression line of y on x.

x		y		
2	17	15	20	18
4	15	14	16	15
6	12	14	10	12
8	8	10	10	12

Estimate the mean value of y when $x = 3$.   (MEI)

10 The marks of 100 pupils in English language and English literature examinations were graded from 1 to 9 and their results are given in the frequency table below:

		English language (x)								
		1	2	3	4	5	6	7	8	9
English literature (y)	1	6	4	1						
	2	2	1	1		1	1			
	3	3	3	3	7	2	1			
	4	1	1	4	1	1	1		1	
	5		2	1	3	1		2		
	6			2	11	6	4			
	7				2	2	3	1		
	8				1		3	2		1
	9				2	3	1		1	

Find the equations of the lines of regression of **a)** y on x, and **b)** x on y.

11 The following table gives, in coded form, the breadths and lengths of 1306 human heads.

frequencies of heads
X (length)

	−4	−3	−2	−1	0	1	2	3	4	5	total
−4		2	2	1	1	2		1			9
−3	1	5	5	15	8	4	1	1			40
−2	2	3	21	34	48	41	18	3		1	171
Y −1		4	21	57	92	110	57	19	5	1	366
(breadth) 0	1	2	17	53	93	116	62	27	12	1	384
1		2	3	12	39	65	57	40	12	2	232
2			1	4	9	17	24	12	9	1	77
3				1	1	5	10	6		2	25
4							2				2
total	4	18	70	177	291	360	231	109	38	8	1306

The length X is coded with 18.85 cm as origin and the breadth Y is coded with 15.15 cm as origin. The interval of grouping is 0.4 cm in both cases. Calculate the average length and the average breadth of the heads, and the equation of the line of regression of breadth on length. (MEI)

12 The assets of a certain company have grown over the period 1850 to 1966 as shown in the table:

year	X	1850	1870	1900	1930	1960	1964	1965	1966
assets/£10^5	Y	5	56	161	464	1498	2250	2527	2905

Calculate the line of regression of $\log_{10} Y$ on X. (MEI)

13 Six pairs of values of x and y are given in the following table:

x	1	2	3	4	5	6
y	1.5	3.1	4.9	7.0	9.4	12.2

It is thought that these values are connected by a relation of the form $y = ax + bx^3$. By writing $Y = y/x$, and $X = x^2$, reduce the relation to a linear one and calculate:
a) the means of X and Y;
b) the variance of X;
c) the covariance of X and Y;
d) estimated values of a and b on the assumption that there is a linear relationship between X and Y. (OC)

The correlation coefficient

Let us consider further the general line of regression of y on x:

$$y - \bar{y} = \frac{s_{xy}}{s_x^2}(x - \bar{x})$$

If we divide through by s_y, the standard deviation of the y values, the equation can be written

$$\frac{y - \bar{y}}{s_y} = \frac{s_{xy}}{s_x s_y} \times \frac{x - \bar{x}}{s_x}$$

Now in this equation $(y - \bar{y})/s_y$ and $(x - \bar{x})/s_x$ are standardised versions of the variables x and y, and the quantity $s_{xy}/s_x s_y$, which is symmetrical in x and y, is known as the *coefficient of correlation* of x and y; or, more exactly, the *product–moment correlation coefficient*.

The correlation coefficient is normally denoted by the letter r, and if the standardised variables are written as x' and y', the regression lines become:

$$y' = rx' \quad \text{and} \quad x' = ry' \quad \text{respectively}$$

The limits on the value of r can now be determined, using a result known as *Cauchy's inequality*: If a_i, b_i are real numbers (for $i = 1, 2, \ldots, n$) then

$$\left(\sum a_i b_i\right)^2 \leqslant \left(\sum a_i^2\right)\left(\sum b_i^2\right)$$

[For consider the expression $\sum(\lambda a_i + b_i)^2$, where λ is any real number. Since each term of the summation is squared, and hence positive, it follows that

$$\sum(\lambda a_i + b_i)^2 \geqslant 0 \quad \text{for all } \lambda$$

$$\Rightarrow \quad \sum(\lambda^2 a_i^2 + 2\lambda a_i b_i + b_i^2) \geqslant 0$$

$$\Rightarrow \quad \lambda^2 \sum a_i^2 + 2\lambda \sum a_i b_i + \sum b_i^2 \geqslant 0 \quad \text{for all } \lambda$$

Now we know that in general if $ax^2 + 2bx + c \geqslant 0$ for all x, then $b^2 \leqslant ac$.

So if we regard $\lambda^2 \sum a_i^2 + 2\lambda \sum a_i b_i + \sum b_i^2$ as a quadratic function in λ, it follows that

$$\left(\sum a_i b_i\right)^2 \leqslant \left(\sum a_i^2\right)\left(\sum b_i^2\right) \quad]$$

Now for the correlation coefficient, r,

$$r^2 = \frac{s_{xy}^2}{s_x^2 s_y^2} = \frac{[\sum(x_i - \bar{x})(y_i - \bar{y})]^2}{\sum(x_i - \bar{x})^2 \sum(y_i - \bar{y})^2}$$

So, using Cauchy's inequality with $(x_i - \bar{x}) = a_i$ and $(y_i - \bar{y}) = b_i$, we see that $r^2 \leqslant 1$.

Hence $-1 \leqslant r \leqslant 1$

Summarising,

$$r = \frac{s_{xy}}{s_x s_y} \quad \text{and} \quad -1 \leqslant r \leqslant 1$$

Let us now look at three particular values of r:

a) $r = 1$ In this case, the regression lines of y on x and x on y coincide, and if standardised they have gradient 1; this means that all the points of the scatter diagram are collinear, and we say that there is exact *positive (or direct) correlation* between the variables.

b) $r = 0$ The regression line of y on x is horizontal, and that of x on y is vertical. This indicates a complete lack of linear association between the variables, which are said to be *uncorrelated.*

c) $r = -1$ The regression lines coincide, and if standardised have gradient -1; small values of x correspond to large values of y and vice versa, all points being collinear. In this case we say that there is exact *negative (or inverse) correlation* between the variables.

Significance of the correlation coefficient

If, in a particular example we obtain a value of r close to 1 or a value which is nearly zero, then little further analysis is necessary; but how do we interpret a correlation coefficient of (say) 0.5? We wish to know its significance, that is the probability of obtaining such a high value if the variables are uncorrelated (i.e. with the null hypothesis that $r = 0$ for the background population). This will depend on the size of the sample from which r is calculated: the more data we use, the more confident we can feel about the value of r. The mathematics involved in the relevant significance test is too involved for this text, but we can (without at present being concerned about the mathematical justification) use a graph like the one on p. 290. This shows the size of samples required to give 5%, 1%, and 0.1% significance levels for values of the correlation coefficient r (using a one-tailed significance test).

So, using the graph, we can say that a correlation coefficient of 0.5 obtained from a bivariate sample of 15 pairs is significantly different from zero at the 5% level, but if obtained from a sample of only 10 pairs is not significant. Similarly a value of $r = 0.8$, obtained from a sample of size 9, is significant at the 1% level, but not at the 0.1% level.

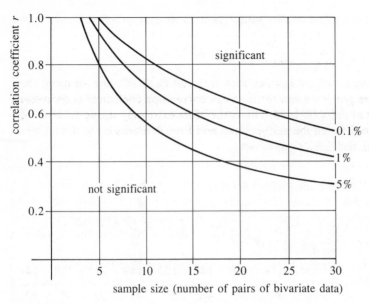

sample size (number of pairs of bivariate data)

Warning

A significant correlation coefficient does not necessarily imply causal connection between two quantities. There is, for instance, a correlation coefficient of 0.75, which is significant at the 5% level, between the number of indictable crimes committed and life expectation in Great Britain between 1900 and 1960 (measured at 10-yearly intervals). But we would hardly accept that an extended life-span increased a person's criminal tendencies; nor even vice-versa!

Example 2

Calculate the product–moment correlation coefficient for the mathematics and French marks given at the beginning of the section, and investigate the significance of this coefficient.

We have already calculated $s_{xy} = 25.90$

and $s_x^2 = 48.23 \Rightarrow s_x = 6.94$

Now for the data, $\sum y_i^2 = 8159$

$\Rightarrow$ $\dfrac{\sum y_i^2}{n} = 407.95$

$\Rightarrow$ $s_y^2 = 407.95 - (19.65)^2$

$\qquad\qquad\qquad\qquad = 407.95 - 386.12$

$\qquad\qquad\qquad\qquad = 21.83$

$\Rightarrow$ $$s_y = 4.67$$

Therefore $r = \dfrac{s_{xy}}{s_x s_y} = \dfrac{25.90}{6.94 \times 4.67} = 0.80$

Now this value of r was calculated from 20 pairs of data, so using the significance graph we can see that this correlation coefficient is definitely significant at the 0.1% level. This is therefore extremely strong evidence of correlation between the mathematics and French marks in the population from which the sample was drawn.

Example 3

The heights and masses of a random sample of 10 first-formers from a secondary school are given below.

height/cm	156	151	152	160	146	157	149	142	158	141
mass/kg	47	38	44	55	46	39	45	30	45	32

On the basis of this sample, how significant is the correlation between height and mass?

Calculation will be considerably simplified if we subtract 150 cm from each height and 40 kg from each mass. This is equivalent to translating the axes of the scatter diagram, so does not affect the value of the correlation coefficient. The data therefore becomes:

height (x)	6	1	2	10	-4	7	-1	-8	8	-9
weight (y)	7	-2	4	15	6	-1	5	-10	5	-8

and the calculations are set out in tabular form:

x	y	x^2	y^2	xy
6	7	36	49	42
1	-2	1	4	-2
2	4	4	16	8
10	15	100	225	150
-4	6	16	36	-24
7	-1	49	1	-7
-1	5	1	25	-5
-8	-10	64	100	80
8	5	64	25	40
-9	-8	81	64	72
12	21	416	545	354

Therefore $\bar{x} = 1.2$ and $\bar{y} = 2.1$.

So, working correct to 3 significant figures,

$$s_x^2 = 41.6 - 1.2^2 \qquad s_y^2 = 54.5 - 2.1^2 \qquad s_{xy} = 35.4 - 1.2 \times 2.1$$

$$ = 40.2 \qquad\qquad = 50.1 \qquad\qquad \phantom{s_{xy}} = 32.9$$

$$\Rightarrow \quad s_x = 6.34 \qquad\qquad s_y = 7.08$$

Hence $\quad r = \dfrac{32.9}{6.34 \times 7.08} = 0.73$

and using the significance graph, we find that this correlation coefficient is significant at the 1% level.

Exercise 15.7b

1–6   Calculate the correlation coefficients for the data in exercise 15.7a, nos. **1–6**, and comment on the significance of the coefficient in each case.

7   Calculate the coefficient of correlation between the mass of the heart and mass of the liver in mice, using the following data.

heart/10^{-2} g	mass of organ							
	20	16	20	21	26	24	18	18
liver/10^{-2} g	230	126	203	241	159	230	140	242

What do you conclude from this analysis?   (MEI)

8   The following table gives the number of goals scored at home and away by ten teams in the First Division of the Football League, part way through the 1972–73 season:

home	25	20	12	15	14	24	20	13	14	10
away	10	14	12	13	6	10	8	5	6	7

Calculate the product–moment correlation coefficient for home and away goals.

9   The intelligence quotients of 8 students were assessed, and the time (in minutes) it took them to complete a certain piece of work was measured. Their IQs and times are given below:

IQ	121	118	132	141	140	137	124	130
time	28	26	23	20	16	17	22	19

Is the negative correlation between these significant?

10 Calculate the correlation coefficient for the data of exercise 15.7a, no. **10**.

11 Calculate the correlation coefficient for the data of exercise 15.7a, no. **11**.

12 Calculate the product–moment correlation coefficient for the bivariate data given in the following frequency table:

y \ x	-3	-2	-1	0	1	2	3	4
-2	4	6	2	1				
-1	1	8	4	3		2		
0		2	6	7	2	3	2	1
1		1	4	3	9	4	1	1
2			2		1	6	6	3	5

13 Ten sets of readings for the variables x, y, and z are given below.

x	2	3	5	7	8	11	13	16	18	19
y	8	5	12	11	22	9	24	17	12	25
z	8	10	10	7	12	14	15	21	28	26

Calculate the coefficients of correlation between   **a)** x and y,   **b)** y and z, and   **c)** z and x, and investigate the significance of each.

Rank correlation: Spearman's rank correlation coefficient

Six athletes competed in both a 1500 m race and a 400 m race, and finished in the following positions:

athlete	A	B	C	D	E	F
1500 m	1	2	3	4	5	6
400 m	2	3	5	1	4	6

Can anything be said about correlation of performance at these two distances?

If the times for the athletes were given, then the product–moment correlation coefficient for these times could be calculated, as in previous examples. Here, however, the only information is about position or rank. Therefore the product–moment correlation coefficient of the *ranks* is calculated, and is referred to as the *coefficient of rank correlation*.

Since ranks (or positions) are less informative than actual measurements, the rank coefficient is correspondingly less informative than the product–moment coefficient. However, since only integers are involved, the rank coefficient does have the advantage of being easily calculated. It is evaluated by means of the formula:

$$\rho = 1 - \frac{6\sum d_i^2}{n(n^2 - 1)}$$

where n is the total number of ranks involved, and d_i the ith rank difference. The derivation of this formula follows the worked examples.

This rank correlation coefficient was first defined by Spearman in 1906, and until recently has been the most widely used. But there are others, so it is usually called *Spearman's rank correlation coefficient* and its significance graph is given below:

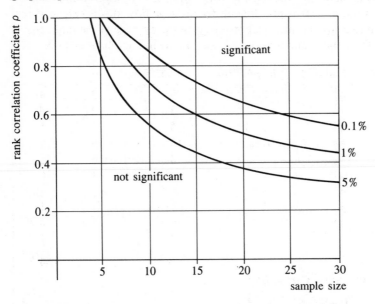

Returning to the data for the two races, ρ is evaluated as follows:

athlete	A	B	C	D	E	F	$n = 6$
1500 m	1	2	3	4	5	6	
400 m	2	3	5	1	4	6	
difference d	-1	-1	-2	3	1	0	
d^2	1	1	4	9	1	0	$\sum d^2 = 16$

Hence $\rho = 1 - \dfrac{6 \times 16}{6(36 - 1)}$

$= 1 - \frac{16}{35}$

$= \frac{19}{35}$

$= 0.54$

which, using the significance graph, is not significantly different from zero at the 5% level.

Example 4

What is the rank correlation coefficient for the heights and masses of schoolboys given in example 3, and how significant is this coefficient?

The first step is to rank (in ascending order) the heights and masses. But the problem arises of how to deal with the two equal masses of 45 kg; between them, they must occupy the 6th and 7th positions, so by convention each is given the rank 6.5. (It is not difficult to envisage how this is extended to more than two equal ranks, each being given the average of the ranks they replace.) The heights and masses are now ranked in the following table:

height/cm	mass/kg	height rank	mass rank	d	d^2
156	47	7	9	2	4
151	38	5	3	2	4
152	44	6	5	1	1
160	55	10	10	0	0
146	46	3	8	5	25
157	39	8	4	4	16
149	45	4	6.5	2.5	6.25
142	30	2	1	1	1
158	45	9	6.5	2.5	6.25
141	32	1	2	1	1
					64.5

So $n = 10$ and $\sum d^2 = 64.5$

$$\Rightarrow \quad \rho = 1 - \frac{6 \times 64.5}{10 \times 99}$$

$$= 1 - 0.39$$

$$= 0.61$$

and using the significance graph, we find that this coefficient is significant at the 5% level. (As compared with the product–moment coefficient which was significant at the 1% level.)

Derivation of the formula for Spearman's rank correlation coefficient

The product-moment correlation coefficient is required where the two variables, x and y, are ranks, each taking values $1, 2, \ldots, n$ in some order or other.

We shall assume the standard results (see chapter 6)

$$\sum_1^n i = \tfrac{1}{2}n(n+1)$$

$$\sum_1^n i^2 = \tfrac{1}{6}n(n+1)(2n+1)$$

Since we need to calculate $\rho = s_{xy}/s_x s_y$, it is necessary first of all to evaluate s_x, s_y and s_{xy}.

Now $\displaystyle\sum_1^n x_i = \sum_1^n y_i = \sum_1^n i = \tfrac{1}{2}n(n+1)$

$\Rightarrow$ $\bar{x} = \bar{y} = \tfrac{1}{2}(n+1)$

Also $\displaystyle\sum_1^n x_i^2 = \sum_1^n y_i^2 = \sum_1^n i^2 = \tfrac{1}{6}n(n+1)(2n+1)$

$\Rightarrow$ $\displaystyle s_x^2 = \frac{1}{n}\sum_1^n x_i^2 - \bar{x}^2$

$\qquad\qquad = \tfrac{1}{6}(n+1)(2n+1) - \tfrac{1}{4}(n+1)^2$

$\qquad\qquad = \tfrac{1}{12}(n+1)(4n+2-3n-3)$

$\qquad\qquad = \tfrac{1}{12}(n+1)(n-1) = \tfrac{1}{12}(n^2-1)$

$\Rightarrow$ $s_x = \sqrt{[\tfrac{1}{12}(n^2-1)]}, \quad$ and similarly $\quad s_y = \sqrt{[\tfrac{1}{12}(n^2-1)]}$

$\Rightarrow$ $s_x s_y = \tfrac{1}{12}(n^2-1)$

We now use the identity:

$$x_i y_i = \tfrac{1}{2}x_i^2 + \tfrac{1}{2}y_i^2 - \tfrac{1}{2}(x_i - y_i)^2$$

$\Rightarrow$ $\displaystyle\sum x_i y_i = \tfrac{1}{2}\sum x_i^2 + \tfrac{1}{2}\sum y_i^2 - \tfrac{1}{2}\sum(x_i - y_i)^2$

$\qquad\qquad = \tfrac{1}{6}n(n+1)(2n+1) - \tfrac{1}{2}\sum d_i^2$

$\Rightarrow$ $\displaystyle\frac{1}{n}\sum x_i y_i = \tfrac{1}{6}(n+1)(2n+1) - \frac{\sum d_i^2}{2n}$

Now $\displaystyle s_{xy} = \frac{1}{n}\sum x_i y_i - \bar{x}\bar{y}$

$\Rightarrow$ $\displaystyle s_{xy} = \tfrac{1}{6}(n+1)(2n+1) - \frac{\sum d_i^2}{2n} - \tfrac{1}{4}(n+1)^2$

$\qquad\qquad = \tfrac{1}{12}(n^2-1) - \frac{\sum d_i^2}{2n}$

Therefore $\rho = \dfrac{s_{xy}}{s_x s_y}$

$$= \frac{\frac{1}{12}(n^2 - 1) - \sum d_i^2/2n}{\frac{1}{12}(n^2 - 1)} = 1 - \frac{6\sum d_i^2}{n(n^2 - 1)}$$

It should be noted that we have assumed in these calculations that the ranks are all different. If, however, equal ranks occur, $\sum x_i^2$ and/or $\sum y_i^2$ are no longer equal to $\frac{1}{6}n(n + 1)(2n + 1)$, and the above formula is no longer strictly valid; but the error involved is generally very small.

Kendall's rank correlation coefficient

Another coefficient of rank correlation, known as Kendall's rank correlation coefficient (τ), is derived quite differently, but has the same range of values $(-1$ to $1)$ as the other correlation coefficients, and values of τ equal to $-1, 0$, and 1 have the same meaning as r and ρ. It is a measure based on a simple principle of agreements and disagreements between the orders in which the objects are placed.

If we have a total of n objects $(A, B, \ldots)$ ranked in two different ways, then we can choose $\dbinom{n}{2} = \frac{1}{2}n(n - 1)$ different pairs, such as A and B. Now for the two rankings, the relative positions (A before B or B before A) can agree or disagree. So the maximum number of agreements, and the maximum number of disagreements of these pairs are both equal to $\frac{1}{2}n(n - 1)$. Kendall's coefficient is calculated by subtracting the total number of such disagreements from the total number of agreements, and then dividing by the maximum possible number of agreements, i.e.

$$\tau = \frac{\text{no. of agreements} - \text{no. of disagreements}}{\frac{1}{2}n(n - 1)} = \frac{\delta}{\frac{1}{2}n(n - 1)}$$

Clearly if both rankings are exactly the same, all pairs will agree in relative position, so that the number of agreements will be $\frac{1}{2}n(n - 1)$ and the number of disagreements will be zero; hence τ will equal 1. Similarly for total disagreement τ will equal -1.

The significance graph for τ is given below.

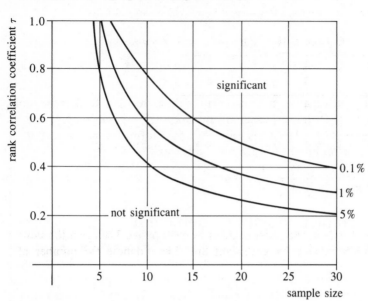

At first glance it would seem a very lengthy procedure to calculate τ, but in practice it can be calculated very quickly and conveniently, by either of the two methods in the following example.

Example 5

In a brass-band competition, two judges place the eight bands in the following order:

band	A	B	C	D	E	F	G	H
first judge	3	6	7	2	4	8	1	5
second judge	5	7	4	1	3	8	2	6

Calculate Kendall's rank correlation coefficient for the two judgements.

Method 1

Re-order the bands so that they are in the correct order of merit according to the first judge. Take the rank given by the second judge to the first band, G, that is 2. Now count $+1$ for every number to the right which is greater than 2 and -1 for each number which is less than 2; these totals are given below the rank. Repeat this procedure for the next rank in the second row, remembering to count only numbers to the *right* of it. And so on for each rank in the second row. The positive numbers so obtained represent agreements, and the negative numbers disagreements, so that their sum gives δ, the total number of agreements less the disagreements.

G	D	A	E	H	B	C	F	$n = 8$
1	2	3	4	5	6	7	8	
2	1	5	3	6	7	4	8	
6	6	3	4	2	1	1		
-1	0	-2	0	-1	-1	0		
5	6	1	4	1	0	1		$\delta = 18$

So $\quad \tau = \dfrac{\delta}{\frac{1}{2}n(n-1)} = \dfrac{18}{28} = \dfrac{9}{14} = 0.64$

Method 2

Set out the bands in the order decided by each judge. Then join the same letter in each ranking by a straight line. Let c denote the number of crossings.

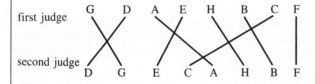

first judge G D A E H B C F

second judge D G E C A H B F

Then $\quad \delta = \frac{1}{2}n(n-1) - 2c$

$\Rightarrow \quad \tau = 1 - \dfrac{4c}{n(n-1)}$

In this case, $\quad c = 5$

So $\quad \tau = 1 - \dfrac{20}{8 \times 7}$

$\qquad = 1 - \frac{5}{14}$

$\qquad = \frac{9}{14} = 0.64$

(The justification for this second method is left as an exercise for the reader.)

So Kendall's rank correlation coefficient is in this case equal to 0.64; which, using the graph, is found to be significant at the 5% level.

Example 6

Calculate τ for the race positions given earlier in this section.

athlete	A	B	C	D	E	F	$n = 6$
1500 m	1	2	3	4	5	6	
400 m	2	3	5	1	4	6	

	4	3	1	2	1	
	-1	-1	-2	0	0	
	3	2	-1	2	1	$\delta = 7$

Hence $\tau = \dfrac{7}{\frac{1}{2} \times 6 \times 5} = \dfrac{7}{15} = 0.47$

and this value of τ compares with the value (already calculated) of Spearman's coefficient, ρ, equal to 0.54.

Exercise 15.7c

1   Six varieties of chocolate, denoted by A, B, C, D, E, F are ranked for taste in the following order (best first): F C A D B E. The ranking on price (most expensive first) is C F B A E D. Calculate a rank correlation coefficient, and comment on the result.   (c)

2   At a village fete the fruit cake competition was judged by the vicar and the local squire. They placed the entries as follows:

cake	A	B	C	D	E	F	G	H	I	J
vicar	5	1	9	2	3	8	7	4	10	6
squire	3	5	4	7	10	8	2	1	6	9

Calculate a rank correlation coefficient. Can you deduce anything about the vicar's and squire's tastes in fruit cake?

3   Twelve chemistry students were each given a theory and a practical examination. Their positions in these two examinations were as follows:

theory	1	2	3	4	5	6	7	8	9	10	11	12
practical	1	4	7	3	5	2	9	8	10	6	12	11

Calculate   a) Spearman's and   b) Kendall's rank correlation coefficient for the data, and test the significance of each coefficient.

4   Rank the data of exercise 15.7a, no. 4, and calculate Spearman's rank correlation coefficient.

5 Rank the data of exercise 15.7a, no. **5**, and calculate Spearman's rank correlation coefficient for the lengths and breadths of leaves.

6 Rank the data of exercise 15.7a, no. **6**, and calculate Spearman's coefficient. Is the rank correlation coefficient between parity and yield significant?

7 Rank the data of exercise 15.7a, no. **7**, and calculate **a)** Spearman's and **b)** Kendall's rank correlation coefficient for the chronological and reading ages of the children.

8 The batting and bowling averages of seven of the members of a cricket team are put in order as follows:

player	A	B	C	D	E	F	G
batting order	1	2	3	4	5	6	7
bowling order	7	4	6	5	3	1	2

a) Calculate Kendall's rank correlation coefficient between the two orders.
b) Is this significant at the 1% level?
c) If the positions of A and C in the bowling order were reversed, show that Kendall's r.c.c. would be significant at the 5% level. (SMP)

9 A, B, C, D, E was the order of merit for the five marrows entered in a local vegetable show. One admirer put them in the order B, A, C, D, E; show that he got 0.8 as a Kendall coefficient of rank correlation on comparing his order with the official result. List all other possible orders which would give the same coefficient when compared with A B C D E. How many different orders could there be for the five marrows? How likely is he to obtain a coefficient as high, or higher than he did, purely by chance? (SMP)

10 At the end of a particular season, the goals scored by the ten teams forming a football league, as compared with their league positions were as follows:

position	1	2	3	4	5	6	7	8	9	10
goals scored	34	19	31	24	26	20	22	18	19	15

By ranking the number of goals scored (in descending order), calculate a rank correlation coefficient between league position and goal score. Is this coefficient significant?

11 Eleven pupils from the same class took part in a sponsored walk for a charity. The distance each walked, and the amount per kilometre that they each earned are given below. Calculate the Kendall rank correlation coefficient for these two sets of figures.

distance/km	25	23	22	21	19	18	12	10	9	5	3
rate/p km^{-1}	14	16	12	13	11	7	9	6	10	4	8

Comment briefly on the result.   (SMP)

Miscellaneous problems

1   In a game of Ludo a six has to be thrown with a die before each of a player's four counters can be moved. Find an expression for the probability that a player will be able to move his fourth counter with his tenth throw.

  Find the probability, for a binomial distribution with parameters n, p and q, that r trials are required to obtain k successes.

2   A certain device, used for measuring Earth tremors in a seismological station, can only maintain the required degree of sensitivity for a total of ten tremors, and after this is disposed of. If such tremors occur randomly, but averaging 4 per day, find the probability density function for the useful lifetime (in days) of such a device, and use this to show that the mean lifetime is $2\frac{1}{2}$ days. What is the variance of the lifetime?

  Find the p.d.f. if the device can be used for k tremors and tremors average λ per day; the probability distribution with this p.d.f. is referred to as the *gamma distribution*.

3   For a certain type of bacterium, the time x from birth to death is a random variable with p.d.f. $(1 + x)^{-2}, (x \geqslant 0)$. A culture of such bacteria is routinely inspected at regular intervals of time.

  One bacterium, inspected at time t after birth, is found to be dead already. Find the probability that this bacterium has been dead for at least time kt $(0 < k < 1)$.

  Another bacterium of the same type is alive at time t after birth, but is found to be dead by the end of a further time t. Find the probability that this bacterium had been dead for at least time kt $(0 < k < 1)$.   (MEI)

4   When a patient A arrives at a doctor's consulting room he will be seen at once, and will occupy the doctor for 30 minutes; he is equally likely to arrive at any time between 2.00 and 3.30 p.m. Patient B, who comes independently, is equally likely to arrive at any time between 2.00 and 4.00 p.m.
  a) What is the probability that B will arrive while the doctor is seeing A?
  b) What is B's expected waiting time?   (OS)

5   In a certain large population of men, heights are distributed Normally about a mean of 180 cm with standard deviation 5 cm. Random samples

are taken with three men in each sample and their heights are arranged in increasing order. In 1000 such samples, approximately how many will have:

a) the middle height under 175 cm?

b) the least height less than 175 cm?

c) the least height between 175 and 180 cm?   (SMP)

6   The number of eggs laid by a farmyard hen in a week is a Poisson variable with mean λ. The chance that an egg is fertile is p. Write down the probability that in one week n eggs are laid, and r of these are fertile.

By summing this probability over appropriate values of n find the probability that r fertile eggs are collected in one week.

Show that if this number r is given, then the distribution of $(n - r)$ conditional on it is of Poisson form, and that the mean of n is $r + \lambda(1 - p)$.   (C)

7   By means of probability generators, calculate the expected value and variance of the number of throws of a normal unbiased die required to obtain the sequence 1, 2, 3, 4, 5, 6 (not necessarily consecutively).

8   Matches are put into a box five at a time until the mass of the box and matches combined reaches M g when the box is said to be full. The mass of an individual match is Normally distributed with mean m g and standard deviation σ g. The mass of an empty box is Normally distributed with mean $5m$ g and standard deviation 2σ g. Find the value of M such that there is only chance in a hundred that a full match-box contains fewer than 50 matches.   (CS)

9   Each of four players is dealt 13 cards from a pack of 52 which contains 4 aces. Player A looks at her hand and winks at her partner, Player B which is a pre-arranged signal that her hand has at least one ace. Player B winks back to show that he has at least one ace as well. Player C looks at her hand and sees that she has just one ace. From Player C's point of view what is the probability that her partner, Player D, has at least one ace if:

a) she saw the winks and understood their meaning;

b) she knows nothing about her opponents' signals.   (CS)

10   R is the number of successes in a binomial distribution with parameters n, p; S is the number of failures.

a) Find the covariance of R and S.

b) Use the expectation and variance of $(R - S)$, together with a suitable approximation to the distribution of $(R - S)$ when $p = \frac{1}{2}$, to find λ such that

$$P(|R - S| \geqslant \lambda) = 0.05$$

c) Calculate the probability $P(|R - S| \geqslant 8)$ when $n = 16$ and $p = \frac{1}{2}$, to two significant figures, using the exact distribution.   (MEI)

11 a) If $A_1, A_2, A_3, \ldots, A_n$ are certain events, we know (from chapter 7) that
$$P(A_1 \cup A_2) = P(A_1) + P(A_2) - P(A_1 \cap A_2)$$
Find corresponding expressions for $P(A_1 \cup A_2 \cup A_3)$ and $P(A_1 \cup A_2 \cup \cdots \cup A_n)$.
b) n boys go to a riotous party, each with a girl; and though the party ends in complete confusion, each boy also leaves with a girl. What is the probability that no boy leaves with the girl he came with, and to what value does this probability tend for a large party?

12   Initially a machine is in good running order but it is subsequently liable to break down. As soon as a breakdown occurs repairs begin. If the machine is in good order at time t then the probability that a breakdown occurs in a small interval $(t, t + \delta t)$ is $\alpha \, \delta t$, and if it is under repair at the time t the probability that the repair is completed in time $(t, t + \delta t)$ is $\beta \, \delta t$. Let $P(t)$ be the probability that the machine is under repair at time t.

 Write down an equation relating $P(t + \delta t)$ to $P(t)$ and hence show that $P(t)$ is $\alpha/(\alpha + \beta) \left\{1 - e^{-(\alpha + \beta)t}\right\}$.   (cs)

13   Two jars, one white and the other black, contain $a + b$ balls each; in the white jar there are a white and b black balls and in the black jar b white and a black. Single draws are made as follows: at the rth draw a ball is drawn from the white or black jar according as the $(r - 1)$th ball drawn was white or black, the colour of the ball noted, and then returned. If p_n is the probability that the nth ball is white, show that

$$(a + b)p_n = b + (a - b)p_{n-1}$$

By means of the substitution

$$p_n = \tfrac{1}{2} + q_n$$

or otherwise, determine p_n, when the jar from which the first draw is made is **a)** chosen at random, **b)** white.   (os)

14   Three deadly enemies A, B, and C take part in a three-cornered duel with pistols. The probability of A hitting his target is 0.6, that of B 0.8, and C 1. They draw lots for the order of firing and each uses his best strategy (they all know one another's capabilities). If they take it in turns to keep on firing until just one is left, who is most likely to survive, and what is his probability of survival?

15   In a game between two players both players have an equal chance of winning each point. The game continues until one player has scored N points. Find the probability p_r that the winning player has a lead of exactly r points when the game is complete. Deduce that

$$(2N - r - 1)p_{r+1} = 2(N - r)p_r \quad (r = 1, 2, \ldots, N)$$

and hence find the expected value of the lead at the end of the game. (cs)

16 Matrices and transformations, determinants and linear equations

16.1 Matrices

A *matrix* is any rectangular table, or array, of numbers. One very simple example is the table of postage charges (in pence) for inland mail in the British Isles in 1985.

weight not over	1st class	2nd class
60 g	17	13
100 g	24	18
150 g	31	22
200 g	38	28
250 g	45	34
300 g	53	40

Matrices are often used for administrative purposes. For instance, the distribution of pupils from contributory primary schools into the four first form classes of a secondary school can be summarised by the matrix

	1A	1B	1C	1D
St Michael's	15	11	13	10
Woodville Junior	5	10	6	11
Belton Junior	8	9	10	7

If the information is set out in exactly the same format each year then there is no need to give headings to rows and columns, and the matrix can just be given as

$$\begin{pmatrix} 15 & 11 & 13 & 10 \\ 5 & 10 & 6 & 11 \\ 8 & 9 & 10 & 7 \end{pmatrix}$$

The *order* of this matrix is said to be 3 by 4 (usually written 3×4) since it has 3 rows and 4 columns.

In quite a different context, a route matrix can be used to describe a road system, or network. The numbers, or *elements*, of the matrix in this case give the number of routes connecting any two junctions.

$$\begin{array}{c} \quad\; A\; B\; C\; D \\ \begin{array}{c} A \\ B \\ C \\ D \end{array} \left(\begin{array}{cccc} 0 & 2 & 1 & 0 \\ 2 & 0 & 1 & 0 \\ 1 & 1 & 0 & 1 \\ 0 & 0 & 1 & 0 \end{array} \right) \end{array}$$

This is a 4×4 matrix, and for obvious reasons is called a *square* matrix.

However, the principal area in which we shall be using matrices is that of transformations and sets of linear equations, and it is in this context that the ideas of matrix addition and multiplication most naturally arise.

Matrix addition

A column vector is a particularly simple and important example of a matrix. We already know that vector addition is achieved by the addition of equivalent elements of the vectors,

e.g. $\begin{pmatrix} 1 \\ 5 \end{pmatrix} + \begin{pmatrix} 3 \\ 2 \end{pmatrix} = \begin{pmatrix} 4 \\ 7 \end{pmatrix}$

and it is natural to extend this to matrices in general.

So $\begin{pmatrix} 5 & 1 & 3 \\ 2 & 4 & 2 \end{pmatrix} + \begin{pmatrix} 2 & 1 & 1 \\ 3 & 8 & 0 \end{pmatrix} = \begin{pmatrix} 7 & 2 & 4 \\ 5 & 12 & 2 \end{pmatrix}$

Similarly for subtraction,

$\begin{pmatrix} 5 & 1 & 3 \\ 2 & 4 & 2 \end{pmatrix} - \begin{pmatrix} 2 & 1 & 1 \\ 3 & 8 & 0 \end{pmatrix} = \begin{pmatrix} 3 & 0 & 2 \\ -1 & -4 & 2 \end{pmatrix}$

The multiplication of a matrix by a scalar is also defined in the obvious way, by multiplying each element of the matrix by the scalar:

If matrix $\mathbf{A} = \begin{pmatrix} 5 & 1 & 3 \\ 2 & 4 & 2 \end{pmatrix}$ then $3\mathbf{A} = \begin{pmatrix} 15 & 3 & 9 \\ 6 & 12 & 6 \end{pmatrix}$

Matrix multiplication

As we have already seen in chapter 1 of Book 1, matrix multiplication arises from the consideration of linear transformations and their combination.

The transformation given by the equations

$x' = ax + by$
$y' = cx + dy$

is defined by the elements of the matrix $\begin{pmatrix} a & b \\ c & d \end{pmatrix}$. A second transformation

$$x'' = \alpha x' + \beta y'$$
$$y'' = \gamma x' + \delta y'$$

is defined by the matrix $\begin{pmatrix} \alpha & \beta \\ \gamma & \delta \end{pmatrix}$.

Therefore the combined transformation is

$$x'' = \alpha(ax + by) + \beta(cx + dy) = (\alpha a + \beta c)x + (\alpha b + \beta d)y$$
$$y'' = \gamma(ax + by) + \delta(cx + dy) = (\gamma a + \delta c)x + (\gamma b + \delta d)y$$

with matrix $\begin{pmatrix} \alpha a + \beta c & \alpha b + \beta d \\ \gamma a + \delta c & \gamma b + \delta d \end{pmatrix}$

What we have done here is to combine the rows of matrix $\left(\begin{array}{c} \alpha \ \ \beta \\ \hline \gamma \ \ \delta \end{array} \right)$ with the columns of matrix $\left(\begin{array}{c|c} a & b \\ c & d \end{array} \right)$, and the result is defined as the product of the two matrices:

$$\begin{pmatrix} \alpha & \beta \\ \gamma & \delta \end{pmatrix}\begin{pmatrix} a & b \\ c & d \end{pmatrix} = \begin{pmatrix} \alpha a + \beta c & \alpha b + \beta d \\ \gamma a + \delta c & \gamma b + \delta d \end{pmatrix}$$

With this definition of matrix multiplication, the transformation equations can also be written in matrix form:

$$\begin{pmatrix} x' \\ y' \end{pmatrix} = \begin{pmatrix} a & b \\ c & d \end{pmatrix}\begin{pmatrix} x \\ y \end{pmatrix} \qquad \begin{pmatrix} x'' \\ y'' \end{pmatrix} = \begin{pmatrix} \alpha & \beta \\ \gamma & \delta \end{pmatrix}\begin{pmatrix} x' \\ y' \end{pmatrix}$$

And by combining the two,

$$\begin{pmatrix} x'' \\ y'' \end{pmatrix} = \begin{pmatrix} \alpha & \beta \\ \gamma & \delta \end{pmatrix}\begin{pmatrix} a & b \\ c & d \end{pmatrix}\begin{pmatrix} x \\ y \end{pmatrix}$$

the reason for writing the matrix product as $\begin{pmatrix} \alpha & \beta \\ \gamma & \delta \end{pmatrix}\begin{pmatrix} a & b \\ c & d \end{pmatrix}$ and not $\begin{pmatrix} a & b \\ c & d \end{pmatrix}\begin{pmatrix} \alpha & \beta \\ \gamma & \delta \end{pmatrix}$ becomes evident.

This principle of combining rows and columns can now be extended to matrices in general. We define the product **AB** of two matrices, **A** and **B**, to be the matrix whose element in the ith row and jth column is the *inner product* (as defined above) of the ith row of **A** and the jth column of **B**. This of course necessitates that the number of columns of **A** is equal to the number of rows of **B**, such matrices being called *conformable* or *compatible* for multiplication. Hence if **A** is an $m \times n$ matrix and **B** is an $n \times p$ matrix, then **AB** will be an $m \times p$ matrix.

Example 1

Evaluate the product **AB** when

$$A = \begin{pmatrix} 2 & 0 & 1 \\ 1 & 3 & 8 \end{pmatrix} \quad \text{and} \quad B = \begin{pmatrix} 1 & 3 & 4 & 3 \\ 5 & 1 & 2 & 0 \\ 2 & 0 & 3 & 1 \end{pmatrix}$$

Writing out the orders of the matrices, we can check that the product can be evaluated. This also gives the order of **AB**:

A   ×   **B**   =   **AB**

2 ×③—③× 4   2 × 4

Matrix **A** is divided into rows and **B** into columns, and we can determine the shape and size of the product **AB**:

$$\begin{pmatrix} 2 & 0 & 1 \\ 1 & 3 & 8 \end{pmatrix} \begin{pmatrix} 1 & 3 & 4 & 3 \\ 5 & 1 & 2 & 0 \\ 2 & 0 & 3 & 1 \end{pmatrix} = \left(\begin{array}{c} \\ 2 \times 4 \end{array} \right)$$

Each element of **AB** is then evaluated as the inner product of the equivalent row from **A** and column from **B**. So for the entry in row 2 and column 3 of **AB**,

$$\begin{pmatrix} \\ 1 & 3 & 8 \end{pmatrix} \begin{pmatrix} & & 4 & \\ & & 2 & \\ & & 3 & \end{pmatrix} = \begin{pmatrix} & & & \\ & & 34 & \end{pmatrix}$$

The final result is then

$$AB = \begin{pmatrix} 2 & 0 & 1 \\ 1 & 3 & 8 \end{pmatrix} \begin{pmatrix} 1 & 3 & 4 & 3 \\ 5 & 1 & 2 & 0 \\ 2 & 0 & 3 & 1 \end{pmatrix} = \begin{pmatrix} 4 & 6 & 11 & 7 \\ 32 & 6 & 34 & 11 \end{pmatrix}$$

It should be noted that **BA** cannot be evaluated since rows and columns in this order do not correspond:

B   **A**

3 × 4   2 × 3

Example 2

Find **AB**, **BA**, **IA** and **BI** when

$$\mathbf{A} = \begin{pmatrix} 0 & 1 & 2 \\ 0 & -1 & -2 \end{pmatrix}, \quad \mathbf{B} = \begin{pmatrix} 3 & -3 \\ 4 & -4 \\ 5 & -5 \end{pmatrix}, \quad \mathbf{I} = \begin{pmatrix} 1 & 0 \\ 0 & 1 \end{pmatrix}$$

$$\mathbf{AB} = \begin{pmatrix} 0 & 1 & 2 \\ 0 & -1 & -2 \end{pmatrix} \begin{pmatrix} 3 & -3 \\ 4 & -4 \\ 5 & -5 \end{pmatrix} = \begin{pmatrix} 14 & -14 \\ -14 & 14 \end{pmatrix}$$

$$2 \times \textcircled{3} \quad\quad \textcircled{3} \times 2 \quad\quad\quad\quad 2 \times 2$$

$$\mathbf{BA} = \begin{pmatrix} 3 & -3 \\ 4 & -4 \\ 5 & -5 \end{pmatrix} \begin{pmatrix} 0 & 1 & 2 \\ 0 & -1 & -2 \end{pmatrix} = \begin{pmatrix} 0 & 6 & 12 \\ 0 & 8 & 16 \\ 0 & 10 & 20 \end{pmatrix}$$

$$3 \times \textcircled{2} \quad\quad \textcircled{2} \times 3 \quad\quad\quad\quad 3 \times 3$$

$$\mathbf{IA} = \begin{pmatrix} 1 & 0 \\ 0 & 1 \end{pmatrix} \begin{pmatrix} 0 & 1 & 2 \\ 0 & -1 & -2 \end{pmatrix} = \begin{pmatrix} 0 & 1 & 2 \\ 0 & -1 & -2 \end{pmatrix}$$

$$2 \times \textcircled{2} \quad\quad \textcircled{2} \times 3 \quad\quad\quad\quad 2 \times 3$$

$$\mathbf{BI} = \begin{pmatrix} 3 & -3 \\ 4 & -4 \\ 5 & -5 \end{pmatrix} \begin{pmatrix} 1 & 0 \\ 0 & 1 \end{pmatrix} = \begin{pmatrix} 3 & -3 \\ 4 & -4 \\ 5 & -5 \end{pmatrix}$$

$$3 \times \textcircled{2} \quad\quad \textcircled{2} \times 2 \quad\quad\quad\quad 3 \times 2$$

Examples 1 and 2 illustrate the fact that in general matrix multiplication is not commutative. However, it is associative:

$$\mathbf{A(BC)} = \mathbf{(AB)C}$$

and it is distributive over addition:

$$\mathbf{A(B + C)} = \mathbf{AB} + \mathbf{AC}$$
$$\mathbf{(A + B)C} = \mathbf{AC} + \mathbf{BC}$$

Furthermore, the matrix $\mathbf{I} = \begin{pmatrix} 1 & 0 \\ 0 & 1 \end{pmatrix}$, which leaves the matrix it multiplies unaltered, is called the *identity* or *unit* matrix. This will clearly commute with any other 2×2 matrix. Similarly the 3×3 matrix $\begin{pmatrix} 1 & 0 & 0 \\ 0 & 1 & 0 \\ 0 & 0 & 1 \end{pmatrix}$ is also an identity matrix.

If the rows and columns of a matrix, $\mathbf{A}$, are exchanged, then the resulting matrix is called the *transpose* of $\mathbf{A}$, and is written $\mathbf{A}^T$.

So if $\quad \mathbf{A} = \begin{pmatrix} 3 & 1 & 4 \\ 2 & 6 & 0 \end{pmatrix} \quad$ then $\quad \mathbf{A}^T = \begin{pmatrix} 3 & 2 \\ 1 & 6 \\ 4 & 0 \end{pmatrix}$

Example 3

For the matrices $\quad \mathbf{A} = \begin{pmatrix} 1 & 4 \\ 2 & 0 \\ -1 & 3 \end{pmatrix} \quad$ and $\quad \mathbf{B} = \begin{pmatrix} 5 & 2 \\ 1 & -3 \end{pmatrix}$

evaluate **a)** $(\mathbf{AB})^T$ and **b)** $\mathbf{B}^T\mathbf{A}^T$

a) $\quad \mathbf{AB} = \begin{pmatrix} 1 & 4 \\ 2 & 0 \\ -1 & 3 \end{pmatrix}\begin{pmatrix} 5 & 2 \\ 1 & -3 \end{pmatrix} = \begin{pmatrix} 9 & -10 \\ 10 & 4 \\ -2 & -11 \end{pmatrix}$

$\Rightarrow \quad (\mathbf{AB})^T = \begin{pmatrix} 9 & 10 & -2 \\ -10 & 4 & -11 \end{pmatrix}$

b) $\quad \mathbf{B}^T\mathbf{A}^T = \begin{pmatrix} 5 & 1 \\ 2 & -3 \end{pmatrix}\begin{pmatrix} 1 & 2 & -1 \\ 4 & 0 & 3 \end{pmatrix} = \begin{pmatrix} 9 & 10 & -2 \\ -10 & 4 & -11 \end{pmatrix}$

Example 3 illustrates a general result, that if matrices $\mathbf{A}$, $\mathbf{B}$ are compatible for multiplication, then

$$(\mathbf{AB})^T = \mathbf{B}^T\mathbf{A}^T$$

Exercise 16.1

1 If $\mathbf{A} = \begin{pmatrix} 3 & 1 \\ 0 & 2 \\ -1 & 4 \end{pmatrix}$ and $\mathbf{B} = \begin{pmatrix} 2 & 3 \\ -4 & 1 \\ 2 & -5 \end{pmatrix}$ find

$\mathbf{A} + \mathbf{B}, \quad 2\mathbf{A}, \quad \mathbf{A} - \mathbf{B}, \quad \mathbf{A} + 3\mathbf{B}, \quad 3\mathbf{A} - 2\mathbf{B}.$

2 $\mathbf{A} = \begin{pmatrix} 4 & 1 \\ 3 & 2 \end{pmatrix}$, $\mathbf{B} = \begin{pmatrix} 1 & 0 \\ 2 & 1 \end{pmatrix}$, $\mathbf{C} = \begin{pmatrix} 1 & 1 \\ 2 & 0 \\ 0 & 4 \end{pmatrix}$, $\mathbf{D} = \begin{pmatrix} -1 & 0 & 2 \\ 2 & -3 & 1 \end{pmatrix}$.

Find where possible $\mathbf{AB}$, $\mathbf{BA}$, $\mathbf{CD}$, $\mathbf{DC}$, $\mathbf{DB}$, $\mathbf{BD}$, $\mathbf{DA}$, $\mathbf{AD}$, $\mathbf{CA}$, $\mathbf{AC}$, $\mathbf{BC}$, $\mathbf{CB}$.

3 Find the missing elements:

$$\begin{pmatrix} 1 & -3 \\ 1 & . \\ 1 & . \\ . & 1 \end{pmatrix} \begin{pmatrix} -1 & . & 7 & . \\ . & 1 & . & 0 \end{pmatrix} = \begin{pmatrix} -25 & -1 & 1 & 3 \\ -1 & . & . & . \\ . & . & 5 & . \\ . & . & . & 0 \end{pmatrix}$$

4 Find the products:

a) $\begin{pmatrix} 2 & -1 & 0 \\ 3 & 1 & -2 \\ 0 & 4 & -3 \end{pmatrix} \begin{pmatrix} 5 & -3 & 2 \\ 9 & -6 & 4 \\ 12 & -8 & 5 \end{pmatrix}$

b) $\begin{pmatrix} 3 - \sqrt{2} & 4 \\ -4 & 3 + \sqrt{2} \end{pmatrix} \begin{pmatrix} 2 & 5 - 2\sqrt{2} \\ -5 - 2\sqrt{2} & 2 \end{pmatrix}$

5 Pre-multiply each of the matrices $\begin{pmatrix} 2 & 6 & 3 \\ 4 & 0 & 1 \\ -1 & 2 & 3 \end{pmatrix}$ and $\begin{pmatrix} 1 & 4 & 2 \\ 2 & 3 & 0 \\ 3 & 5 & 6 \end{pmatrix}$

by $\mathbf{M} = \begin{pmatrix} 0 & 1 & 0 \\ 1 & 0 & 0 \\ 0 & 0 & 1 \end{pmatrix}$. What appears to be the effect of pre-multiplying

a 3×3 matrix by $\mathbf{M}$? Check by pre-multiplying $\begin{pmatrix} a_1 & b_1 & c_1 \\ a_2 & b_2 & c_2 \\ a_3 & b_3 & c_3 \end{pmatrix}$ by $\mathbf{M}$.

What happens when you post-multiply by $\mathbf{M}$?

6 Pre-multiply $\begin{pmatrix} 2 & 6 & 3 \\ 4 & 0 & 1 \\ -1 & 2 & 3 \end{pmatrix}$ and $\begin{pmatrix} 1 & 4 & 2 \\ 2 & 3 & 0 \\ 3 & 5 & 6 \end{pmatrix}$ by $\mathbf{N} = \begin{pmatrix} 1 & 0 & 0 \\ 0 & 3 & 0 \\ 0 & 0 & 1 \end{pmatrix}$.

What is the effect on a 3×3 matrix of pre-multiplication by $\mathbf{N}$? What is the effect of post-multiplication by $\mathbf{N}$?

7 If $\mathbf{A} = \begin{pmatrix} 6 & -3 \\ -4 & 2 \end{pmatrix}$, $\mathbf{B} = \begin{pmatrix} 4 & 3 \\ 8 & 6 \end{pmatrix}$, find $\mathbf{AB}$ and $\mathbf{BA}$. Find other counter-examples to illustrate that $\mathbf{AB} = \mathbf{0} \not\Rightarrow \mathbf{A} = \mathbf{0}$ or $\mathbf{B} = \mathbf{0}$, where $\mathbf{0}$ is the null matrix.

8 If $\mathbf{A} = \begin{pmatrix} -1 & 1 \\ 2 & -1 \end{pmatrix}$, $\mathbf{B} = \begin{pmatrix} 1 & 0 \\ 4 & -2 \end{pmatrix}$ and $\mathbf{C} = \begin{pmatrix} 2 & 1 \\ 4 & 2 \end{pmatrix}$, find $\mathbf{AC}$ and $\mathbf{BC}$.

Find other counter-examples to illustrate that

$$\mathbf{AC} = \mathbf{BC} \not\Rightarrow \mathbf{A} = \mathbf{B}$$

9 For the matrices $A = \begin{pmatrix} 2 & 3 \\ 1 & 4 \end{pmatrix}$, $B = \begin{pmatrix} 3 & 0 \\ 2 & -1 \end{pmatrix}$, $C = \begin{pmatrix} 1 & 1 \\ -2 & 5 \end{pmatrix}$ verify that

a) $(AB)C = A(BC)$
b) $A(B + C) = AB + AC$
c) $(A + B)C = AC + BC$

By writing $A = \begin{pmatrix} a_1 & a_3 \\ a_2 & a_4 \end{pmatrix}$, $B = \begin{pmatrix} b_1 & b_3 \\ b_2 & b_4 \end{pmatrix}$, $C = \begin{pmatrix} c_1 & c_3 \\ c_2 & c_4 \end{pmatrix}$ prove that results a), b), c) are true for all 2×2 matrices.

10 For the matrix $A = \begin{pmatrix} 3 & 0 & -1 \\ 1 & 2 & 1 \end{pmatrix}$ evaluate

a) AA^T and b) A^TA

11 For the matrices $A = \begin{pmatrix} a & b \\ c & d \end{pmatrix}$ and $B = \begin{pmatrix} \alpha & \beta \\ \gamma & \delta \end{pmatrix}$, show that $(AB)^T = B^TA^T$.

12 If A is a 2×2 matrix such that $A^2 = 0$, find the general form of A. Repeat when $A^2 = I$.

13 Show that the only 2×2 matrices which commute with *every* 2×2 matrix are those of the form $\begin{pmatrix} k & 0 \\ 0 & k \end{pmatrix}$. Extend the result to 3×3 matrices.

14 Given that $D = \begin{pmatrix} 1 & -2 \\ -3 & 2 \end{pmatrix}$, express $D^2 - 4I$ in terms of D and hence find the matrix given by $(D - 4I)(D + I)$. (AEB)

15 If $A = \begin{pmatrix} a_1 & a_2 \\ -a_2 & a_1 \end{pmatrix}$, $B = \begin{pmatrix} b_1 & b_2 \\ -b_2 & b_1 \end{pmatrix}$, find AB and BA. What does this result remind you of?

16 If $A = \begin{pmatrix} 2 & -3 \\ -3 & 5 \end{pmatrix}$, find $A^2, A^3, A^4, \ldots$. If $AX = XA$ where X is a 2×2 matrix, show that X may be expressed in the form $\begin{pmatrix} p & q \\ q & p - q \end{pmatrix}$.

17 Let $A = \begin{pmatrix} 1 & 0 & 0 \\ 1 & 0 & 1 \\ 0 & 1 & 0 \end{pmatrix}$. Show, by induction, or otherwise, that for all integers n greater than 2,

$$A^n = A^{n-2} + A^2 - I$$

Hence, calculate A^{20}. (MEI)

18 If $M = \begin{pmatrix} 1 & 0 & 0 \\ 0 & 0 & -1 \\ 0 & -1 & 0 \end{pmatrix}$, prove that $M^{2n+1} = M$.

19 If $\mathbf{C} = \begin{pmatrix} 0 & i \\ -i & i \end{pmatrix}$ find $\mathbf{C}^2, \mathbf{C}^3, \ldots$ and show that $\mathbf{C}^{12} = \mathbf{I}$.

20 Express $\begin{pmatrix} 1 & 2 & -1 \\ 1 & 1 & 0 \\ 4 & 8 & -1 \end{pmatrix}$ in the form $\begin{pmatrix} a & 0 & 0 \\ b & c & 0 \\ d & e & f \end{pmatrix} \begin{pmatrix} 1 & g & h \\ 0 & 1 & k \\ 0 & 0 & 1 \end{pmatrix}$.

16.2 Linear transformations in two dimensions

We have seen that linear transformations of the plane, which keep the origin invariant, are defined by equations of the form

$$x' = ax + by$$
$$y' = cx + dy$$

These equations can be expressed in matrix form as

$$\begin{pmatrix} x' \\ y' \end{pmatrix} = \begin{pmatrix} a & b \\ c & d \end{pmatrix} \begin{pmatrix} x \\ y \end{pmatrix}$$

or more briefly as $\mathbf{x}' = \mathbf{M}\mathbf{x}$, where $\mathbf{x}, \mathbf{x}'$ are the column vectors $\begin{pmatrix} x \\ y \end{pmatrix}, \begin{pmatrix} x' \\ y' \end{pmatrix}$ and $\mathbf{M}$ is the matrix $\begin{pmatrix} a & b \\ c & d \end{pmatrix}$.

The effect of this transformation is to take the grid of unit squares in the x, y plane into a grid of parallelograms in the x', y' plane, so that the unit square OIHJ transforms into the parallelogram OI'H'J'.

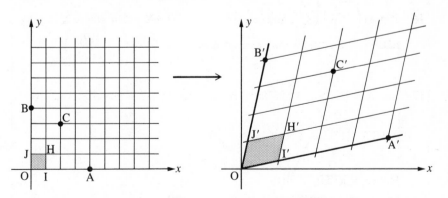

Since $\begin{pmatrix} a & b \\ c & d \end{pmatrix} \begin{pmatrix} 1 \\ 0 \end{pmatrix} = \begin{pmatrix} a \\ c \end{pmatrix}$ and $\begin{pmatrix} a & b \\ c & d \end{pmatrix} \begin{pmatrix} 0 \\ 1 \end{pmatrix} = \begin{pmatrix} b \\ d \end{pmatrix}$

it follows that

$$\mathbf{OI'} = \begin{pmatrix} a \\ c \end{pmatrix}, \quad \mathbf{OJ'} = \begin{pmatrix} b \\ d \end{pmatrix}$$

and therefore that

$$\mathbf{OH'} = \begin{pmatrix} a + b \\ c + d \end{pmatrix}$$

The area of the parallelogram OI'H'J' is $ad - bc$, (the cross product of $\begin{pmatrix} a \\ c \end{pmatrix}$ and $\begin{pmatrix} b \\ d \end{pmatrix}$—see section 11.3). This quantity, $ad - bc$, which represents the area scale factor of the transformation, is called the *determinant* of the matrix $\begin{pmatrix} a & b \\ c & d \end{pmatrix}$, written

$$\det \mathbf{M} \quad \text{or} \quad \begin{vmatrix} a & b \\ c & d \end{vmatrix}.$$

Example 1

Find the images of the points A (2, 1), B (5, 1) and C (4, 3) under the linear transformation with matrix $\begin{pmatrix} 2 & -1 \\ -1 & 3 \end{pmatrix}$. What is the area of the transformed triangle A'B'C'?

$$\begin{pmatrix} 2 & -1 \\ -1 & 3 \end{pmatrix}\begin{pmatrix} 2 \\ 1 \end{pmatrix} = \begin{pmatrix} 3 \\ 1 \end{pmatrix}, \quad \begin{pmatrix} 2 & -1 \\ -1 & 3 \end{pmatrix}\begin{pmatrix} 5 \\ 1 \end{pmatrix} = \begin{pmatrix} 9 \\ -2 \end{pmatrix}$$

and $\begin{pmatrix} 2 & -1 \\ -1 & 3 \end{pmatrix}\begin{pmatrix} 4 \\ 3 \end{pmatrix} = \begin{pmatrix} 5 \\ 5 \end{pmatrix}$

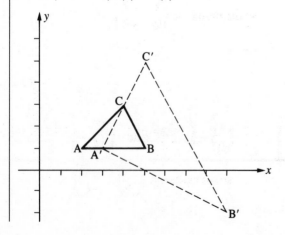

Area of triangle ABC $= \frac{1}{2}(3 \times 2) = 3$ square units

$$\begin{vmatrix} 2 & -1 \\ -1 & 3 \end{vmatrix} = (2 \times 3) - (-1 \times -1) = 5$$

So the area of triangle A'B'C' is 5 times the area of triangle ABC

$\qquad = 5 \times 3 = 15$ square units

The fact that the first column of the matrix $\begin{pmatrix} a & b \\ c & d \end{pmatrix}$ gives the transform of the base (or unit) vector $\begin{pmatrix} 1 \\ 0 \end{pmatrix}$, whilst the second column gives the transform of the base vector $\begin{pmatrix} 0 \\ 1 \end{pmatrix}$, enables us to write down immediately the 2×2 matrix representing any transformation. We do this by considering just the unit square, though all such transformations apply to the whole plane.

Example 2

Find the matrices which represent the following transformations:
a) reflection in the line $y = 0$;
b) rotation of $120°$ about the origin;
c) one-way stretch parallel to the x-axis, factor 3;
d) shear with the x-axis invariant, and $(0, 1) \rightarrow (2, 1)$.

a)

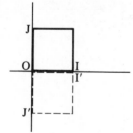

$$\begin{pmatrix} 1 \\ 0 \end{pmatrix} \rightarrow \begin{pmatrix} 1 \\ 0 \end{pmatrix} \quad \text{and} \quad \begin{pmatrix} 0 \\ 1 \end{pmatrix} \rightarrow \begin{pmatrix} 0 \\ -1 \end{pmatrix}$$

so the matrix is $\begin{pmatrix} 1 & 0 \\ 0 & -1 \end{pmatrix}$

b)

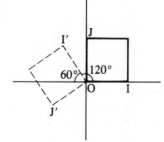

$$\begin{pmatrix} 1 \\ 0 \end{pmatrix} \rightarrow \begin{pmatrix} -\frac{1}{2} \\ \frac{\sqrt{3}}{2} \end{pmatrix} \quad \text{and} \quad \begin{pmatrix} 0 \\ 1 \end{pmatrix} \rightarrow \begin{pmatrix} -\frac{\sqrt{3}}{2} \\ -\frac{1}{2} \end{pmatrix}$$

so the matrix is $\begin{pmatrix} -\frac{1}{2} & -\frac{\sqrt{3}}{2} \\ \frac{\sqrt{3}}{2} & -\frac{1}{2} \end{pmatrix}$

c)

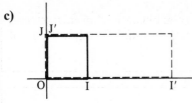

$$\begin{pmatrix}1\\0\end{pmatrix}\to\begin{pmatrix}3\\0\end{pmatrix} \quad\text{and}\quad \begin{pmatrix}0\\1\end{pmatrix}\to\begin{pmatrix}0\\1\end{pmatrix}$$

so the matrix is $\begin{pmatrix}3 & 0\\0 & 1\end{pmatrix}$

d)

$$\begin{pmatrix}1\\0\end{pmatrix}\to\begin{pmatrix}1\\0\end{pmatrix} \quad\text{and}\quad \begin{pmatrix}0\\1\end{pmatrix}\to\begin{pmatrix}2\\1\end{pmatrix}$$

so the matrix is $\begin{pmatrix}1 & 2\\0 & 1\end{pmatrix}$

The determinants of these matrices are
a) $(1\times-1)-(0\times0)=-1$
b) $(-\frac{1}{2}\times-\frac{1}{2})-(-\frac{\sqrt3}{2}\times\frac{\sqrt3}{2})=\frac{1}{4}+\frac{3}{4}=1$
c) $(3\times1)-(0\times0)=3$
d) $(1\times1)-(2\times0)=1$
Determinants of 1 for the rotation and shear show, as expected, that areas are unchanged by these transformations. In the one-way stretch the area is clearly three times greater. A determinant of -1 for the reflection, however, indicates that though the area is unchanged, the order of vertices of the shape is reversed (OIJ anti-clockwise becomes OI′J′ clockwise).

Example 3

Identify the transformation defined by the matrix $\begin{pmatrix}3 & -2\\2 & 3\end{pmatrix}$.

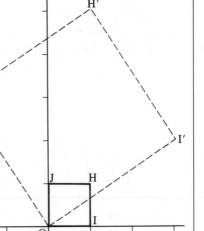

We can write down immediately that

$$\begin{pmatrix}1\\0\end{pmatrix}\to\begin{pmatrix}3\\2\end{pmatrix} \quad\text{and}\quad \begin{pmatrix}0\\1\end{pmatrix}\to\begin{pmatrix}-2\\3\end{pmatrix}$$

so the vectors **OI′** and **OJ′** can be marked on the diagram, and the parallelogram (in fact in this case, square) OI′H′J′ completed.

The transformation can be identified as a combination of an enlargement, with scale factor $\sqrt{(2^2 + 3^2)} = \sqrt{13}$, and a rotation about $(0,0)$ of $\tan^{-1}\frac{2}{3} \approx 33.7°$.

Example 4

Investigate the transformation defined by the matrix $\begin{pmatrix} 4 & 2 \\ 2 & 1 \end{pmatrix}$.

Since

$$\begin{pmatrix} 1 \\ 0 \end{pmatrix} \rightarrow \begin{pmatrix} 4 \\ 2 \end{pmatrix} \quad \text{and} \quad \begin{pmatrix} 0 \\ 1 \end{pmatrix} \rightarrow \begin{pmatrix} 2 \\ 1 \end{pmatrix}$$

the unit square (and the whole plane) is squashed on to the line $y = \frac{1}{2}x$.

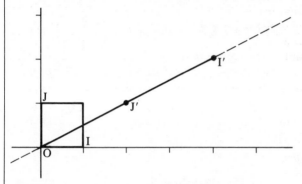

The area scale factor must therefore be zero, as can be checked by evaluating the determinant

$$\begin{vmatrix} 4 & 2 \\ 2 & 1 \end{vmatrix} = (4 \times 1) - (2 \times 2) = 0$$

Such a matrix, whose determinant is zero, is said to be *singular*.

Eigenvectors

Any vector, whose direction is unchanged after a transformation, is said to be an *eigenvector* of the transformation (or of its associated matrix). The factor by which its length is increased is called the *eigenvalue* of the vector.

So the matrix $\begin{pmatrix} 1 & 0 \\ 0 & -1 \end{pmatrix}$, which describes a reflection in the x-axis, has eigenvectors $\begin{pmatrix} 1 \\ 0 \end{pmatrix}$ and $\begin{pmatrix} 0 \\ 1 \end{pmatrix}$ with eigenvalues 1 and -1 respectively. (If $\begin{pmatrix} 1 \\ 0 \end{pmatrix}$ is an eigenvector, then clearly any multiple of it, $\begin{pmatrix} k \\ 0 \end{pmatrix}$, will also be an eigenvector.)

The matrix $\begin{pmatrix} 3 & 0 \\ 0 & 1 \end{pmatrix}$ also has eigenvectors $\begin{pmatrix} 1 \\ 0 \end{pmatrix}$ and $\begin{pmatrix} 0 \\ 1 \end{pmatrix}$, this time with eigenvalues 3 and 1.

However, a rotation of $120°$ changes the direction of all vectors, so matrix $\begin{pmatrix} -\frac{1}{2} & -\frac{\sqrt{3}}{2} \\ \frac{\sqrt{3}}{2} & -\frac{1}{2} \end{pmatrix}$ has no real eigenvectors.

Most 2×2 matrices do not however describe such simple transformations, and their eigenvectors and eigenvalues have to be obtained in a more methodical way, as in the following example.

Example 5

Find the eigenvectors and eigenvalues of the matrix $\begin{pmatrix} -\frac{3}{5} & \frac{4}{5} \\ \frac{4}{5} & \frac{3}{5} \end{pmatrix}$. Hence describe the corresponding transformation.

If $\begin{pmatrix} \alpha \\ \beta \end{pmatrix}$ is an eigenvector of the matrix, with corresponding eigenvalue λ, then

$$\begin{pmatrix} -\frac{3}{5} & \frac{4}{5} \\ \frac{4}{5} & \frac{3}{5} \end{pmatrix} \begin{pmatrix} \alpha \\ \beta \end{pmatrix} = \lambda \begin{pmatrix} \alpha \\ \beta \end{pmatrix}$$

$$\Rightarrow \quad \begin{matrix} -\frac{3}{5}\alpha + \frac{4}{5}\beta = \lambda\alpha \\ \frac{4}{5}\alpha + \frac{3}{5}\beta = \lambda\beta \end{matrix} \quad \Rightarrow \quad \begin{matrix} (-\frac{3}{5} - \lambda)\alpha + \frac{4}{5}\beta = 0 \\ \frac{4}{5}\alpha + (\frac{3}{5} - \lambda)\beta = 0 \end{matrix}$$

which can be written in the form

$$\begin{pmatrix} -\frac{3}{5} - \lambda & \frac{4}{5} \\ \frac{4}{5} & \frac{3}{5} - \lambda \end{pmatrix} \begin{pmatrix} \alpha \\ \beta \end{pmatrix} = \begin{pmatrix} 0 \\ 0 \end{pmatrix} \tag{1}$$

So the matrix $\begin{pmatrix} -\frac{3}{5} - \lambda & \frac{4}{5} \\ \frac{4}{5} & \frac{3}{5} - \lambda \end{pmatrix}$ 'squashes' the vector $\begin{pmatrix} \alpha \\ \beta \end{pmatrix}$ on to the origin. This can only occur if the matrix is singular, i.e. if its determinant is zero.

Hence $\begin{vmatrix} -\frac{3}{5} - \lambda & \frac{4}{5} \\ \frac{4}{5} & \frac{3}{5} - \lambda \end{vmatrix} = 0$

$$\Rightarrow \quad (-\tfrac{3}{5} - \lambda)(\tfrac{3}{5} - \lambda) - \tfrac{16}{25} = 0$$

$$\Rightarrow \quad \lambda^2 - \tfrac{9}{25} - \tfrac{16}{25} = 0$$

$$\Rightarrow \quad \lambda^2 - 1 = 0$$

$$\Rightarrow \quad \lambda = \pm 1$$

When $\lambda = 1$, equation (1) takes the form

$$\begin{pmatrix} -\frac{8}{5} & \frac{4}{5} \\ \frac{4}{5} & -\frac{2}{5} \end{pmatrix} \begin{pmatrix} \alpha \\ \beta \end{pmatrix} = \begin{pmatrix} 0 \\ 0 \end{pmatrix}$$

so $\quad -\frac{8}{5}\alpha + \frac{4}{5}\beta = 0 \quad \Rightarrow \quad -2\alpha + \beta = 0$

and an eigenvector is therefore $\begin{pmatrix} 1 \\ 2 \end{pmatrix}$.

When $\lambda = -1$, equation (1) take the form

$$\begin{pmatrix} \frac{2}{5} & \frac{4}{5} \\ \frac{4}{5} & \frac{8}{5} \end{pmatrix} \begin{pmatrix} \alpha \\ \beta \end{pmatrix} = \begin{pmatrix} 0 \\ 0 \end{pmatrix}$$

so $\quad \frac{2}{5}\alpha + \frac{4}{5}\beta = 0 \quad \Rightarrow \quad \alpha + 2\beta = 0$

and a second eigenvector is $\begin{pmatrix} 2 \\ -1 \end{pmatrix}$.

If we now look at the effect of this transformation on the unit square, it is clear that the matrix defines a reflection.

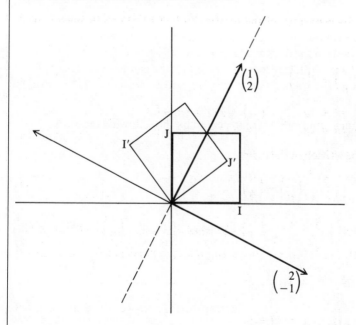

Since the eigenvector $\begin{pmatrix} 1 \\ 2 \end{pmatrix}$ has an eigenvalue of 1, this vector must lie on the mirror line (or axis) of the reflection. So the matrix defines a reflection in the line $y = 2x$.

The second eigenvector $\begin{pmatrix} 2 \\ -1 \end{pmatrix}$ is perpendicular to the mirror line, so its direction is reversed by the transformation, and this is confirmed by its eigenvalue of -1.

In general therefore, to find the eigenvectors and eigenvalues of a matrix $\begin{pmatrix} a & b \\ c & d \end{pmatrix}$, we first find the eigenvalues from the equation

$$\begin{vmatrix} a - \lambda & b \\ c & d - \lambda \end{vmatrix} = 0$$

Once the eigenvalues are known, the corresponding eigenvectors can easily be found from the equation

$$\begin{pmatrix} a & b \\ c & d \end{pmatrix} \begin{pmatrix} \alpha \\ \beta \end{pmatrix} = \lambda \begin{pmatrix} \alpha \\ \beta \end{pmatrix}$$

Exercise 16.2a

1   Write down the matrices representing the following transformations:
   a) rotation about the origin through $\pi, \frac{1}{2}\pi.$ $-\frac{1}{2}\pi, \frac{1}{3}\pi, -\frac{1}{3}\pi, \frac{3}{4}\pi$;
   b) reflections in the lines $x = 0, y = x, y = -x, y = \frac{1}{2}x$;
   c) an enlargement, centre $(0, 0)$, scale factor 3; a one-way stretch, factor 2, parallel to the x-axis; a shear with the y-axis invariant, which moves the point $(1, 1)$ to $(1, 0)$.

2   What transformations are represented by the following matrices:

   a) $\begin{pmatrix} 2 & 0 \\ 0 & 3 \end{pmatrix}$
   b) $\begin{pmatrix} -3 & 0 \\ 0 & -3 \end{pmatrix}$
   c) $\begin{pmatrix} 3 & 4 \\ -4 & 3 \end{pmatrix}$

   d) $\begin{pmatrix} 1 & 2 \\ 0 & 1 \end{pmatrix}$
   e) $\begin{pmatrix} 1 & 0 \\ -3 & 1 \end{pmatrix}$
   f) $\begin{pmatrix} 0 & -2 \\ -2 & 0 \end{pmatrix}$

   g) $\begin{pmatrix} 2 & -1 \\ 0 & 2 \end{pmatrix}$
   h) $\begin{pmatrix} \cos\alpha & \sin\alpha \\ \sin\alpha & -\cos\alpha \end{pmatrix}$
   i) $\begin{pmatrix} -\sin\alpha & \cos\alpha \\ \cos\alpha & \sin\alpha \end{pmatrix}$

   j) $\begin{pmatrix} \cos 40° & -\sin 40° \\ \sin 40° & \cos 40° \end{pmatrix}$
   k) $\begin{pmatrix} 2 & 4 \\ 1 & 2 \end{pmatrix}$
   l) $\begin{pmatrix} 1 & 0 \\ 1 & 0 \end{pmatrix}$?

3   In no. 2 check the effects of the transformations on areas, and compare the values of the determinant $ad - bc$ in each case.

4   Describe the transformations represented by the eight matrices

$$\begin{pmatrix} \pm 1 & 0 \\ 0 & \pm 1 \end{pmatrix}, \quad \begin{pmatrix} 0 & \pm 1 \\ \pm 1 & 0 \end{pmatrix}$$

5 For each of the matrices **M** and **N** where

$$\mathbf{M} = \begin{pmatrix} \dfrac{1}{\sqrt{2}} & \dfrac{1}{\sqrt{2}} \\ -\dfrac{1}{\sqrt{2}} & \dfrac{1}{\sqrt{2}} \end{pmatrix} \quad \mathbf{N} = \begin{pmatrix} \dfrac{1}{2} & \dfrac{\sqrt{3}}{2} \\ \dfrac{\sqrt{3}}{2} & -\dfrac{1}{2} \end{pmatrix}$$

describe fully the geometrical transformation to which it corresponds.
Explain the geometrical significance of the value of the determinant of each of these matrices.   (JMB)

6 Obtain the matrix for reflection in the line $y = x \tan \theta$, by first rotating the coordinate axes so that the x-axis coincides with the reflection axis (operation **R**), then performing the reflection in this axis, and finally restoring the axes to their original position ($\mathbf{R}^{-1}$).

7 A triangle has vertices   A $\begin{pmatrix} 2 \\ 1 \end{pmatrix}$, B $\begin{pmatrix} -1 \\ 2 \end{pmatrix}$, C $\begin{pmatrix} 0 \\ 3 \end{pmatrix}$. What is its image under the transformations represented by the matrices:

a) $\begin{pmatrix} 2 & 3 \\ -1 & -2 \end{pmatrix}$ **b)** $\begin{pmatrix} 3 & 1 \\ 2 & 0 \end{pmatrix}$ **c)** $\begin{pmatrix} 4 & -6 \\ -2 & 3 \end{pmatrix}$?

8 Investigate the effect of the transformation represented by $\begin{pmatrix} 1 & 0 \\ 1 & 1 \end{pmatrix}$ on the rectangle with vertices $\begin{pmatrix} 0 \\ -1 \end{pmatrix}, \begin{pmatrix} 0 \\ 1 \end{pmatrix}, \begin{pmatrix} 1 \\ 1 \end{pmatrix}, \begin{pmatrix} 1 \\ -1 \end{pmatrix}$.

9 What is the image of the point $\begin{pmatrix} 3 \\ -2 \end{pmatrix}$ under the matrix which moves $\begin{pmatrix} 1 \\ 0 \end{pmatrix}$ to $\begin{pmatrix} -2 \\ 2 \end{pmatrix}$ and $\begin{pmatrix} 1 \\ 1 \end{pmatrix}$ to $\begin{pmatrix} 0 \\ 3 \end{pmatrix}$?

10 In a parallelogram OABC the coordinates of O, A and C are $(0, 0)$, $(5, 2)$ and $(1, 3)$. Under a shear transformation, with matrix

$$\begin{pmatrix} 1 & k \\ 0 & 1 \end{pmatrix}$$

this parallelogram is transformed into a parallelogram OA′B′C′. Show that there are two values of k for which OA′B′C′ is a rectangle, and find them. Prove that one of these rectangles is a square.   (SMP)

11 Show that $\mathbf{M} = \begin{pmatrix} 2 & -1 \\ -3 & 4 \end{pmatrix}$ operating on *any* vector in the direction $\begin{pmatrix} 1 \\ 1 \end{pmatrix}$ leaves it invariant. Show that vectors in the direction $\begin{pmatrix} -3 \\ 1 \end{pmatrix}$ are also unchanged in direction, but are enlarged by a factor 5. Repeat using the matrix $\mathbf{M}^2$.

12 Find the eigenvalues and eigenvectors of the following matrices:

a) $\begin{pmatrix} \frac{4}{5} & \frac{3}{5} \\ \frac{3}{5} & -\frac{4}{5} \end{pmatrix}$ **b)** $\begin{pmatrix} 2 & 4 \\ 1 & 5 \end{pmatrix}$ **c)** $\begin{pmatrix} 1 & 3 \\ 3 & 1 \end{pmatrix}$

13 The transformation with matrix $\mathbf{T}$, where $\mathbf{T} = \begin{pmatrix} 2 & 9 \\ 1 & 2 \end{pmatrix}$, maps (x, y) on to (x', y'), where

$$\begin{pmatrix} x' \\ y' \end{pmatrix} = \mathbf{T} \begin{pmatrix} x \\ y \end{pmatrix}$$

Find the possible values of the constant λ for which there exist non-zero x and y such that

$$\begin{pmatrix} x' \\ y' \end{pmatrix} = \begin{pmatrix} \lambda x \\ \lambda y \end{pmatrix}$$

Use these values of λ to find the equations of the lines through the origin, each of which is mapped on to itself by the transformation.   (JMB)

14 The matrix M, where

$$\mathbf{M} = \begin{pmatrix} p & q \\ r & s \end{pmatrix}$$

is such that
a) all its elements are real and non-zero,
b) $\det \mathbf{M} = 1$,
c) $\mathbf{M}\mathbf{M}^T = \mathbf{I}$, where $\mathbf{M}^T$ is the transpose of $\mathbf{M}$ and $\mathbf{I}$ is the unit matrix.
By showing that

$$s(p^2 + q^2) = p$$

or otherwise, express each of q, r and s in terms of p.

Given now that $p = \frac{3}{5}$ write down the two possible forms of the matrix $\mathbf{M}$. State the relation between the geometrical transformations represented by these forms.   (JMB)

15 Show that, if a 2×2 matrix represents an *isometry*, i.e. a distance-preserving transformation, then it must be either of the form

$$\begin{pmatrix} \cos\theta & \sin\theta \\ -\sin\theta & \cos\theta \end{pmatrix} \quad \text{or else of the form} \quad \begin{pmatrix} \cos\theta & \sin\theta \\ \sin\theta & -\cos\theta \end{pmatrix}$$

i.e. a rotation or a reflection.

Combined transformations

In section 16.1 we saw that matrix multiplication was defined in such a way that a transformation with matrix $\mathbf{A}$ followed by a transformation with matrix $\mathbf{B}$ is equivalent to a transformation with matrix $\mathbf{BA}$.

Example 6

Use matrices to find the single transformation which is equivalent to a rotation of 90° about the origin, followed by a reflection in $y = 0$.

Rotation of 90° about $(0, 0)$:

Matrix $\mathbf{A} = \begin{pmatrix} 0 & -1 \\ 1 & 0 \end{pmatrix}$

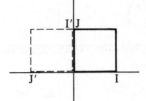

Reflection in $y = 0$:

Matrix $\mathbf{B} = \begin{pmatrix} 1 & 0 \\ 0 & -1 \end{pmatrix}$

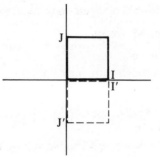

Combined matrix $= \mathbf{BA} = \begin{pmatrix} 1 & 0 \\ 0 & -1 \end{pmatrix} \begin{pmatrix} 0 & -1 \\ 1 & 0 \end{pmatrix} = \begin{pmatrix} 0 & -1 \\ -1 & 0 \end{pmatrix}$

and the matrix $\begin{pmatrix} 0 & -1 \\ -1 & 0 \end{pmatrix}$

defines a reflection in the

line $x + y = 0$

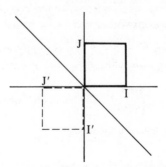

So a rotation of 90° about the origin followed by a reflection in $y = 0$ is equivalent to a reflection in $x + y = 0$.

Example 7

By considering its geometrical effect, write the matrix $\begin{pmatrix} 0 & -1 \\ 1 & 2 \end{pmatrix}$ as a product of two other 2×2 matrices.

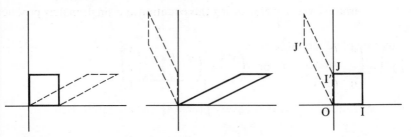

From the diagram it can be seen that the transformation associated with this matrix is equivalent to a shear parallel to the x-axis, followed by a rotation of $90°$ about the origin. The matrix describing the shear is $\begin{pmatrix} 1 & 2 \\ 0 & 1 \end{pmatrix}$ and the matrix describing the rotation is $\begin{pmatrix} 0 & -1 \\ 1 & 0 \end{pmatrix}$.

So $\begin{pmatrix} 0 & -1 \\ 1 & 2 \end{pmatrix} = \begin{pmatrix} 0 & -1 \\ 1 & 0 \end{pmatrix} \begin{pmatrix} 1 & 2 \\ 0 & 1 \end{pmatrix}$

Translations

In all the transformations considered so far, defined by a matrix $\begin{pmatrix} a & b \\ c & d \end{pmatrix}$, the origin has been invariant, since $\begin{pmatrix} a & b \\ c & d \end{pmatrix} \begin{pmatrix} 0 \\ 0 \end{pmatrix} = \begin{pmatrix} 0 \\ 0 \end{pmatrix}$

However there are other *non-linear* transformations, where the origin is not invariant, and which cannot be expressed in this form.

The simplest example of a non-linear transformation is a translation, which can be represented by

$$\begin{pmatrix} x' \\ y' \end{pmatrix} = \begin{pmatrix} x \\ y \end{pmatrix} + \begin{pmatrix} e \\ f \end{pmatrix}$$

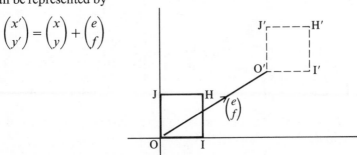

Any *affine* transformation (one in which straight lines remain straight and parallel lines remain parallel) can be expressed in the form

$$\begin{pmatrix} x' \\ y' \end{pmatrix} = \begin{pmatrix} a & b \\ c & d \end{pmatrix}\begin{pmatrix} x \\ y \end{pmatrix} + \begin{pmatrix} e \\ f \end{pmatrix}$$

which is a combination of a linear transformation and a translation.

Two alternative ways of expressing this equation as a single matrix product are

$$\begin{pmatrix} x' \\ y' \\ 1 \end{pmatrix} = \begin{pmatrix} a & b & e \\ c & d & f \\ 0 & 0 & 1 \end{pmatrix}\begin{pmatrix} x \\ y \\ 1 \end{pmatrix} \quad \text{or} \quad \begin{pmatrix} x' \\ y' \end{pmatrix} = \begin{pmatrix} a & b & e \\ c & d & f \end{pmatrix}\begin{pmatrix} x \\ y \\ 1 \end{pmatrix}$$

but these representations are not very practical, and are rarely used.

Example 8

The images of the points $(0,0)$, $(1,0)$, $(1,1)$ and $(0,1)$ after a certain transformation are $(3,1)$, $(5,1)$, $(6,3)$ and $(4,3)$ respectively. Represent the transformation in the form

$$\begin{pmatrix} x' \\ y' \end{pmatrix} = \begin{pmatrix} a & b \\ c & d \end{pmatrix}\begin{pmatrix} x \\ y \end{pmatrix} + \begin{pmatrix} e \\ f \end{pmatrix}$$

Translate the image of the unit square to $O''I''H''J''$ so that O'' coincides with the origin.

OIHJ is transformed on to $O''I''H''J''$ by the matrix $\begin{pmatrix} 2 & 1 \\ 0 & 2 \end{pmatrix}$

i.e. $\begin{pmatrix} x'' \\ y'' \end{pmatrix} = \begin{pmatrix} 2 & 1 \\ 0 & 2 \end{pmatrix}\begin{pmatrix} x \\ y \end{pmatrix}$

$O''I''H''J''$ is translated on to $O'I'H'J'$ by the vector $\begin{pmatrix} 3 \\ 1 \end{pmatrix}$

i.e. $\begin{pmatrix} x' \\ y' \end{pmatrix} = \begin{pmatrix} x'' \\ y'' \end{pmatrix} + \begin{pmatrix} 3 \\ 1 \end{pmatrix}$

So $\begin{pmatrix} x' \\ y' \end{pmatrix} = \begin{pmatrix} 2 & 1 \\ 0 & 2 \end{pmatrix}\begin{pmatrix} x \\ y \end{pmatrix} + \begin{pmatrix} 3 \\ 1 \end{pmatrix}$

Exercise 16.2b

1 What transformations are represented by the matrices

$$\mathbf{A} = \begin{pmatrix} 0 & -1 \\ 1 & 0 \end{pmatrix}, \quad \mathbf{B} = \begin{pmatrix} -1 & 0 \\ 0 & 1 \end{pmatrix}, \quad \mathbf{C} = \begin{pmatrix} 2 & 0 \\ 0 & 2 \end{pmatrix}, \quad \mathbf{D} = \begin{pmatrix} 1 & 1 \\ 0 & 1 \end{pmatrix}?$$

Evaluate the products $\mathbf{AB}, \mathbf{BA}, \mathbf{AC}, \mathbf{CB}, \mathbf{A}^2, \mathbf{B}^2, \mathbf{C}^2, \mathbf{D}^2$. Interpret each of these products geometrically.

2 For the matrices $\mathbf{A} = \begin{pmatrix} 1 & 0 \\ 0 & -1 \end{pmatrix}$ and $\mathbf{B} = \begin{pmatrix} 0 & 1 \\ 1 & 0 \end{pmatrix}$, find the transformations associated with each of the products $\mathbf{AB}, \mathbf{BA}, \mathbf{ABA}, \mathbf{BAB}, \mathbf{ABAB}$.

3 Decompose the following as the products of two matrices representing 'simple' two-dimensional transformations such as rotations, reflections, enlargements, etc.:

a) $\begin{pmatrix} -3 & 0 \\ 0 & -3 \end{pmatrix}$ b) $\begin{pmatrix} 0 & -3 \\ -3 & 0 \end{pmatrix}$ c) $\begin{pmatrix} 1 & \sqrt{3} \\ \sqrt{3} & -1 \end{pmatrix}$ d) $\begin{pmatrix} 5 & -12 \\ 12 & 5 \end{pmatrix}$

e) $\begin{pmatrix} 2 & -3 \\ 0 & 2 \end{pmatrix}$ f) $\begin{pmatrix} 0 & -2 \\ 3 & 0 \end{pmatrix}$ g) $\begin{pmatrix} 1 & 1 \\ -1 & 3 \end{pmatrix}$

4 Write down the matrices which describe reflection in the line $y = 0$, and reflection in the line $y = x \tan \theta$. By multiplying these matrices together find the single transformation equivalent to a reflection in the line $y = 0$ followed by a reflection in the line $y = x \tan \theta$.

5 Prove that, if a 2×2 real matrix $\mathbf{A}$ is such that $\mathbf{A}^T\mathbf{A} = \mathbf{I}$, then $\mathbf{A}$ has one of the forms

$$\mathbf{A}_1(\alpha) = \begin{pmatrix} \cos \alpha & -\sin \alpha \\ \sin \alpha & \cos \alpha \end{pmatrix} \quad \text{or} \quad \mathbf{A}_2(\alpha) = \begin{pmatrix} \cos \alpha & \sin \alpha \\ \sin \alpha & -\cos \alpha \end{pmatrix}$$

for some real α.

A transformation with matrix $\mathbf{A}$ maps the point P with position vector $\begin{pmatrix} x \\ y \end{pmatrix}$ in the plane relative to the origin O to a point P' with position vector $\begin{pmatrix} x' \\ y' \end{pmatrix}$ where $\begin{pmatrix} x' \\ y' \end{pmatrix} = \mathbf{A}\begin{pmatrix} x \\ y \end{pmatrix}$. By considering the images of $\begin{pmatrix} 1 \\ 0 \end{pmatrix}$ and $\begin{pmatrix} 0 \\ 1 \end{pmatrix}$, or otherwise, determine the geometric transformations represented by the matrices $\mathbf{A}_1(\alpha)$ and $\mathbf{A}_2(\alpha)$.

Verify that $\mathbf{A}_2(\alpha)\mathbf{A}_2(\beta) = \mathbf{A}_1(\alpha - \beta)$ and interpret this result geometrically. (OC)

6 By considering rotations about the origin, find 2×2 matrices $\mathbf{A}, \mathbf{B}, \mathbf{C}, \mathbf{D}, \mathbf{E}$ such that $\mathbf{A}^3 = \mathbf{I}$, $\mathbf{B}^4 = \mathbf{I}$, $\mathbf{C}^6 = \mathbf{I}$, $\mathbf{D}^8 = \mathbf{I}$, $\mathbf{E}^{12} = \mathbf{I}$.

7   By interpreting geometrically each of the matrices, $\overset{\checkmark}{\mathbf{M}}$, below, in each case find a second matrix, $\mathbf{N}$, such that $\mathbf{MN} = \mathbf{I}$.

a) $\begin{pmatrix} 0 & -1 \\ 1 & 0 \end{pmatrix}$ b) $\begin{pmatrix} 2 & 0 \\ 0 & 2 \end{pmatrix}$ c) $\begin{pmatrix} 1 & 3 \\ 0 & 1 \end{pmatrix}$ d) $\begin{pmatrix} -1 & 0 \\ 0 & 1 \end{pmatrix}$

8   A non-singular matrix $\begin{pmatrix} a & b \\ c & d \end{pmatrix}$, where $d \neq 0$, can be expressed as $\mathbf{UL}$ where $\mathbf{U}$ is $\begin{pmatrix} 1 & p \\ 0 & 1 \end{pmatrix}$ and $\mathbf{L}$ is $\begin{pmatrix} q & 0 \\ r & s \end{pmatrix}$ for suitable values of p, q, r and s. Find p, q, r and s in terms of a, b, c and d.

The transformation, T, is associated with $\mathbf{U}$ when $p = 2$. Given that O is the origin and A, B and C have coordinates $(1, 0)$, $(1, 1)$ and $(0, 1)$ respectively, sketch the quadrilateral into which OABC is mapped by this transformation. Describe the transformation.   (AEB, 1981)

9   The images of the points $(0, 0)$, $(1, 0)$, $(1, 1)$ and $(0, 1)$ (in order) after a number of transformations are given below. Express each transformation in the form

$$\begin{pmatrix} x' \\ y' \end{pmatrix} = \begin{pmatrix} a & b \\ c & d \end{pmatrix} \begin{pmatrix} x \\ y \end{pmatrix} + \begin{pmatrix} e \\ f \end{pmatrix}$$

a) $(1, 0)$, $(3, 0)$, $(3, 2)$, $(1, 2)$
b) $(0, -2)$, $(2, 0)$, $(0, 2)$, $(-2, 0)$
c) $(-2, 1)$, $(-2, 4)$, $(-3, 4)$, $(-3, 1)$

10   Express each of the following transformations of the plane in the form

$$\begin{pmatrix} x' \\ y' \end{pmatrix} = \begin{pmatrix} a & b \\ c & d \end{pmatrix} \begin{pmatrix} x \\ y \end{pmatrix} + \begin{pmatrix} p \\ q \end{pmatrix}$$

giving the values of a, b, c, d, p and q in each case.
a) $\mathbf{T}_1$:   reflection in the line $x + y = 0$;
b) $\mathbf{T}_2$:   reflection in the line $x - y = 2$;
c) $\mathbf{T}_3$:   rotation through $90°$ anticlockwise about the point $(2, -1)$;
d) $\mathbf{T}_4$: $\mathbf{T}_4 = \mathbf{T}_2 \mathbf{T}_3 \mathbf{T}_1$;
e) $\mathbf{T}_5$: $\mathbf{T}_5 = \mathbf{T}_1 \mathbf{T}_3^{-1} \mathbf{T}_2 \mathbf{T}_3$.

Show that $\mathbf{T}_5$ has no invariant points, and that $\mathbf{T}_4$ has a single invariant point. Give a simple geometric description of $\mathbf{T}_4$.   (O)

16.3 Linear transformations in three dimensions

In three dimensions a linear transformation is defined by equations of the form

$$x' = a_1 x + b_1 y + c_1 z$$
$$y' = a_2 x + b_2 y + c_2 z$$
$$z' = a_3 x + b_3 y + c_3 z$$

which can be written in matrix form as

$$\begin{pmatrix} x' \\ y' \\ z' \end{pmatrix} = \begin{pmatrix} a_1 & b_1 & c_1 \\ a_2 & b_2 & c_2 \\ a_3 & b_3 & c_3 \end{pmatrix} \begin{pmatrix} x \\ y \\ z \end{pmatrix}$$

The geometrical effect of this is to transform the 3-dimensional lattice of cubes into a lattice of parallelepipeds. The effect on the unit cube is illustrated in the diagram.

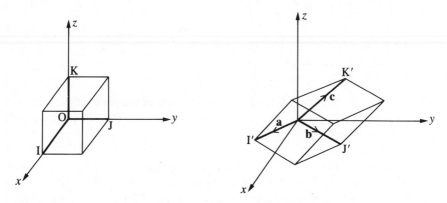

In a simple extension of the 2-dimensional case,

$$\begin{pmatrix} 1 \\ 0 \\ 0 \end{pmatrix} \rightarrow \begin{pmatrix} a_1 \\ a_2 \\ a_3 \end{pmatrix}, \quad \begin{pmatrix} 0 \\ 1 \\ 0 \end{pmatrix} \rightarrow \begin{pmatrix} b_1 \\ b_2 \\ b_3 \end{pmatrix} \quad \text{and} \quad \begin{pmatrix} 0 \\ 0 \\ 1 \end{pmatrix} \rightarrow \begin{pmatrix} c_1 \\ c_2 \\ c_3 \end{pmatrix}$$

So the columns of the matrix give the images of the base vectors **OI**, **OJ** and **OK** (or **i**, **j** and **k** as they are often written). In this sense the matrix may be thought of as a collection of three column vectors **a**, **b**, **c**.

Conversely, to find the matrix representing a particular 3-dimensional transformation, all we have to do is to find the images of the points I, J, K, and their coordinates will give the three columns of the required matrix. So we concentrate our attention on the unit cube.

Example 1

Find the matrices which represent the following 3-dimensional transformations:

a) reflection in the plane $z = 0$;   **b)** rotation of $90°$ about the y-axis, with the point $(1,0,0)$ moving to $(0,0,1)$;   **c)** enlargement, centre the origin, scale factor 2.

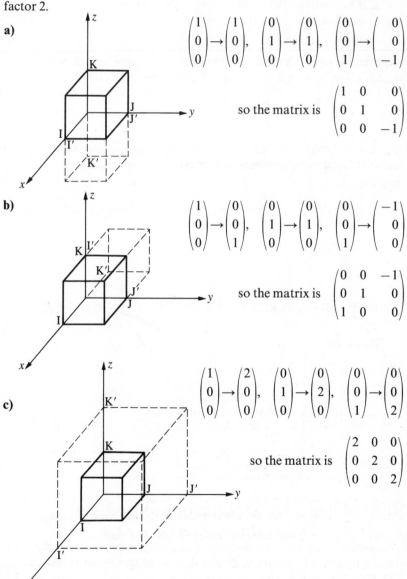

a)

$$\begin{pmatrix} 1 \\ 0 \\ 0 \end{pmatrix} \rightarrow \begin{pmatrix} 1 \\ 0 \\ 0 \end{pmatrix}, \quad \begin{pmatrix} 0 \\ 1 \\ 0 \end{pmatrix} \rightarrow \begin{pmatrix} 0 \\ 1 \\ 0 \end{pmatrix}, \quad \begin{pmatrix} 0 \\ 0 \\ 1 \end{pmatrix} \rightarrow \begin{pmatrix} 0 \\ 0 \\ -1 \end{pmatrix}$$

so the matrix is $\begin{pmatrix} 1 & 0 & 0 \\ 0 & 1 & 0 \\ 0 & 0 & -1 \end{pmatrix}$

b)

$$\begin{pmatrix} 1 \\ 0 \\ 0 \end{pmatrix} \rightarrow \begin{pmatrix} 0 \\ 0 \\ 1 \end{pmatrix}, \quad \begin{pmatrix} 0 \\ 1 \\ 0 \end{pmatrix} \rightarrow \begin{pmatrix} 0 \\ 1 \\ 0 \end{pmatrix}, \quad \begin{pmatrix} 0 \\ 0 \\ 1 \end{pmatrix} \rightarrow \begin{pmatrix} -1 \\ 0 \\ 0 \end{pmatrix}$$

so the matrix is $\begin{pmatrix} 0 & 0 & -1 \\ 0 & 1 & 0 \\ 1 & 0 & 0 \end{pmatrix}$

c)

$$\begin{pmatrix} 1 \\ 0 \\ 0 \end{pmatrix} \rightarrow \begin{pmatrix} 2 \\ 0 \\ 0 \end{pmatrix}, \quad \begin{pmatrix} 0 \\ 1 \\ 0 \end{pmatrix} \rightarrow \begin{pmatrix} 0 \\ 2 \\ 0 \end{pmatrix}, \quad \begin{pmatrix} 0 \\ 0 \\ 1 \end{pmatrix} \rightarrow \begin{pmatrix} 0 \\ 0 \\ 2 \end{pmatrix}$$

so the matrix is $\begin{pmatrix} 2 & 0 & 0 \\ 0 & 2 & 0 \\ 0 & 0 & 2 \end{pmatrix}$

Example 2

What transformation is defined by the matrix $\begin{pmatrix} 0 & 0 & 1 \\ 1 & 0 & 0 \\ 0 & 1 & 0 \end{pmatrix}$?

Since $\begin{pmatrix} 1 \\ 0 \\ 0 \end{pmatrix} \rightarrow \begin{pmatrix} 0 \\ 1 \\ 0 \end{pmatrix}$, $\begin{pmatrix} 0 \\ 1 \\ 0 \end{pmatrix} \rightarrow \begin{pmatrix} 0 \\ 0 \\ 1 \end{pmatrix}$ and $\begin{pmatrix} 0 \\ 0 \\ 1 \end{pmatrix} \rightarrow \begin{pmatrix} 1 \\ 0 \\ 0 \end{pmatrix}$

the unit cube occupies the same position after the transformation, but it has rotated about an axis through the origin and the point $(1, 1, 1)$.

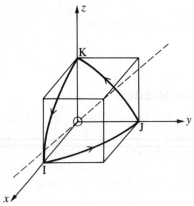

From the symmetry of the diagram, it is clear that the angle of rotation is 120°.

Since $\begin{pmatrix} 0 & 0 & 1 \\ 1 & 0 & 0 \\ 0 & 1 & 0 \end{pmatrix} \begin{pmatrix} 1 \\ 1 \\ 1 \end{pmatrix} = \begin{pmatrix} 1 \\ 1 \\ 1 \end{pmatrix}$

it follows that $\begin{pmatrix} 1 \\ 1 \\ 1 \end{pmatrix}$ is an eigenvector of the matrix, with eigenvalue 1. This confirms the position of the axis of rotation.

Effect of linear transformations on volume

We have seen that the linear transformation, whose matrix is

$$\mathbf{M} = \begin{pmatrix} a_1 & b_1 & c_1 \\ a_2 & b_2 & c_2 \\ a_3 & b_3 & c_3 \end{pmatrix}$$

transforms the unit cube into the parallelepiped, whose adjacent edges are given by the vectors

$$\mathbf{a} = \begin{pmatrix} a_1 \\ a_2 \\ a_3 \end{pmatrix}, \quad \mathbf{b} = \begin{pmatrix} b_1 \\ b_2 \\ b_3 \end{pmatrix}, \quad \mathbf{c} = \begin{pmatrix} c_1 \\ c_2 \\ c_3 \end{pmatrix}$$

As with areas in two dimensions, let the corresponding volume scale factor be called det $\mathbf{M}$ (or Δ).

Now we know (from section 11.4) that this scale factor is $\mathbf{a} \cdot (\mathbf{b} \times \mathbf{c})$.

So $\det \mathbf{M} = \mathbf{a} \cdot (\mathbf{b} \times \mathbf{c})$

$$\Rightarrow \quad \begin{vmatrix} a_1 & b_1 & c_1 \\ a_2 & b_2 & c_2 \\ a_3 & b_3 & c_3 \end{vmatrix} = a_1(b_2 c_3 - b_3 c_2) + a_2(b_3 c_1 - b_1 c_3) + a_3(b_1 c_2 - b_2 c_1)$$

Singular matrices: linear dependence

As in the 2×2 case, if the determinant of a 3×3 matrix is zero, the matrix is said to be singular. In this case the volume scale factor is zero, which means that the 3-dimensional lattice has been squashed on to a plane (or in exceptional cases a single line or point). So the vectors $\mathbf{a}$, $\mathbf{b}$ and $\mathbf{c}$ are coplanar, and one of the vectors can be written as a linear combination of the other two.

e.g. $\mathbf{c} = s\mathbf{a} + t\mathbf{b}$

Or equivalently there exist scalars λ, μ, ν (not all zero) such that

$$\lambda\mathbf{a} + \mu\mathbf{b} + \nu\mathbf{c} = 0$$

If this is the case then $\mathbf{a}$, $\mathbf{b}$, $\mathbf{c}$ are said to be *linearly dependent*. Otherwise the vectors are *independent*.

Example 3

What is the geometric effect of the transformation with matrix

$$\begin{pmatrix} 1 & 2 & 1 \\ 1 & 2 & 1 \\ 3 & 1 & 8 \end{pmatrix}?$$

Since $\begin{pmatrix} 1 \\ 1 \\ 8 \end{pmatrix} = 3\begin{pmatrix} 1 \\ 1 \\ 3 \end{pmatrix} - \begin{pmatrix} 2 \\ 2 \\ 1 \end{pmatrix}$

the columns of the matrix are linearly dependent, and the matrix is singular, with zero determinant. Hence the transformation is a 'squash' of the 3-dimensional space on to a plane.

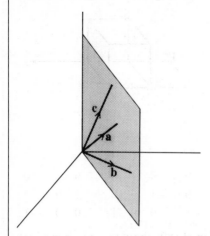

Inspection of the vectors shows the equation of the plane to be $x = y$.

Combined transformations

We may now consider the combined effect of successive 3-dimensional transformations (which are often difficult to draw or to visualise) by simply multiplying their corresponding matrices.

Example 4

What single transformation is equivalent to a rotation of 90° about the z-axis (in the direction which takes $(1, 0, 0)$ to $(0, -1, 0)$), followed by a rotation of 180° about the x-axis? Is the result different if the order of the two transformations is reversed?

Rotation of 90° about the z-axis:

$$\text{Matrix } \mathbf{A} = \begin{pmatrix} 0 & 1 & 0 \\ -1 & 0 & 0 \\ 0 & 0 & 1 \end{pmatrix}$$

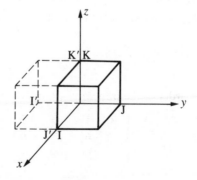

Rotation of 180° about the x-axis:

Matrix $\mathbf{B} = \begin{pmatrix} 1 & 0 & 0 \\ 0 & -1 & 0 \\ 0 & 0 & -1 \end{pmatrix}$

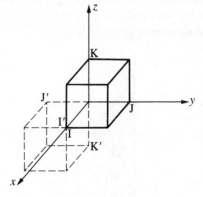

Combined matrix $= \mathbf{BA} = \begin{pmatrix} 1 & 0 & 0 \\ 0 & -1 & 0 \\ 0 & 0 & -1 \end{pmatrix}\begin{pmatrix} 0 & 1 & 0 \\ -1 & 0 & 0 \\ 0 & 0 & 1 \end{pmatrix} = \begin{pmatrix} 0 & 1 & 0 \\ 1 & 0 & 0 \\ 0 & 0 & -1 \end{pmatrix}$

And the matrix $\mathbf{BA}$ represents a rotation of 180° about the axis shown in the diagram, which has equation $\quad x = y, \quad z = 0$.

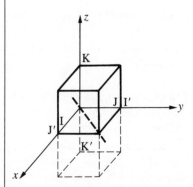

So a rotation of 90° about the z-axis, followed by a rotation of 180° about the x-axis is equivalent to a single rotation of 180° about an axis in the direction of the vector $\begin{pmatrix} 1 \\ 1 \\ 0 \end{pmatrix}$.

If the order of the transformations is reversed,

$$\mathbf{AB} = \begin{pmatrix} 0 & 1 & 0 \\ -1 & 0 & 0 \\ 0 & 0 & 1 \end{pmatrix}\begin{pmatrix} 1 & 0 & 0 \\ 0 & -1 & 0 \\ 0 & 0 & -1 \end{pmatrix} = \begin{pmatrix} 0 & -1 & 0 \\ -1 & 0 & 0 \\ 0 & 0 & -1 \end{pmatrix}$$

This again represents a rotation of 180°, but this time about an axis in the direction of the vector $\begin{pmatrix} 1 \\ -1 \\ 0 \end{pmatrix}$.

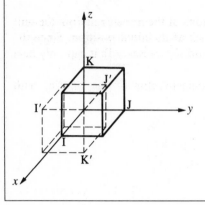

Example 5

By evaluating $\mathbf{A}^3$, describe the transformation represented by the matrix

$$\mathbf{A} = \begin{pmatrix} 0 & 1 & 0 \\ 0 & 0 & -1 \\ -1 & 0 & 0 \end{pmatrix}.$$

Even with a diagram showing the effect of the transformation on the unit square, it is difficult to visualise just what the transformation is.

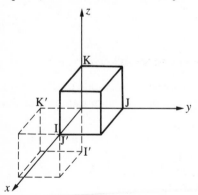

Now $\mathbf{A}^2 = \begin{pmatrix} 0 & 1 & 0 \\ 0 & 0 & -1 \\ -1 & 0 & 0 \end{pmatrix}\begin{pmatrix} 0 & 1 & 0 \\ 0 & 0 & -1 \\ -1 & 0 & 0 \end{pmatrix} = \begin{pmatrix} 0 & 0 & -1 \\ 1 & 0 & 0 \\ 0 & -1 & 0 \end{pmatrix}$

$$\text{so} \quad \mathbf{A}^3 = \begin{pmatrix} 0 & 1 & 0 \\ 0 & 0 & -1 \\ -1 & 0 & 0 \end{pmatrix} \begin{pmatrix} 0 & 0 & -1 \\ 1 & 0 & 0 \\ 0 & -1 & 0 \end{pmatrix} = \begin{pmatrix} 1 & 0 & 0 \\ 0 & 1 & 0 \\ 0 & 0 & 1 \end{pmatrix}$$

which is the 3×3 identity matrix.

This means that after three applications of the transformation, the unit cube (and any other shape) arrives back at its initial position. Since the transformation is an isometry (shape and size preserved), it can only be a rotation of 120°.

The axis of rotation must be an eigenvector of the matrix, with eigenvalue equal to 1:

$$\begin{pmatrix} 0 & 1 & 0 \\ 0 & 0 & -1 \\ -1 & 0 & 0 \end{pmatrix} \begin{pmatrix} x \\ y \\ z \end{pmatrix} = \begin{pmatrix} x \\ y \\ z \end{pmatrix} \quad \Rightarrow \quad \begin{pmatrix} y \\ -z \\ -x \end{pmatrix} = \begin{pmatrix} x \\ y \\ z \end{pmatrix}$$

So this vector is of the form $\begin{pmatrix} k \\ k \\ -k \end{pmatrix}$.

Therefore the transformation is a rotation of 120° about an axis in the direction of the vector $\begin{pmatrix} 1 \\ 1 \\ -1 \end{pmatrix}$.

Exercise 16.3

1 Write down the 3×3 matrices representing the following:
   a) reflection in the plane $y = 0$;
   b) reflection in the plane $y = x$;
   c) reflection in the plane $y = x \tan \alpha$;
   d) reflection in the plane $x + y + z = 0$;
   e) half-turn about the y-axis;
   f) half-turn about the line $x = z$, $y = 0$;
   g) half-turn about the line $x = y = z$;
   h) quarter-turns about the x-axis;
   i) quarter-turns about the line $x = 0 = y + z$;
   j) rotation through θ about the y-axis;
   k) an enlargement, centre the origin, scale factor 5;
   l) an orthogonal projection on to the plane $y = 0$.

2 What transformations are represented by the following matrices?

a) $\begin{pmatrix} 1 & 0 & 1 \\ 0 & -1 & 0 \\ 0 & 0 & 1 \end{pmatrix}$ b) $\begin{pmatrix} 1 & 0 & 0 \\ 0 & 1 & 0 \\ 0 & 0 & 0 \end{pmatrix}$ c) $\begin{pmatrix} 0 & 1 & 0 \\ 0 & 0 & 1 \\ 1 & 0 & 0 \end{pmatrix}$

d) $\begin{pmatrix} 1 & 0 & 0 \\ 0 & 0 & -1 \\ 0 & 1 & 0 \end{pmatrix}$ e) $\begin{pmatrix} 2 & 0 & 0 \\ 0 & 3 & 0 \\ 0 & 0 & -1 \end{pmatrix}$ f) $\begin{pmatrix} -3 & 0 & 0 \\ 0 & -3 & 0 \\ 0 & 0 & -3 \end{pmatrix}$

g) $\begin{pmatrix} 0.8 & -0.6 & 0 \\ 0.6 & 0.8 & 0 \\ 0 & 0 & 1 \end{pmatrix}$ h) $\begin{pmatrix} 1 & 0 & 0 \\ 0 & 0.6 & 0.8 \\ 0 & 0.8 & -0.6 \end{pmatrix}$

3 Consider the effects of the above transformations on volumes and check by evaluating the determinants in each case.

4 Show that the matrix $\begin{pmatrix} 6 & -4 & 3 \\ 29 & -20 & 13 \\ 33 & -23 & 14 \end{pmatrix}$ transforms the vector $\begin{pmatrix} 1 \\ 2 \\ 1 \end{pmatrix}$ into itself.

5 Check that the vectors $\begin{pmatrix} 2 \\ 1 \\ 0 \end{pmatrix}$, $\begin{pmatrix} 1 \\ 0 \\ 1 \end{pmatrix}$, $\begin{pmatrix} 0 \\ -1 \\ 2 \end{pmatrix}$ are coplanar, and investigate the effect of the matrix

$$\begin{pmatrix} 2 & 1 & 0 \\ 1 & 0 & -1 \\ 0 & 1 & 2 \end{pmatrix}$$

6 Find the effect on the cube with vertices $\begin{pmatrix} \pm 1 \\ \pm 1 \\ \pm 1 \end{pmatrix}$ of the matrix

$$\begin{pmatrix} 0 & -1 & 0 \\ 1 & 0 & 0 \\ 0 & 0 & -1 \end{pmatrix}$$

7 Given that

$$p = \begin{pmatrix} 1 \\ 1 \\ 1 \end{pmatrix}, \quad A = \begin{pmatrix} 1 & -1 & 1 \\ -1 & 2 & -1 \\ 2 & -1 & 1 \end{pmatrix}, \quad B = \begin{pmatrix} -1 & 0 & 1 \\ 1 & 1 & 0 \\ 3 & 1 & -1 \end{pmatrix}$$

show that $AB = BA$, and that the column vectors of A are linearly independent.

Express p as a linear combination of the column vectors of A.

Find the point which maps on to the point $(1, 1, 1)$ under the transformation with matrix $\mathbf{A}$, and the point on to which the point $(1, 1, 1)$ is mapped under the transformation with matrix $\mathbf{B}$.   (L)

8 A linear transformation T represented by the equation $\mathbf{y} = \mathbf{A}\mathbf{x}$, where $\mathbf{y}$ and $\mathbf{x}$ are column vectors with 3 elements and $\mathbf{A}$ is a 3×3 matrix, maps $\mathbf{x}_1, \mathbf{x}_2, \mathbf{x}_3$, where

$$\mathbf{x}_1 = \begin{pmatrix} 2 \\ 1 \\ 1 \end{pmatrix}, \quad \mathbf{x}_2 = \begin{pmatrix} 1 \\ -1 \\ 2 \end{pmatrix}, \quad \mathbf{x}_3 = \begin{pmatrix} 1 \\ 0 \\ 2 \end{pmatrix}$$

into $\begin{pmatrix} 3 \\ 6 \\ 1 \end{pmatrix}, \begin{pmatrix} 3 \\ 0 \\ -2 \end{pmatrix}, \begin{pmatrix} 1 \\ 3 \\ 2 \end{pmatrix}$ respectively.

Find the images of the unit vectors $\begin{pmatrix} 1 \\ 0 \\ 0 \end{pmatrix}, \begin{pmatrix} 0 \\ 1 \\ 0 \end{pmatrix}$ and $\begin{pmatrix} 0 \\ 0 \\ 1 \end{pmatrix}$ under the transformation T.

Find also the matrix $\mathbf{A}$.   (L)

9 Describe the transformations represented by the matrices

$$\mathbf{A} = \begin{pmatrix} 0 & 0 & 1 \\ 1 & 0 & 0 \\ 0 & 1 & 0 \end{pmatrix}, \quad \mathbf{B} = \begin{pmatrix} -1 & 0 & 0 \\ 0 & -1 & 0 \\ 0 & 0 & 1 \end{pmatrix}, \quad \mathbf{C} = \begin{pmatrix} 0 & 0 & 1 \\ 0 & 1 & 0 \\ 1 & 0 & 0 \end{pmatrix},$$

$$\mathbf{D} = \begin{pmatrix} 2 & 0 & 0 \\ 0 & 2 & 0 \\ 0 & 0 & 2 \end{pmatrix}$$

Evaluate the products $\mathbf{AB}, \mathbf{BA}, \mathbf{AC}, \mathbf{CB}, \mathbf{A}^2, \mathbf{B}^2, \mathbf{C}^2, \mathbf{D}^2$. Interpret each of these products geometrically.

10 Decompose the following as products of two simpler matrices, representing simple 3-dimensional transformations such as rotations, reflections, enlargements, etc.:

a) $\begin{pmatrix} -1 & 0 & 0 \\ 0 & 0 & 1 \\ 0 & 3 & 0 \end{pmatrix}$ b) $\begin{pmatrix} 0 & 2 & 0 \\ 2 & 0 & 0 \\ 0 & 0 & 2 \end{pmatrix}$ c) $\begin{pmatrix} 0 & 1 & 0 \\ 1 & 0 & 0 \\ 0 & 0 & 0 \end{pmatrix}$

11 By interpreting each of the matrices $\mathbf{M}$ below, geometrically, find a second matrix, $\mathbf{N}$ such that $\mathbf{MN} = \mathbf{I}$:

a) $\begin{pmatrix} 3 & 0 & 0 \\ 0 & 3 & 0 \\ 0 & 0 & 3 \end{pmatrix}$ b) $\begin{pmatrix} 0 & 1 & 0 \\ -1 & 0 & 0 \\ 0 & 0 & 1 \end{pmatrix}$ c) $\begin{pmatrix} 1 & 0 & 0 \\ 0 & 1 & 0 \\ 0 & 0 & -1 \end{pmatrix}$

12 Use geometrical considerations to find 3×3 matrices **A**, **B** such that $\mathbf{A}^4 = \mathbf{I}, \mathbf{B}^8 = \mathbf{I}$.

13 Show that if $\mathbf{M} = \begin{pmatrix} 6 & -4 & 3 \\ 29 & -20 & 13 \\ 33 & -23 & 14 \end{pmatrix}$, then $\mathbf{M}\begin{pmatrix} 1 \\ 2 \\ 1 \end{pmatrix} = \begin{pmatrix} 1 \\ 2 \\ 1 \end{pmatrix}$ and $\mathbf{M}^3 = \mathbf{I}$. What do you suspect?

14 Find vectors ('eigenvectors') whose directions are unaltered by

$$\mathbf{N} = \begin{pmatrix} 1 & -2 & 0 \\ 1 & -1 & 0 \\ -1 & 0 & 1 \end{pmatrix}$$

Show that $\mathbf{N}^4 = \mathbf{I}$, but that **N** does *not* represent a quarter-turn.

15 If a 3×3 matrix has the property $\mathbf{M}^2 = \mathbf{I}$ and $\mathbf{MX} = \mathbf{X}$, show that it may well *not* represent a half-turn about the vector **X**. What conditions are needed in addition to the two given in order that it may represent a half-turn?

16 **M** is a certain 3×3 matrix whose elements are 0, 1 or -1, and each row and each column of **M** contains exactly one non-zero element. Prove that $\mathbf{M}^2, \mathbf{M}^3, \ldots, \mathbf{M}^n$ are all of the same form, and deduce that $\mathbf{M}^h = \mathbf{I}$ for some positive integer $h \leqslant 48$. Interpret the action of **M** on a vector $\begin{pmatrix} x \\ y \\ z \end{pmatrix}$ geometrically.

(MEI)

*16.4 Inverse matrices

Suppose that a linear transformation has a matrix **M**, and transforms any point P to P′. Can we find the inverse transformation, whose effect is to move P′ back to P? Or in matrix notation, if $\mathbf{x}' = \mathbf{Mx}$, can we find a matrix **N** such that $\mathbf{x} = \mathbf{Nx}'$?

If so, it will clearly follow that

$$\mathbf{x} = \mathbf{Nx}' = \mathbf{NMx}$$
and $\quad \mathbf{x}' = \mathbf{Mx} = \mathbf{MNx}'$

So, if these are true for all vectors $\mathbf{x}, \mathbf{x}'$, then $\mathbf{MN} = \mathbf{NM} = \mathbf{I}$ (the identity matrix) and our problem is to find **N** which is the *inverse* of **M** (and which we shall write $\mathbf{M}^{-1}$).

In some cases finding the inverse transformation, and its matrix, is quite simple: the inverse of a stretch is a contraction, and that of a clockwise rotation is an anti-clockwise rotation, whilst reflection is clearly its own inverse. In other cases, however, the answer is not so simple. In example 4 of section 16.2 for instance, we saw that the whole plane was squashed on to the line $y = \frac{1}{2}x$. So it appears that if P′ is on this line there are many points P, from which P′ originated, whilst if P′ is not on the line, there is no point P. It was for this reason that we called it a singular transformation, and we see that such a transformation has no inverse.

We now consider the general procedure for finding inverse matrices and transformations.

Two-dimensional case

For the transformation with matrix $\mathbf{M} = \begin{pmatrix} a & b \\ c & d \end{pmatrix}$,

$$x' = ax + by$$
$$y' = cx + dy$$

$$\Rightarrow \quad \begin{aligned} dx' &= adx + bdy \\ by' &= bcx + bdy \end{aligned} \quad \Rightarrow \quad (ad - bc)x = dx' - by'$$

$$\text{and} \quad \begin{aligned} cx' &= acx + bcy \\ ay' &= acx + ady \end{aligned} \quad \Rightarrow \quad (ad - bc)y = -cx' + ay'$$

The quantity $ad - bc$ is of course the determinant of the matrix, and for convenience we shall refer to it as Δ.

If $\Delta = 0$ clearly x and y cannot in general be evaluated, the transformation is singular, and $\mathbf{M}^{-1}$ does not exist. The fact that such a situation is determined by $\Delta = 0$ is the reason why Δ is called the 'determinant' of the matrix.

We now consider the case where $\Delta \neq 0$. The above equations can then be written

$$x = \frac{d}{\Delta}x' - \frac{b}{\Delta}y'$$

$$y = -\frac{c}{\Delta}x' + \frac{a}{\Delta}y'$$

so that the inverse transformation has matrix

$$\begin{pmatrix} d/\Delta & -b/\Delta \\ -c/\Delta & a/\Delta \end{pmatrix}$$

We readily see that

$$\begin{pmatrix} a & b \\ c & d \end{pmatrix}\begin{pmatrix} d/\Delta & -b/\Delta \\ -c/\Delta & a/\Delta \end{pmatrix} = \begin{pmatrix} (ad - bc)/\Delta & 0 \\ 0 & (ad - bc)/\Delta \end{pmatrix} = \begin{pmatrix} 1 & 0 \\ 0 & 1 \end{pmatrix} = \mathbf{I}$$

and $\begin{pmatrix} d/\Delta & -b/\Delta \\ -c/\Delta & a/\Delta \end{pmatrix} \begin{pmatrix} a & b \\ c & d \end{pmatrix} = \begin{pmatrix} (ad-bc)/\Delta & 0 \\ 0 & (ad-bc)/\Delta \end{pmatrix}$

$$= \begin{pmatrix} 1 & 0 \\ 0 & 1 \end{pmatrix} = I$$

So $\begin{pmatrix} d/\Delta & -b/\Delta \\ -c/\Delta & a/\Delta \end{pmatrix}$ is called the *inverse* of $\begin{pmatrix} a & b \\ c & d \end{pmatrix}$ and we write

$$\mathbf{M}^{-1} = \begin{pmatrix} d/\Delta & -b/\Delta \\ -c/\Delta & a/\Delta \end{pmatrix} = \frac{1}{\Delta} \begin{pmatrix} d & -b \\ -c & a \end{pmatrix}$$

Example 1

For the transformation with matrix $\begin{pmatrix} 3 & 2 \\ 1 & 4 \end{pmatrix}$, find the point from which $(1, 2)$ originates.

We need to find the point (x, y) such that

$$\begin{pmatrix} 3 & 2 \\ 1 & 4 \end{pmatrix} \begin{pmatrix} x \\ y \end{pmatrix} = \begin{pmatrix} 1 \\ 2 \end{pmatrix}$$

The inverse of matrix $\begin{pmatrix} 3 & 2 \\ 1 & 4 \end{pmatrix}$ is $\frac{1}{10} \begin{pmatrix} 4 & -2 \\ -1 & 3 \end{pmatrix}$

so $\begin{pmatrix} x \\ y \end{pmatrix} = \frac{1}{10} \begin{pmatrix} 4 & -2 \\ -1 & 3 \end{pmatrix} \begin{pmatrix} 1 \\ 2 \end{pmatrix} = \frac{1}{10} \begin{pmatrix} 0 \\ 5 \end{pmatrix} = \begin{pmatrix} 0 \\ \frac{1}{2} \end{pmatrix}$

and $(1, 2)$ originates from the point $(0, \frac{1}{2})$.

Example 2

Use an inverse matrix to find the point of intersection of the lines $x - y = 6$ and $2x + 3y = 4$.

$\begin{matrix} x - y = 6 \\ 2x + 3y = 4 \end{matrix}$ can be written $\begin{pmatrix} 1 & -1 \\ 2 & 3 \end{pmatrix} \begin{pmatrix} x \\ y \end{pmatrix} = \begin{pmatrix} 6 \\ 4 \end{pmatrix}$

The inverse of $\begin{pmatrix} 1 & -1 \\ 2 & 3 \end{pmatrix}$ is $\begin{pmatrix} \frac{3}{5} & \frac{1}{5} \\ -\frac{2}{5} & \frac{1}{5} \end{pmatrix}$

So $\begin{pmatrix} x \\ y \end{pmatrix} = \begin{pmatrix} \frac{3}{5} & \frac{1}{5} \\ -\frac{2}{5} & \frac{1}{5} \end{pmatrix} \begin{pmatrix} 6 \\ 4 \end{pmatrix} = \begin{pmatrix} \frac{22}{5} \\ -\frac{8}{5} \end{pmatrix}$

and the required point is $(\frac{22}{5}, -\frac{8}{5})$.

Exercise 16.4a

1 Calculate, if possible, A^{-1} and check that $AA^{-1} = A^{-1}A = I$, when A is

a) $\begin{pmatrix} 2 & 0 \\ 0 & 1 \end{pmatrix}$ b) $\begin{pmatrix} 1 & 2 \\ 0 & 1 \end{pmatrix}$ c) $\begin{pmatrix} -1 & 0 \\ 0 & 1 \end{pmatrix}$ d) $\begin{pmatrix} 0 & 1 \\ -2 & 0 \end{pmatrix}$

e) $\begin{pmatrix} 1 & -1 \\ -1 & 1 \end{pmatrix}$ f) $\begin{pmatrix} 0 & 1 \\ 1 & 0 \end{pmatrix}$ g) $\begin{pmatrix} 2 & 1 \\ 3 & 5 \end{pmatrix}$ h) $\begin{pmatrix} 4 & 0 \\ -2 & 1 \end{pmatrix}$

2 Simplify: **a)** I^{-1} **b)** $(A^{-1})^{-1}$

3 Find the points which are mapped on to **a)** $(1, 0)$ **b)** $(1, 6)$
   c) $(3, -2)$ **d)** $(-4, 1)$ by the transformation with matrix $\begin{pmatrix} 2 & -1 \\ 2 & 4 \end{pmatrix}$.

4 Find the inverse of $\begin{pmatrix} 3 & -2 \\ -2 & 2 \end{pmatrix}$ and hence find the quadrilateral which is mapped by this matrix into the square $\begin{pmatrix} \pm 1 \\ \pm 1 \end{pmatrix}$.

5 For each of the following, write down the inverse transformation. Then write down the matrix for the transformation and the matrix for its inverse; check that the product of the two matrices gives the identity matrix:
   a) rotation of $90°$ about the origin;
   b) reflection in $y = 0$;
   c) enlargement, centre $(0, 0)$, scale factor 3.

6 Calculate the inverse of $\begin{pmatrix} 4 & 5 \\ 2 & 3 \end{pmatrix}$ and use your result to solve the equations

   a) $4x + 5y = 2$ **b)** $4x + 5y = 3$
      $2x + 3y = 4$ $2x + 3y = 1$

   c) $4x + 5y = -1$ **d)** $4x + 5y = -4$
      $2x + 3y = 2$ $2x + 3y = -3$

7 Use inverse matrices where possible to find the points of intersection of the pairs of lines:

   a) $2x + 5y = 7$ **b)** $3x - y = -3$
      $x + 3y = 11$ $x + 2y = 4$

   c) $x + 2y = -6$ **d)** $-7x - 3y = 2$
      $2x + 4y = 5$ $x + y = -3$

8   If $A = \begin{pmatrix} 1 & 3 \\ 2 & 4 \end{pmatrix}$, $B = \begin{pmatrix} 3 & 2 \\ -1 & 0 \end{pmatrix}$ and X is a 2 × 2 matrix,
    solve the equations   **a)** $AX = B$ **b)** $XA = B$

9   If $M = \begin{pmatrix} a & b \\ c & d \end{pmatrix}$, under what conditions does $M = M^{-1}$?
    What types of transformations may such matrices represent?

10 A 2 × 2 matrix M has the property that

$$M\begin{pmatrix} a \\ b \end{pmatrix} = \begin{pmatrix} p \\ q \end{pmatrix} \quad \text{and} \quad M\begin{pmatrix} c \\ d \end{pmatrix} = \begin{pmatrix} r \\ s \end{pmatrix}$$

Prove that, provided a, b, c, d satisfy a certain restriction, then
$M = VU^{-1}$, where $U = \begin{pmatrix} a & c \\ b & d \end{pmatrix}$ and $V = \begin{pmatrix} p & r \\ q & s \end{pmatrix}$
State the restriction on a, b, c, d.   (SMP)

11 Given that $A = \begin{pmatrix} 1 & 2 \\ 1 & 3 \end{pmatrix}$, $B = \begin{pmatrix} 2 & 5 \\ 1 & 2 \end{pmatrix}$, find matrices X, Y such that

    a) $AXB = I$,   **b)** $AY = Y + B$.   (OC)

12 Identify the symmetry transformations represented by the matrices

$$P = \begin{pmatrix} -1 & 0 \\ 0 & 1 \end{pmatrix} \quad \text{and} \quad Q = \begin{pmatrix} -\frac{1}{2} & -\frac{1}{2}\sqrt{3} \\ \frac{1}{2}\sqrt{3} & -\frac{1}{2} \end{pmatrix}$$

Describe geometrically the transformation given by the matrix R where

    $R = Q^{-1}PQ$

Verify your answer by evaluating the components of R and investigating
the eigenvalues and eigenvectors of R.   (SMP)

Three-dimensional case

Let us now consider the transformation

$$\left.\begin{aligned} x' &= a_1 x + b_1 y + c_1 z \\ y' &= a_2 x + b_2 y + c_2 z \\ z' &= a_3 x + b_3 y + c_3 z \end{aligned}\right\} \quad \text{or} \quad \begin{pmatrix} x' \\ y' \\ z' \end{pmatrix} = \begin{pmatrix} a_1 & b_1 & c_1 \\ a_2 & b_2 & c_2 \\ a_3 & b_3 & c_3 \end{pmatrix} \begin{pmatrix} x \\ y \\ z \end{pmatrix}$$

Our problem is to find the vector $\begin{pmatrix} x \\ y \\ z \end{pmatrix}$ from which the vector $\begin{pmatrix} x' \\ y' \\ z' \end{pmatrix}$
originated.

We start by asking the simpler question of where the vector $\begin{pmatrix} 1 \\ 0 \\ 0 \end{pmatrix}$ originated,

i.e., we try to find $\begin{pmatrix} x \\ y \\ z \end{pmatrix}$ such that

$$a_1 x + b_1 y + c_1 z = 1 \tag{1}$$

$$a_2 x + b_2 y + c_2 z = 0 \tag{2}$$

$$a_3 x + b_3 y + c_3 z = 0 \tag{3}$$

Looking at the last two of these equations, and writing them in the form

$$\begin{pmatrix} a_2 \\ b_2 \\ c_2 \end{pmatrix} \cdot \begin{pmatrix} x \\ y \\ z \end{pmatrix} = 0 \quad \text{and} \quad \begin{pmatrix} a_3 \\ b_3 \\ c_3 \end{pmatrix} \cdot \begin{pmatrix} x \\ y \\ z \end{pmatrix} = 0$$

we see that we need to find a vector $\begin{pmatrix} x \\ y \\ z \end{pmatrix}$ which is perpendicular to both $\begin{pmatrix} a_2 \\ b_2 \\ c_2 \end{pmatrix}$

and $\begin{pmatrix} a_3 \\ b_3 \\ c_3 \end{pmatrix}$.

Now we know (see section 11.4) that their *vector product* is precisely such a vector, and this we abbreviate by writing the vector product as:

$$\begin{pmatrix} b_2 c_3 - b_3 c_2 \\ c_2 a_3 - c_3 a_2 \\ a_2 b_3 - a_3 b_2 \end{pmatrix} = \begin{pmatrix} A_1 \\ B_1 \\ C_1 \end{pmatrix}$$

Similarly, we write

$$\begin{pmatrix} b_3 c_1 - b_1 c_3 \\ c_3 a_1 - c_1 a_3 \\ a_3 b_1 - a_1 b_3 \end{pmatrix} = \begin{pmatrix} A_2 \\ B_2 \\ C_2 \end{pmatrix} \quad \text{and} \quad \begin{pmatrix} b_1 c_2 - b_2 c_1 \\ c_1 a_2 - c_2 a_1 \\ a_1 b_2 - a_2 b_1 \end{pmatrix} = \begin{pmatrix} A_3 \\ B_3 \\ C_3 \end{pmatrix}$$

These $A_1, A_2, \ldots, C_3$ are referred to as the *cofactors* of $a_1, a_2, \ldots, c_3$.

We now consider the matrix $\mathbf{M}^*$ formed from these cofactors, with the As forming the first row:

$$\mathbf{M}^* = \begin{pmatrix} A_1 & A_2 & A_3 \\ B_1 & B_2 & B_3 \\ C_1 & C_2 & C_3 \end{pmatrix}$$

This is called the *adjoint matrix*, and we now consider the product

$$\mathbf{MM^*} = \begin{pmatrix} a_1 & b_1 & c_1 \\ a_2 & b_2 & c_2 \\ a_3 & b_3 & c_3 \end{pmatrix} \begin{pmatrix} A_1 & A_2 & A_3 \\ B_1 & B_2 & B_3 \\ C_1 & C_2 & C_3 \end{pmatrix}$$

$$= \begin{pmatrix} a_1A_1 + b_1B_1 + c_1C_1 & a_1A_2 + b_1B_2 + c_1C_2 & a_1A_3 + b_1B_3 + c_1C_3 \\ a_2A_1 + b_2B_1 + c_2C_1 & a_2A_2 + b_2B_2 + c_2C_2 & a_2A_3 + b_2B_3 + c_2C_3 \\ a_3A_1 + b_3B_1 + c_3C_1 & a_3A_2 + b_3B_2 + c_3C_2 & a_3A_3 + b_3B_3 + c_3C_3 \end{pmatrix}$$

Now $\begin{pmatrix} A_1 \\ B_1 \\ C_1 \end{pmatrix}$ is, by definition, perpendicular to $\begin{pmatrix} a_2 \\ b_2 \\ c_2 \end{pmatrix}$ and $\begin{pmatrix} a_3 \\ b_3 \\ c_3 \end{pmatrix}$.

So $a_2A_1 + b_2B_1 + c_2C_1 = 0$ and $a_3A_1 + b_3B_1 + c_3C_1 = 0$ (as may be verified by direct substitution). Indeed, the only terms in $\mathbf{MM^*}$ which do not vanish are those on the leading diagonal:

$$a_1A_1 + b_1B_1 + c_1C_1, \quad a_2A_2 + b_2B_2 + c_2C_2, \quad a_3A_3 + b_3B_3 + c_3C_3$$

But each of these is quickly seen to be equal to the expression

$$\Delta = a_1b_2c_3 - a_1b_3c_2 + b_1c_2a_3 - b_1c_3a_2 + c_1a_2b_3 - c_1a_3b_2$$

which we have already called the *determinant* of $\mathbf{M}$.

So $\mathbf{MM^*} = \begin{pmatrix} \Delta & 0 & 0 \\ 0 & \Delta & 0 \\ 0 & 0 & \Delta \end{pmatrix} = \Delta \mathbf{I}, \quad$ where $\Delta = \det \mathbf{M}$

and it can similarly be proved that

$$\mathbf{M^*M} = \Delta \mathbf{I}$$

Now if $\Delta \neq 0$, it follows that

$$\frac{\mathbf{M^*}}{\Delta} \mathbf{M} = \mathbf{M} \frac{\mathbf{M^*}}{\Delta} = \mathbf{I}$$

so that $\mathbf{M^*}/\Delta$ is an inverse of $\mathbf{M}$ for both pre- and post-multiplication and we can write

$$\mathbf{M^{-1}} = \frac{\mathbf{M^*}}{\Delta} \quad \text{(provided } \Delta \neq 0\text{)}$$

It is immediately clear that $\mathbf{M^{-1}}$ is unique, for if we suppose that $\mathbf{L}$ is another inverse for pre-multiplication (or *left inverse*) such that $\mathbf{LM} = \mathbf{I}$, then $\mathbf{M^{-1}} = \mathbf{IM^{-1}} = (\mathbf{LM})\mathbf{M^{-1}} = \mathbf{L}(\mathbf{MM^{-1}}) = \mathbf{LI} = \mathbf{L}$ and $\mathbf{L}$ is identical with $\mathbf{M^{-1}}$.

Similarly, $\mathbf{M^{-1}}$ is the unique inverse for post-multiplication. If, however, $\Delta = 0$, then $\mathbf{M}$ has no such inverse and is singular.

Example 3

Find the adjoints and the inverses of the matrices:

a) $\mathbf{A} = \begin{pmatrix} 1 & 0 & 0 \\ 0 & 0 & 2 \\ 0 & -1 & 0 \end{pmatrix}$ **b)** $\mathbf{B} = \begin{pmatrix} 0 & 1 & 2 \\ 1 & 2 & 3 \\ 2 & 3 & 4 \end{pmatrix}$

a) Since $\mathbf{A} = \begin{pmatrix} 1 & 0 & 0 \\ 0 & 0 & 2 \\ 0 & -1 & 0 \end{pmatrix}$, we see that $\mathbf{A}^* = \begin{pmatrix} 2 & 0 & 0 \\ 0 & 0 & -2 \\ 0 & 1 & 0 \end{pmatrix}$

$\Rightarrow \quad \mathbf{AA}^* = \begin{pmatrix} 1 & 0 & 0 \\ 0 & 0 & 2 \\ 0 & -1 & 0 \end{pmatrix} \begin{pmatrix} 2 & 0 & 0 \\ 0 & 0 & -2 \\ 0 & 1 & 0 \end{pmatrix} = \begin{pmatrix} 2 & 0 & 0 \\ 0 & 2 & 0 \\ 0 & 0 & 2 \end{pmatrix} = 2\mathbf{I}$

$\Rightarrow \quad \mathbf{A}^{-1} = \tfrac{1}{2}\mathbf{A}^* = \begin{pmatrix} 1 & 0 & 0 \\ 0 & 0 & -1 \\ 0 & \tfrac{1}{2} & 0 \end{pmatrix}$

b) Since $\mathbf{B} = \begin{pmatrix} 0 & 1 & 2 \\ 1 & 2 & 3 \\ 2 & 3 & 4 \end{pmatrix}$, we see that $\mathbf{B}^* = \begin{pmatrix} -1 & 2 & -1 \\ 2 & -4 & 2 \\ -1 & 2 & -1 \end{pmatrix}$

$\Rightarrow \quad \mathbf{BB}^* = \begin{pmatrix} 0 & 1 & 2 \\ 1 & 2 & 3 \\ 2 & 3 & 4 \end{pmatrix} \begin{pmatrix} -1 & 2 & -1 \\ 2 & -4 & 2 \\ -1 & 2 & -1 \end{pmatrix} = \begin{pmatrix} 0 & 0 & 0 \\ 0 & 0 & 0 \\ 0 & 0 & 0 \end{pmatrix} = \mathbf{0}$

Hence $\Delta = 0$, so that $\mathbf{B}$ is singular and does not possess an inverse.

Example 4

If $\mathbf{A}$ and $\mathbf{B}$ are both non-singular matrices, show that:
a) $(\mathbf{A}^{-1})^{-1} = \mathbf{A}$ **b)** $(\mathbf{AB})^{-1} = \mathbf{B}^{-1}\mathbf{A}^{-1}$

a) The inverse of $\mathbf{A}$ is $\mathbf{A}^{-1}$,

so $\quad \mathbf{AA}^{-1} = \mathbf{A}^{-1}\mathbf{A} = \mathbf{I}$

Hence the inverse of $\mathbf{A}^{-1}$ is $\mathbf{A}$

and $\quad (\mathbf{A}^{-1})^{-1} = \mathbf{A}$

b) $\quad (\mathbf{AB})(\mathbf{B}^{-1}\mathbf{A}^{-1}) = \mathbf{A}(\mathbf{BB}^{-1})\mathbf{A}^{-1} = \mathbf{AIA}^{-1} = \mathbf{AA}^{-1} = \mathbf{I}$
and $\quad (\mathbf{B}^{-1}\mathbf{A}^{-1})(\mathbf{AB}) = \mathbf{B}^{-1}(\mathbf{A}^{-1}\mathbf{A})\mathbf{B} = \mathbf{B}^{-1}\mathbf{IB} = \mathbf{B}^{-1}\mathbf{B} = \mathbf{I}$

So the inverse of $\mathbf{AB}$ is $\mathbf{B}^{-1}\mathbf{A}^{-1}$

or $\quad (\mathbf{AB})^{-1} = \mathbf{B}^{-1}\mathbf{A}^{-1}$

Example 5

Given the transformation

$$x' = x + 2y + 3z$$
$$y' = 3x + y + 2z$$
$$z' = 2x + 3y + z$$

find: **a)** its inverse transformation;
b) the point from which $(1, 2, 3)$ arises under the given transformation.

a) The matrix of the original transformation is

$$\mathbf{A} = \begin{pmatrix} 1 & 2 & 3 \\ 3 & 1 & 2 \\ 2 & 3 & 1 \end{pmatrix}$$

and we can find its adjoint matrix

$$\mathbf{A}^* = \begin{pmatrix} -5 & 7 & 1 \\ 1 & -5 & 7 \\ 7 & 1 & -5 \end{pmatrix}$$

Furthermore, it is seen that

$$\mathbf{AA}^* = \begin{pmatrix} 1 & 2 & 3 \\ 3 & 1 & 2 \\ 2 & 3 & 1 \end{pmatrix} \begin{pmatrix} -5 & 7 & 1 \\ 1 & -5 & 7 \\ 7 & 1 & -5 \end{pmatrix} = \begin{pmatrix} 18 & 0 & 0 \\ 0 & 18 & 0 \\ 0 & 0 & 18 \end{pmatrix} = 18\mathbf{I}$$

So the determinant Δ of the transformation is 18, the transformation is non-singular and

$$\mathbf{A}^{-1} = \frac{\mathbf{A}^*}{\Delta} = \tfrac{1}{18} \begin{pmatrix} -5 & 7 & 1 \\ 1 & -5 & 7 \\ 7 & 1 & -5 \end{pmatrix}$$

Hence the inverse transformation is

$$x = \tfrac{1}{18}(-5x' + 7y' + z')$$
$$y = \tfrac{1}{18}(x' - 5y' + 7z')$$
$$z = \tfrac{1}{18}(7x' + y' - 5z')$$

b) Finally, putting $x' = 1, y' = 2, z' = 3$ we see that the point $(1, 2, 3)$ arises from $x = \tfrac{2}{3}, y = \tfrac{2}{3}, z = -\tfrac{1}{3}$, i.e. from the point $(\tfrac{2}{3}, \tfrac{2}{3}, -\tfrac{1}{3})$.

Example 6

Solve the equations
$$\begin{aligned} x + y + 2z &= 0 \\ 2x - y + z &= 3 \\ x - 2y + 3z &= -5 \end{aligned}$$

These equations may be written in matrix form as:

$$\begin{pmatrix} 1 & 1 & 2 \\ 2 & -1 & 1 \\ 1 & -2 & 3 \end{pmatrix} \begin{pmatrix} x \\ y \\ z \end{pmatrix} = \begin{pmatrix} 0 \\ 3 \\ -5 \end{pmatrix}$$

The adjoint of $\begin{pmatrix} 1 & 1 & 2 \\ 2 & -1 & 1 \\ 1 & -2 & 3 \end{pmatrix}$ is $\begin{pmatrix} -1 & -7 & 3 \\ -5 & 1 & 3 \\ -3 & 3 & -3 \end{pmatrix}$

and $\begin{pmatrix} 1 & 1 & 2 \\ 2 & -1 & 1 \\ 1 & -2 & 3 \end{pmatrix} \begin{pmatrix} -1 & -7 & 3 \\ -5 & 1 & 3 \\ -3 & 3 & -3 \end{pmatrix} = \begin{pmatrix} -12 & 0 & 0 \\ 0 & -12 & 0 \\ 0 & 0 & -12 \end{pmatrix}$

So the inverse of $\begin{pmatrix} 1 & 1 & 2 \\ 2 & -1 & 1 \\ 1 & -2 & 3 \end{pmatrix}$ is $-\frac{1}{12} \begin{pmatrix} -1 & -7 & 3 \\ -5 & 1 & 3 \\ -3 & 3 & -3 \end{pmatrix}$

and $\begin{pmatrix} x \\ y \\ z \end{pmatrix} = -\frac{1}{12} \begin{pmatrix} -1 & -7 & 3 \\ -5 & 1 & 3 \\ -3 & 3 & -3 \end{pmatrix} \begin{pmatrix} 0 \\ 3 \\ -5 \end{pmatrix}$

$$= -\frac{1}{12} \begin{pmatrix} -36 \\ -12 \\ 24 \end{pmatrix} = \begin{pmatrix} 3 \\ 1 \\ -2 \end{pmatrix}$$

This procedure for solving equations is, however, rather tedious, and although we shall investigate it further in section 16.6 we shall also discover how more practical methods can be employed.

Calculation of the entries in the adjoint matrix from their basic definition can seem a very daunting prospect. But an alternative approach, illustrated below, may make the working simpler. It is used to obtain the adjoint of the matrix

$$\mathbf{M} = \begin{pmatrix} 1 & 1 & 2 \\ 2 & -1 & 1 \\ 1 & -2 & 3 \end{pmatrix} \text{ from example 6.}$$

Step 1

Write out the elements of the matrix four times in a square pattern:

$$
\begin{array}{ccc|ccc}
1 & 1 & 2 & 1 & 1 & 2 \\
2 & -1 & 1 & 2 & -1 & 1 \\
1 & -2 & 3 & 1 & -2 & 3 \\
\hline
1 & 1 & 2 & 1 & 1 & 2 \\
2 & -1 & 1 & 2 & -1 & 1 \\
1 & -2 & 3 & 1 & -2 & 3
\end{array}
$$

Step 2

Strike out the outer rows and columns to leave a 4 by 4 array:

$$
\begin{array}{cccccc}
1 & 1 & 2 & 1 & 1 & 2 \\
2 & -1 & 1 & 2 & -1 & 1 \\
1 & -2 & 3 & 1 & -2 & 3 \\
1 & 1 & 2 & 1 & 1 & 2 \\
2 & -1 & 1 & 2 & -1 & 1 \\
1 & -2 & 3 & 1 & -2 & 3
\end{array}
$$

Step 3

Form a 3×3 matrix by evaluating all possible 2×2 determinants in the array:

e.g. $\begin{vmatrix} -1 & 1 \\ -2 & 3 \end{vmatrix} = -1, \quad \begin{vmatrix} 1 & 2 \\ 3 & 1 \end{vmatrix} = -5, \quad \begin{vmatrix} 2 & -1 \\ 1 & -2 \end{vmatrix} = -3$ give the top row;

$\begin{vmatrix} -2 & 3 \\ 1 & 2 \end{vmatrix} = -7, \quad \ldots$ etc. give the middle row:

$$
\begin{pmatrix}
-1 & -5 & -3 \\
-7 & 1 & 3 \\
3 & 3 & -3
\end{pmatrix}
$$

Step 4

The transpose of this matrix (obtained by exchanging rows and columns) gives the adjoint matrix:

$$
\mathbf{M}^* = \begin{pmatrix}
-1 & -7 & 3 \\
-5 & 1 & 3 \\
-3 & 3 & -3
\end{pmatrix}
$$

In practice steps 3 and 4 are usually carried out at the same time.

By repeating the same steps with the general matrix $\begin{pmatrix} a_1 & b_1 & c_1 \\ a_2 & b_2 & c_2 \\ a_3 & b_3 & c_3 \end{pmatrix}$, the above procedure can be readily verified.

Exercise 16.4b

In nos **1–8** use the adjoint method to find the inverse of the given matrix.

1 $\begin{pmatrix} 1 & 4 & -3 \\ -2 & -5 & 7 \\ 3 & 3 & -10 \end{pmatrix}$
2 $\begin{pmatrix} 1 & 2 & 3 \\ 0 & 1 & 2 \\ 0 & 0 & 1 \end{pmatrix}$
3 $\begin{pmatrix} -15 & 11 & 7 \\ 11 & -8 & -5 \\ -1 & 2 & 1 \end{pmatrix}$

4 $\begin{pmatrix} 6 & 0 & -1 \\ 3 & -3 & -2 \\ -3 & 6 & 3 \end{pmatrix}$
5 $\begin{pmatrix} 2 & -1 & 0 \\ -1 & 2 & -1 \\ 0 & -1 & 1 \end{pmatrix}$
6 $\begin{pmatrix} 1 & a & b \\ 0 & 1 & c \\ 0 & 0 & 1 \end{pmatrix}$

7 $\begin{pmatrix} 1 & a & a \\ a & 1 & a \\ a & a & 1 \end{pmatrix}$
8 $\begin{pmatrix} 1 & a & b \\ a & 1 & c \\ b & c & 1 \end{pmatrix}$

9 Use geometrical considerations to write down the inverses of the following matrices:

a) $\begin{pmatrix} -1 & 0 & 0 \\ 0 & 1 & 0 \\ 0 & 0 & 1 \end{pmatrix}$
b) $\begin{pmatrix} 1 & 0 & 0 \\ 0 & 0 & -1 \\ 0 & 1 & 0 \end{pmatrix}$
c) $\begin{pmatrix} 1 & 0 & 0 \\ 0 & 2 & 0 \\ 0 & 0 & 1 \end{pmatrix}$

Check your results by the adjoint method.

10 Find the adjoint and inverse of $\mathbf{M} = \begin{pmatrix} 1 & 2 & 0 \\ -1 & 3 & 1 \\ 0 & 4 & 2 \end{pmatrix}$ and use them to find the column vector which, when transformed by $\mathbf{M}$, becomes $\begin{pmatrix} -2 \\ 1 \\ 1 \end{pmatrix}$.

11 Find the adjoint and the inverse of $\begin{pmatrix} 2 & -1 & 4 \\ 4 & 0 & 2 \\ 3 & -2 & 7 \end{pmatrix}$ and hence solve the equations

$$2x - y + 4z = 1$$
$$4x + 2z = -1$$
$$3x - 2y + 7z = -3$$

12 Solve
$$\begin{aligned} a + b &= 3 \\ b + c &= 4 \\ c + a &= 5 \end{aligned}$$
by first inverting the matrix $\begin{pmatrix} 1 & 1 & 0 \\ 0 & 1 & 1 \\ 1 & 0 & 1 \end{pmatrix}$.

13 Determine the value of k for which the matrix $\begin{pmatrix} k & 1 & 2 \\ 0 & 1 & 1 \\ 1 & -1 & 0 \end{pmatrix}$ has no multiplicative inverse.

Given that $k = \frac{3}{2}$, obtain the multiplicative inverse.   (AEB, 1982)

14 Invert the matrix $\mathbf{M} = \begin{pmatrix} 1 & 1 & 0 \\ 5 & 1 & -3 \\ 2 & 7 & 4 \end{pmatrix}$ and also the transposed matrix $\mathbf{M}^{\mathsf{T}} = \begin{pmatrix} 1 & 5 & 2 \\ 1 & 1 & 7 \\ 0 & -3 & 4 \end{pmatrix}$.

Can you draw any general conclusion about the inverse of the transpose of a 3×3 matrix? Deduce that the inverse of a symmetric matrix (i.e., one for which $\mathbf{M}^{\mathsf{T}} = \mathbf{M}$) is itself symmetric.

15 Find the inverse of $\begin{pmatrix} a & b & c \\ c & a & b \\ b & c & a \end{pmatrix}$ and state under what circumstances it is singular.

16 A matrix which satisfies $\mathbf{M}\mathbf{M}^{\mathsf{T}} = \mathbf{M}^{\mathsf{T}}\mathbf{M} = \mathbf{I}$, i.e., whose inverse is its transpose, is called an *orthogonal* matrix. Investigate such matrices **a)** in the 2×2 case,   **b)** in the 3×3 case, and give examples of each.

17 Show that if $\mathbf{M}$ is an orthogonal 3×3 matrix, and $\mathbf{M}\mathbf{X} = \mathbf{X}'$, where $\mathbf{X}, \mathbf{X}'$ are column 3-vectors, then these vectors have the same *length*, i.e.

if $\mathbf{X} = \begin{pmatrix} x \\ y \\ z \end{pmatrix}$ and $\mathbf{X}' = \begin{pmatrix} x' \\ y' \\ z' \end{pmatrix}$, then $x^2 + y^2 + z^2 = x'^2 + y'^2 + z'^2$

18 Find the inverse of $\begin{pmatrix} \cos\alpha\cos\beta & \sin\alpha\cos\beta & -\sin\beta \\ \cos\alpha\sin\beta & \sin\alpha\sin\beta & \cos\beta \\ \sin\alpha & -\cos\alpha & 0 \end{pmatrix}$.

How do you account for the result geometrically?

19 If $\mathbf{A}$ is a matrix of period n (i.e. $\mathbf{A}^r = \mathbf{I}$ when $r = n$, but $\mathbf{A}^r \neq \mathbf{I}$ when $r < n$), show that $\mathbf{B}\mathbf{A}\mathbf{B}^{-1}$ is also of period n, where $\mathbf{B}$ is another square matrix of the same order as $\mathbf{A}$.

* 16.5   Determinants

We have already met instances of both 2×2 and 3×3 determinants, which have occurred as area and volume scale factors in linear transformations, and also in the evaluation of inverse matrices.

Unlike a matrix, a determinant is not just a table of numerical data, but is (quite differently) a shorthand way of writing an algebraic expression. If the elements of the determinant are known, then its numerical value can be calculated.

Although we shall be principally interested in 3×3 determinants, it is interesting first of all to consider 2×2 determinants. This is because the rules for 2×2 determinants, which can be easily demonstrated, are imitated by those for larger determinants.

2×2 determinants

A number of properties can easily be established.

a) Transposing:

$$\begin{vmatrix} a & c \\ b & d \end{vmatrix} = ad - bc = \begin{vmatrix} a & b \\ c & d \end{vmatrix}$$

b)
$$\begin{vmatrix} ka & b \\ kc & d \end{vmatrix} = kad - kbc = k(ad - bc) = k\begin{vmatrix} a & b \\ c & d \end{vmatrix}$$

and
$$\begin{vmatrix} ka & kb \\ c & d \end{vmatrix} = kad - kbc = k\begin{vmatrix} a & b \\ c & d \end{vmatrix}$$

It therefore follows that

$$\begin{vmatrix} ka & kb \\ kc & kd \end{vmatrix} = k^2 \begin{vmatrix} a & b \\ c & d \end{vmatrix}$$

which should be contrasted with the equivalent result for matrices:

$$\begin{pmatrix} ka & kb \\ kc & kd \end{pmatrix} = k\begin{pmatrix} a & b \\ c & d \end{pmatrix}$$

c) Interchanging rows or columns,

$$\begin{vmatrix} b & a \\ d & c \end{vmatrix} = bc - ad = -(ad - bc) = -\begin{vmatrix} a & b \\ c & d \end{vmatrix}$$

d)
$$\begin{vmatrix} a & a \\ c & c \end{vmatrix} = ac - ac = 0$$

and
$$\begin{vmatrix} a & ka \\ c & kc \end{vmatrix} = k \begin{vmatrix} a & a \\ c & c \end{vmatrix} = 0$$

e) Addition of matrices which differ in just one row or column:

$$\begin{vmatrix} a & b \\ c & d \end{vmatrix} + \begin{vmatrix} x & b \\ y & d \end{vmatrix} = (ad - bc) + (xd - by)$$

$$= (a + x)d - b(c + y)$$

$$= \begin{vmatrix} a+x & b \\ c+y & d \end{vmatrix}$$

f) Addition of a scalar multiple of one row (or column) to the other row (or column):

$$\begin{vmatrix} a + kc & b + kd \\ c & d \end{vmatrix} = (a + kc)d - (b + kd)c$$

$$= ad + kcd - bc - kcd$$
$$= ad - bc$$
$$= \begin{vmatrix} a & b \\ c & d \end{vmatrix}$$

Exercise 16.5a

1 Evaluate:

a) $\begin{vmatrix} 3 & -4 \\ -2 & 1 \end{vmatrix}$
b) $\begin{vmatrix} 8 & -3 \\ 2 & 6 \end{vmatrix}$
c $\begin{vmatrix} 8 & 5 \\ 80 & 50 \end{vmatrix}$

d) $\begin{vmatrix} 173 & 163 \\ 94 & 104 \end{vmatrix}$
e) $\begin{vmatrix} 42 & 6 \\ -36 & 48 \end{vmatrix}$
f) $\begin{vmatrix} 13 & 1 \\ 1001 & 77 \end{vmatrix}$

g) $\begin{vmatrix} 300 & -400 \\ -240 & 320 \end{vmatrix}$

2 Show that $\begin{vmatrix} a & b \\ a^2 & b^2 \end{vmatrix}$ has a factor $(a - b)$. Generalise the result.

3 Simplify:

a) $\begin{vmatrix} a+b & a-b \\ a-b & a+b \end{vmatrix}$
b) $\begin{vmatrix} \cos\theta & \sin\theta \\ -\sin\theta & \cos\theta \end{vmatrix}$

c) $\begin{vmatrix} \cos\theta & \sin\theta \\ \sin\theta & \cos\theta \end{vmatrix}$
d) $\begin{vmatrix} \cosh u & \sinh u \\ \sinh u & \cosh u \end{vmatrix}$

e) $\begin{vmatrix} 2t & 1 - t^2 \\ 1 - t^2 & 2t \end{vmatrix}$ f) $\begin{vmatrix} px + qz & rx + sz \\ py + qz & ry + sz \end{vmatrix}$

4 If $\Delta = \begin{vmatrix} a & b \\ c & d \end{vmatrix}$, show that $\begin{vmatrix} a + c & b + d \\ c & d \end{vmatrix}$ and $\begin{vmatrix} a & b + ka \\ c & d + kc \end{vmatrix}$
both have the value Δ, and write down some other determinants which have the same value.

5 Express $\begin{vmatrix} a + x & b + y \\ c + p & d + q \end{vmatrix}$ as the sum of determinants.

6 If $A = \begin{pmatrix} a & b \\ c & d \end{pmatrix}$ and $B = \begin{pmatrix} p & q \\ r & s \end{pmatrix}$ show that $|AB| = |A| \times |B|$.

7 Find the value of λ so that the equations

$$2x + 5y = \lambda x$$
$$3x + 4y = \lambda y$$

may have non-trivial solution.

8 Solve the following equation for x: $\begin{vmatrix} 2 - x & -1 \\ -6 & 3 + x \end{vmatrix} = 0$.

9 Show that, in general, the solutions of the simultaneous equations

$$\begin{aligned} a_1x + b_1y + c_1 &= 0 \\ a_2x + b_2y + c_2 &= 0 \end{aligned} \quad \text{may be written} \quad \frac{x}{\begin{vmatrix} b_1 & c_1 \\ b_2 & c_2 \end{vmatrix}} = \frac{y}{\begin{vmatrix} c_1 & a_1 \\ c_2 & a_2 \end{vmatrix}} = \frac{1}{\begin{vmatrix} a_1 & b_1 \\ a_2 & b_2 \end{vmatrix}}$$

Use this to solve the simultaneous equations $3x - y = -8$, $x + 7y = 1$.

3×3 determinants

The determinant of a 3×3 matrix has been defined (in section 16.3) as

$$\Delta = \begin{vmatrix} a_1 & b_1 & c_1 \\ a_2 & b_2 & c_2 \\ a_3 & b_3 & c_3 \end{vmatrix} = a_1b_2c_3 - a_1b_3c_2 + a_2b_3c_1 - a_2b_1c_3 + a_3b_1c_2 - a_3b_2c_1$$

This expression can be written in a number of different ways, but one of the most useful is

$$\begin{vmatrix} a_1 & b_1 & c_1 \\ a_2 & b_2 & c_2 \\ a_3 & b_3 & c_3 \end{vmatrix} = a_1(b_2c_3 - c_2b_3) - b_1(a_2c_3 - c_2a_3) + c_1(a_2b_3 - b_2a_3)$$

$$= a_1 \begin{vmatrix} b_2 & c_2 \\ b_3 & c_3 \end{vmatrix} - b_1 \begin{vmatrix} a_2 & c_2 \\ a_3 & c_3 \end{vmatrix} + c_1 \begin{vmatrix} a_2 & b_2 \\ a_3 & b_3 \end{vmatrix}$$

Here each element from the top row of the determinant is, in turn, multiplied by the 2 × 2 determinant left when its row and column are deleted, and the resulting products, with alternatively positive, negative, positive signs, are added together.

$$\begin{vmatrix} \textcircled{a_1} & b_1 & c_1 \\ a_2 & b_2 & c_2 \\ a_3 & b_3 & c_3 \end{vmatrix} \quad \begin{vmatrix} a_1 & \textcircled{b_1} & c_1 \\ a_2 & b_2 & c_2 \\ a_3 & b_3 & c_3 \end{vmatrix} \quad \begin{vmatrix} a_1 & b_1 & \textcircled{c_1} \\ a_2 & b_2 & c_2 \\ a_3 & b_3 & c_3 \end{vmatrix}$$

$$+\ ve \qquad\qquad -\ ve \qquad\qquad +\ ve$$

The 2 × 2 determinants obtained in this way are called the *minors* of a_1, b_1, c_1 respectively, and this method of evaluating Δ is usually called 'expansion by the first row'.

In fact the determinant can be expanded by any row:

e.g. $\quad \Delta = -a_2 \begin{vmatrix} b_1 & c_1 \\ b_3 & c_3 \end{vmatrix} + b_2 \begin{vmatrix} a_1 & c_1 \\ a_3 & c_3 \end{vmatrix} - c_2 \begin{vmatrix} a_1 & b_1 \\ a_3 & b_3 \end{vmatrix}$

or by any column

e.g. $\quad \Delta = a_1 \begin{vmatrix} b_2 & c_2 \\ b_3 & c_3 \end{vmatrix} - a_2 \begin{vmatrix} b_1 & c_1 \\ b_3 & c_3 \end{vmatrix} + a_3 \begin{vmatrix} b_1 & c_1 \\ b_2 & c_2 \end{vmatrix}$

where each element in the row or column is multiplied by its minor, appropriate signs being given by the 'chessboard' scheme:

+	−	+
−	+	−
+	−	+

It can now be seen that the cofactor of each element of the matrix $\begin{pmatrix} a_1 & b_1 & c_1 \\ a_2 & b_2 & c_2 \\ a_3 & b_3 & c_3 \end{pmatrix}$, as defined in section 16.4, is in fact just the minor of that element, with the appropriate sign as defined above.

So, for instance, $\quad A_1 = \begin{vmatrix} b_2 & c_2 \\ b_3 & c_3 \end{vmatrix} \quad$ and $\quad B_3 = -\begin{vmatrix} a_1 & c_1 \\ a_2 & c_2 \end{vmatrix}$

Therefore the value of the determinant can be obtained (as we have already seen indirectly in section 16.4) by multiplying the elements of any row or column by their respective cofactors, and adding these products together.

e.g. $\quad a_1 A_1 + b_1 B_1 + c_1 C_1 \quad$ for the first row or $b_1B_1 + b_2B_2 + b_3B_3$ for the second column

These expressions are sometimes referred to as *inner products*.

Example 1

Evaluate $\Delta = \begin{vmatrix} 3 & 2 & 1 \\ 7 & 4 & -2 \\ 1 & -3 & 5 \end{vmatrix}$

Expanding the determinant by the first row,

$$\Delta = 3 \begin{vmatrix} 4 & -2 \\ -3 & 5 \end{vmatrix} - 2 \begin{vmatrix} 7 & -2 \\ 1 & 5 \end{vmatrix} + 1 \begin{vmatrix} 7 & 4 \\ 1 & -3 \end{vmatrix}$$

$$= 3 \times 14 - 2 \times 37 + 1 \times -25$$
$$= 42 - 74 - 25$$
$$= -57$$

Alternatively, we might expand by the first column, to give

$$\Delta = 3 \begin{vmatrix} 4 & -2 \\ -3 & 5 \end{vmatrix} - 7 \begin{vmatrix} 2 & 1 \\ -3 & 5 \end{vmatrix} + 1 \begin{vmatrix} 2 & 1 \\ 4 & -2 \end{vmatrix}$$

$$= 3 \times 14 - 7 \times 13 + 1 \times -8$$
$$= 42 - 91 - 8$$
$$= -57$$

A different, and quite neat, method of evaluating a 3×3 determinant, which follows directly from the original definition of

$$\Delta = a_1b_2c_3 - a_1b_3c_2 + a_2b_3c_1 - a_2b_1c_3 + a_3b_1c_2 - a_3b_2c_1$$

is illustrated below for the determinant of example 1.

Step 1

Set out the determinant, and repeat the first two columns:

$$\begin{vmatrix} 3 & 2 & 1 \\ 7 & 4 & -2 \\ 1 & -3 & 5 \end{vmatrix} \begin{matrix} 3 & 2 \\ 7 & 4 \\ 1 & -3 \end{matrix}$$

Step 2

Evaluate the three diagonal products, working from top left to bottom right. Add these three products.

Step 3

Evaluate the three diagonal products, working from top right to bottom left. Add these three products.

Step 4
Subtract the total for step 3 from that for step 2 to give the determinant.

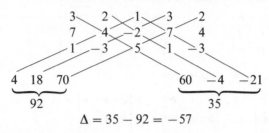

$$\Delta = 35 - 92 = -57$$

Evaluation of determinants of higher order follows the same general pattern as in the 3×3 case. For example, a 4×4 determinant can be expanded by its first row to give it in terms of four 3×3 determinants:

$$\begin{vmatrix} a_1 & b_1 & c_1 & d_1 \\ a_2 & b_2 & c_2 & d_2 \\ a_3 & b_3 & c_3 & d_3 \\ a_4 & b_4 & c_4 & d_4 \end{vmatrix} = a_1 \begin{vmatrix} b_2 & c_2 & d_2 \\ b_3 & c_3 & d_3 \\ b_4 & c_4 & d_4 \end{vmatrix} - b_1 \begin{vmatrix} a_2 & c_2 & d_2 \\ a_3 & c_3 & d_3 \\ a_4 & c_4 & d_4 \end{vmatrix}$$

$$+ c_1 \begin{vmatrix} a_2 & b_2 & d_2 \\ a_3 & b_3 & d_3 \\ a_4 & b_4 & d_4 \end{vmatrix} - d_1 \begin{vmatrix} a_2 & b_2 & c_2 \\ a_3 & b_3 & c_3 \\ a_4 & b_4 & c_4 \end{vmatrix}$$

In this case the signs given to the minors are defined by the table:

+	−	+	−
−	+	−	+
+	−	+	−
−	+	−	+

Exercise 16.5b

1   Evaluate the following determinants and check your answers by expanding them in several different ways:

a) $\begin{vmatrix} 6 & 3 & -7 \\ 5 & 0 & 2 \\ 2 & -1 & 4 \end{vmatrix}$

b) $\begin{vmatrix} 1 & 3 & -1 \\ 2 & 3 & 1 \\ 3 & 3 & 4 \end{vmatrix}$

c) $\begin{vmatrix} -1 & 3 & 1 \\ -6 & 12 & 3 \\ 6 & 1 & 0 \end{vmatrix}$

d) $\begin{vmatrix} 60 & 18 & 49 \\ 50 & 0 & -14 \\ 20 & -6 & -28 \end{vmatrix}$

2 Evaluate the following 4×4 determinants:

a) $\begin{vmatrix} 1 & 0 & 2 & 0 \\ 0 & 3 & 0 & 4 \\ 5 & 0 & 6 & 0 \\ 0 & 7 & 0 & 8 \end{vmatrix}$ **b)** $\begin{vmatrix} 1 & -2 & -3 & 4 \\ -2 & 3 & 4 & -5 \\ 3 & -4 & -5 & 6 \\ -4 & 5 & 6 & -7 \end{vmatrix}$

3 Expand the determinants

$\begin{vmatrix} a & b & c \\ c & a & b \\ b & c & a \end{vmatrix}$, $\begin{vmatrix} a & c & b \\ b & a & c \\ c & b & a \end{vmatrix}$, $\begin{vmatrix} a & b & c \\ b & c & a \\ c & a & b \end{vmatrix}$

and deduce the values of

$\begin{vmatrix} a & b & 0 \\ b & 0 & a \\ 0 & a & b \end{vmatrix}$ and $\begin{vmatrix} 1 & x^2 & x \\ x & 1 & x^2 \\ x^2 & x & 1 \end{vmatrix}$

4 Solve the equation for x:

$\begin{vmatrix} 2-x & -3 & 5 \\ 0 & 5-x & 2 \\ 8 & -6 & 8-x \end{vmatrix} = 0$

5 If $\mathbf{A} = \begin{pmatrix} 2 & -3 & 1 \\ 1 & 4 & 6 \\ 0 & -1 & -1 \end{pmatrix}$, show that $\det \mathbf{A} = 0$,

and that $\det (\mathrm{adj}\, \mathbf{A}) = \begin{vmatrix} 2 & -4 & -22 \\ 1 & -2 & -11 \\ -1 & 2 & 11 \end{vmatrix}$

What do you notice?

6 Expand the determinants:

a) $\begin{vmatrix} a & a & 0 \\ 0 & b & b \\ c & 0 & c \end{vmatrix}$ **b)** $\begin{vmatrix} 1 & 1 & 1 \\ x & y & z \\ y+z & z+x & x+y \end{vmatrix}$ **c)** $\begin{vmatrix} bc^2 + b^2c & bc & 1 \\ ca^2 + c^2a & ca & 1 \\ ab^2 + a^2b & ab & 1 \end{vmatrix}$

Row and column operations

Before dealing with the rules for the manipulation of determinants, we shall introduce a notation which will be found convenient throughout the remainder of this chapter for both matrices and determinants.

We may regard a matrix as a collection of row vectors which we shall call $\mathbf{r}_1$, $\mathbf{r}_2, \mathbf{r}_3$, etc., and similarly the columns will be denoted $\mathbf{c}_1, \mathbf{c}_2, \mathbf{c}_3, \ldots$.

So, in the determinant $\begin{vmatrix} 0 & 3 & -1 \\ 2 & 5 & 1 \\ -1 & 0 & 2 \end{vmatrix}$, $\mathbf{c}_3$ denotes the vector $\begin{pmatrix} -1 \\ 1 \\ 2 \end{pmatrix}$.

Furthermore, we shall use a notation similar to that of transformations to indicate operations of replacing individual rows or columns. For example, $\mathbf{r}_2' = \mathbf{r}_2 - 2\mathbf{r}_1$ simply means 'create a new second row by subtracting from its elements twice the corresponding elements of the first row'; so that the above determinant becomes

$$\begin{vmatrix} 0 & 3 & -1 \\ 2 - 2 \times 0 & 5 - 2 \times 3 & 1 - 2 \times -1 \\ -1 & 0 & 2 \end{vmatrix}, \quad \text{or} \quad \begin{vmatrix} 0 & 3 & -1 \\ 2 & -1 & 3 \\ -1 & 0 & 2 \end{vmatrix}$$

Similarly, the operation $\mathbf{c}_2' = 2\mathbf{c}_2$ has the effect of creating a new second column by doubling the elements of its predecessor.

Rules for manipulation of determinants

The rules for manipulation of 3×3 determinants (and those of higher order) correspond to those for 2×2 determinants, though the proofs are longer and more tedious. We merely state the rules and illustrate them.

a) $\det \mathbf{M} = \det \mathbf{M}^T$, *i.e. the determinant of a matrix is unchanged when the matrix is transposed*

$$\Delta' = \begin{vmatrix} a_1 & a_2 & a_3 \\ b_1 & b_2 & b_3 \\ c_1 & c_2 & c_3 \end{vmatrix} = \Delta \quad \text{(the reader should verify that the expansion of } \Delta' \text{ agrees term by term with that of } \Delta\text{)}$$

b) *Multiple of row (or column)*

e.g. $\Delta' = \begin{vmatrix} ka_1 & b_1 & c_1 \\ ka_2 & b_2 & c_2 \\ ka_3 & b_3 & c_3 \end{vmatrix} = k\Delta; \qquad \Delta' = \begin{vmatrix} a_1 & b_1 & c_1 \\ a_2 & b_2 & c_2 \\ \lambda a_3 & \lambda b_3 & \lambda c_3 \end{vmatrix} = \lambda\Delta, \quad \text{etc.}$

c) *When two rows (or columns) are interchanged, the determinant changes sign*

e.g. $\Delta' = \begin{vmatrix} c_1 & b_1 & a_1 \\ c_2 & b_2 & a_2 \\ c_3 & b_3 & a_3 \end{vmatrix} = -\Delta$, as may be verified by expansion.

d) i) *If two rows (or columns) are identical, the determinant has the value zero;* for in this case, from **c)**, $\Delta = -\Delta \Rightarrow \Delta = 0$

ii) *If two rows (or columns) are proportional, then the determinant is zero*

The two proportional rows (or columns) are, in fact, linearly dependent (see section 16.3) and the determinant is also zero if its three rows (or columns) are linearly dependent, but this cannot usually be spotted at a glance.

e) *Addition of determinants which differ in just one row (or column),*

e.g.
$$\begin{vmatrix} a_1 & b_1 & c_1 \\ a_2 & b_2 & c_2 \\ a_3 & b_3 & c_3 \end{vmatrix} + \begin{vmatrix} x & y & z \\ a_2 & b_2 & c_2 \\ a_3 & b_3 & c_3 \end{vmatrix} = \begin{vmatrix} a_1 + x & b_1 + y & c_1 + z \\ a_2 & b_2 & c_2 \\ a_3 & b_3 & c_3 \end{vmatrix}$$

which is easily verified through expansion by the first row.

f) *A determinant is unchanged when to the elements of any row (or column) are added a constant multiple of the elements of any other row (or column)*

For example, if $\Delta = \begin{vmatrix} a_1 & b_1 & c_1 \\ a_2 & b_2 & c_2 \\ a_3 & b_3 & c_3 \end{vmatrix}$, let $\mathbf{r}_1' = \mathbf{r}_1 + \lambda \mathbf{r}_2$.

Then $\Delta' = \begin{vmatrix} a_1 + \lambda a_2 & b_1 + \lambda b_2 & c_1 + \lambda c_2 \\ a_2 & b_2 & c_2 \\ a_3 & b_3 & c_3 \end{vmatrix}$

$$= \begin{vmatrix} a_1 & b_1 & c_1 \\ a_2 & b_2 & c_2 \\ a_3 & b_3 & c_3 \end{vmatrix} + \lambda \begin{vmatrix} a_2 & b_2 & c_2 \\ a_2 & b_2 & c_2 \\ a_3 & b_3 & c_3 \end{vmatrix}$$

$$= \Delta + \lambda \times 0$$

$$= \Delta$$

g) *For any two 3 × 3 matrices $\mathbf{A}$ and $\mathbf{B}$, $\det(\mathbf{AB}) = \det \mathbf{A} \times \det \mathbf{B}$*
This follows from the interpretation of the determinants as volume scale factors for successive transformations.

Similarly, $\det(\mathbf{BA}) = \det \mathbf{B} \times \det \mathbf{A}$, so $\mathbf{AB}$ and $\mathbf{BA}$ have the same determinant, even though the matrices themselves may well be different.

Example 2

Evaluate $\Delta = \begin{vmatrix} 2 & 4 & 18 \\ -3 & 10 & -27 \\ 1 & 3 & 11 \end{vmatrix}$.

We try to use row and column operations to obtain as many zeros as possible:

$$\Delta = 2 \begin{vmatrix} 1 & 2 & 9 \\ -3 & 10 & -27 \\ 1 & 3 & 11 \end{vmatrix}$$

$$= 2 \begin{vmatrix} 1 & 2 & 9 \\ -3 & 10 & -27 \\ 0 & 1 & 2 \end{vmatrix} \quad (\mathbf{r}'_3 = \mathbf{r}_3 - \mathbf{r}_1)$$

$$= 2 \begin{vmatrix} 1 & 2 & 5 \\ -3 & 10 & -47 \\ 0 & 1 & 0 \end{vmatrix} \quad (\mathbf{c}'_3 = \mathbf{c}_3 - 2\mathbf{c}_2)$$

$$= -2 \begin{vmatrix} 1 & 5 \\ -3 & -47 \end{vmatrix} \quad \text{(expanding by the third row)}$$

$$= -2(-47 + 15)$$

$$= 64$$

Alternatively,

$$\Delta = 2 \begin{vmatrix} 1 & 2 & 9 \\ -3 & 10 & -27 \\ 1 & 3 & 11 \end{vmatrix} = 2 \begin{vmatrix} 1 & 2 & 9 \\ 0 & 16 & 0 \\ 1 & 3 & 11 \end{vmatrix} \quad (\mathbf{r}'_2 = \mathbf{r}_2 + 3\mathbf{r}_1)$$

$$= 32 \begin{vmatrix} 1 & 9 \\ 1 & 11 \end{vmatrix} = 32 \times 2 = 64$$

Example 3

Factorise $\begin{vmatrix} 1 & a & a^3 \\ 1 & b & b^3 \\ 1 & c & c^3 \end{vmatrix}$

Putting $b = c$ gives two identical rows, thus causing the determinant to vanish. Therefore, by the remainder theorem, $(b - c)$ is a factor. Similarly $(c - a)$ and $(a - b)$ are also factors.

Hence $\Delta = (b - c)(c - a)(a - b)F$, and consideration of the degree of the determinant enables us to see that the remaining factor F must have degree 1. Clearly it is symmetric in a, b and c, so must have the form $k(a + b + c)$, where k is a constant.

The 'leading term' $1bc^3$ (the product of the elements in the leading diagonal) has coefficient 1, and this shows us that $k = 1$. Hence

$$\Delta = (b - c)(c - a)(a - b)(a + b + c)$$

Exercise 16.5c

1 Perform the operations specified on the given determinants, and hence evaluate:

a) $\begin{vmatrix} 4 & -18 & 3 \\ 2 & 12 & -4 \\ -4 & 0 & 5 \end{vmatrix}$ $\begin{aligned} \mathbf{r}_1' &= \mathbf{r}_1 + \mathbf{r}_3 \\ \mathbf{r}_3' &= \mathbf{r}_3 + 2\mathbf{r}_2 \end{aligned}$

b) $\begin{vmatrix} 1 & a & b \\ 1 & a^2 & b^2 \\ 1 & a^3 & b^3 \end{vmatrix}$ $\begin{aligned} \mathbf{c}_2' &= \mathbf{c}_2 - \mathbf{c}_3 \\ \mathbf{r}_1'' &= \mathbf{r}_1' - \mathbf{r}_2' \end{aligned}$

c) $\begin{vmatrix} a-b-c & 2a & 2a \\ 2b & b-c-a & 2b \\ 2c & 2c & c-a-b \end{vmatrix}$ $\begin{aligned} \mathbf{c}_2' &= \mathbf{c}_2 - \mathbf{c}_3 \\ \mathbf{r}_2'' &= \mathbf{r}_2' + \mathbf{r}_3' \end{aligned}$

d) $\begin{vmatrix} a & b & c \\ b & c & a \\ c & a & b \end{vmatrix}$ $\mathbf{c}_1' = \mathbf{c}_1 + \omega\mathbf{c}_2 + \omega^2\mathbf{c}_3$, where $\omega^3 = 1$

2 Factorise $\begin{vmatrix} a & b & c \\ c & a & b \\ b & c & a \end{vmatrix}$

3 Use the result of no. 2 above to write down the factors of

$\begin{vmatrix} a & -1 & 0 \\ 0 & a & -1 \\ -1 & 0 & a \end{vmatrix}$ and of $\begin{vmatrix} 1 & a & a^2 \\ a^2 & 1 & a \\ a & a^2 & 1 \end{vmatrix}$

What is the value of the latter determinant when a is a complex cube root of unity?

4 Factorise the following determinants:

a) $\begin{vmatrix} 1 & bc & a \\ 1 & ca & b \\ 1 & ab & c \end{vmatrix}$ b) $\begin{vmatrix} b+c & c+a & a+b \\ bc & ca & ab \\ a & b & c \end{vmatrix}$

c) $\begin{vmatrix} b+c & c+a & a+b \\ c & a & b \\ 1 & 1 & 1 \end{vmatrix}$ d) $\begin{vmatrix} a & bc & a^2 \\ b & ca & b^2 \\ c & ab & c^2 \end{vmatrix}$

e) $\begin{vmatrix} 1 & a & b+c \\ 1 & b & c+a \\ 1 & c & a+b \end{vmatrix}$

5 Show that:

$$\begin{vmatrix} a & b & c \\ a^2 & b^2 & c^2 \\ b+c & c+a & a+b \end{vmatrix} = k(a+b+c)(b-c)(c-a)(a-b)$$

and find the value of the constant k.

6 Evaluate $\begin{vmatrix} b & a & a \\ a & b & a \\ a & a & b \end{vmatrix}$, and find other possible values of a 3×3

determinant which contains six a's and three b's.

7 Show that $x + a + y$ is a factor of the determinant

$$\begin{vmatrix} a & x & y \\ x & a & y \\ x & y & a \end{vmatrix}$$

Express the determinant as a product of three factors.

Hence find all the values of θ in the range $0 \leqslant \theta \leqslant \pi$ which satisfy the equation

$$\begin{vmatrix} 1 & \cos\theta & \cos 2\theta \\ \cos\theta & 1 & \cos 2\theta \\ \cos\theta & \cos 2\theta & 1 \end{vmatrix} = 0$$

(JMB)

8 Show that

$$\begin{vmatrix} 1 & a & a^2 \\ a^2 & 1 & a \\ a & a^2 & 1 \end{vmatrix} = (1 - a^3)^2$$

Hence
a) show that

$$\begin{vmatrix} 1 & -i & -1 \\ -1 & 1 & -i \\ -i & -1 & 1 \end{vmatrix} = ki$$

where k is real,

b) given that ω is a root of the equation $x^3 = 1$, find the numerical value of

$$\begin{vmatrix} 1 & 2\omega & 4\omega^2 \\ 4\omega^2 & 1 & 2\omega \\ 2\omega & 4\omega^2 & 1 \end{vmatrix}$$

(JMB)

9 Find a counter-example to show that, if $\mathbf{X}, \mathbf{Y}, \mathbf{Z}$ are square matrices, such that $\mathbf{Z} = \mathbf{X} + \mathbf{Y}$, then $\det \mathbf{X} + \det \mathbf{Y} \neq \det \mathbf{Z}$.

10 If $D = \begin{vmatrix} a & c & 0 \\ b & 0 & c \\ 0 & a & b \end{vmatrix}$, express D^2 as a 3×3 determinant in two different ways.

11 Express $\begin{vmatrix} a^2 + p^2 & ab + pq & ac + pr \\ ab + pq & b^2 + q^2 & bc + qr \\ ac + pr & bc + qr & c^2 + r^2 \end{vmatrix}$ as the square of a determinant, and hence prove that it is zero.

12 Using the notation $|\mathbf{a} \quad \mathbf{b} \quad \mathbf{c}|$ to denote $\begin{vmatrix} a_1 & b_1 & c_1 \\ a_2 & b_2 & c_2 \\ a_3 & b_3 & c_3 \end{vmatrix}$, where $\mathbf{a}, \mathbf{b}, \mathbf{c}$ are column vectors, prove that

$$|\mathbf{a} + \mathbf{x} \quad \mathbf{b} \quad \mathbf{c}| = |\mathbf{a} \quad \mathbf{b} \quad \mathbf{c}| + |\mathbf{x} \quad \mathbf{b} \quad \mathbf{c}|$$

and that $|\mathbf{a} + \mathbf{x} + \mathbf{y} \quad \mathbf{b} \quad \mathbf{c}| = |\mathbf{a} \quad \mathbf{b} \quad \mathbf{c}| + |\mathbf{x} \quad \mathbf{b} \quad \mathbf{c}| + |\mathbf{y} \quad \mathbf{b} \quad \mathbf{c}|$

and express $|\mathbf{a} + \mathbf{x} \quad \mathbf{b} + \mathbf{y} \quad \mathbf{c} + \mathbf{z}|$ as the sum of eight determinants.

13 Express in factorised form the determinants:

a) $\begin{vmatrix} b^2 + c^2 & ab & ac \\ ab & c^2 + a^2 & bc \\ ca & cb & a^2 + b^2 \end{vmatrix}$

b) $\begin{vmatrix} 1 & a & a & a \\ 1 & b & a & a \\ 1 & a & b & a \\ 1 & a & a & b \end{vmatrix}$

c) $\begin{vmatrix} 1 & 1 & 1 & 1 \\ 1 & -1 & 1 & 1 \\ 1 & -1 & -1 & 1 \\ 1 & -1 & -1 & -1 \end{vmatrix}$

d) $\begin{vmatrix} a & b & c & d \\ b & a & d & c \\ c & d & a & b \\ d & c & b & a \end{vmatrix}$

e) $\begin{vmatrix} a & b & c & d \\ b & a & c & d \\ a & b & d & c \\ b & a & d & c \end{vmatrix}$

14 Evaluate all the cofactors in the determinant $\begin{vmatrix} 1 & 2 & 4 \\ 2 & 4 & 8 \\ 4 & 8 & 16 \end{vmatrix}$.

Write down several other examples of determinants which have all the 2×2 minors equal to zero.

Compose a 4×4 determinant which has every one of its 3×3 minors zero.

* 16.6 Systems of linear equations

In general, a set of three simultaneous linear equations in three unknowns can be written in the form

$$a_1 x + b_1 y + c_1 z = d_1$$
$$a_2 x + b_2 y + c_2 z = d_2$$
$$a_3 x + b_3 y + c_3 z = d_3$$

Each of these can be considered as the equation of a plane in 3-dimensional space, and any set of values (x, y, z) which satisfies all three equations will be a common point, or point of intersection of the three planes.

Alternatively the set of equations can be written in matrix form:

$$\begin{pmatrix} a_1 & b_1 & c_1 \\ a_2 & b_2 & c_2 \\ a_3 & b_3 & c_3 \end{pmatrix} \begin{pmatrix} x \\ y \\ z \end{pmatrix} = \begin{pmatrix} d_1 \\ d_2 \\ d_3 \end{pmatrix} \quad \text{or} \quad \mathbf{MX} = \mathbf{d}$$

In this case the solution $\begin{pmatrix} x \\ y \\ z \end{pmatrix}$ can be interpreted as the vector which is mapped on to $\mathbf{d}$ by the transformation with matrix $\mathbf{M}$.

Yet a third way of representing the equations is as

$$x \begin{pmatrix} a_1 \\ a_2 \\ a_3 \end{pmatrix} + y \begin{pmatrix} b_1 \\ b_2 \\ b_3 \end{pmatrix} + z \begin{pmatrix} c_1 \\ c_2 \\ c_3 \end{pmatrix} = \begin{pmatrix} d_1 \\ d_2 \\ d_3 \end{pmatrix} \quad \text{or} \quad x\mathbf{a} + y\mathbf{b} + z\mathbf{c} = \mathbf{d}$$

Here x, y, and z give the multiples of $\mathbf{a}$, $\mathbf{b}$ and $\mathbf{c}$ which must be combined to produce $\mathbf{d}$.

We now consider each of these three interpretations of the system of equations.

Intersection of planes

Normally three planes will meet in just a single point, and the coordinates of this point represent a unique solution of the equations.

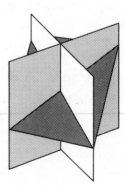

Alternatively, the three planes may have a common line, like pages meeting at the spine of a book. In this case the equations have an infinite number of solutions.

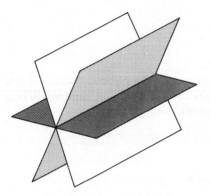

The third possibility is that the three planes will have no points in common. This can arise in three different ways:

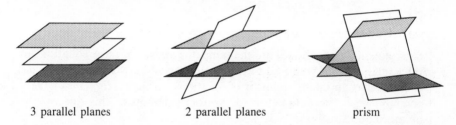

3 parallel planes   2 parallel planes   prism

In each of these cases the system of equations has no solution. The parallel planes can be easily identified from their equations, (e.g. $x + 2y + 3z = 1$, $x + 2y + 3z = 5$, $2x + 4y + 6z = 11$), and so the first two configurations are of little concern. The prism however is not immediately obvious from the form of the equations.

The simplest systematic method of solving a set of three equations is illustrated in the following examples.

Example 1

Solve the equations

$$x + 2y + z = 5 \tag{1}$$
$$3x - y - z = 2 \tag{2}$$
$$x + 5y + 2z = 7 \tag{3}$$

Combine the equations in pairs, to eliminate one of the variables. It does not usually matter which variable is chosen and here we will eliminate z.

$$(1) + (2): \quad 4x + y = 7 \quad\quad\quad\quad\quad\quad (4)$$
$$2(1) - (3): \quad x - y = 3 \quad\quad\quad\quad\quad\quad (5)$$

So the system has been reduced from three equations in three unknowns to two equations in two unknowns.

Now eliminate y:

$$(4) + (5): \quad\quad\quad\quad\quad 5x = 10 \quad \Rightarrow \quad x = 2$$
Substituting in (5): $\quad\quad 2 - y = \;\; 3 \quad \Rightarrow \quad y = -1$
Substituting in (1): $\quad 2 - 2 + z = \;\; 5 \quad \Rightarrow \quad z = 5$

The equations therefore have a unique solution, and the point $(2, -1, 5)$ is the point of intersection of the three planes.

Example 2

Solve the equations $\quad 3x + \;\; y + z = 5 \quad\quad\quad\quad\quad (1)$
$$x + 2y + z = 1 \quad\quad\quad\quad\quad (2)$$
$$x - 3y - z = 3 \quad\quad\quad\quad\quad (3)$$

$(1) - (2): \quad 2x - \;\; y = 4 \quad\quad\quad\quad\quad\quad\quad\quad\quad (4)$
$(1) + (3): \quad 4x - 2y = 8 \quad \Rightarrow \quad 2x - y = 4 \quad\quad (5)$

Since equations (4) and (5) are the same, the system of three equations in three unknowns has reduced to just one equation in two unknowns. This equation has an infinite number of solutions, so the three planes have a common line. The parametric (or vector) equation of this line can be obtained quite easily:

Let $\quad\quad\quad\quad\quad\quad\quad\quad\quad\quad x = \lambda$
therefore from (4) $\quad\quad\quad\quad\quad\quad y = -4 + 2\lambda$
and substituting in (1) $\quad 3\lambda - 4 + 2\lambda + z = 5$
$\Rightarrow \quad\quad\quad\quad\quad\quad\quad\quad\quad\quad z = 9 - 5\lambda$

Hence $\begin{pmatrix} x \\ y \\ z \end{pmatrix} = \begin{pmatrix} \lambda \\ -4 + 2\lambda \\ 9 - 5\lambda \end{pmatrix} = \begin{pmatrix} 0 \\ -4 \\ 9 \end{pmatrix} + \lambda \begin{pmatrix} 1 \\ 2 \\ -5 \end{pmatrix}$

and this represents the parametric equation of the common line (or axis) of the three planes.

Example 3

Solve the equations

$$x + 3y + z = 2 \tag{1}$$
$$2x + y - z = 1 \tag{2}$$
$$x - 2y - 2z = 4 \tag{3}$$

(1) + (2): $\qquad\qquad 3x + 4y = 3 \tag{4}$

2(1) + (3): $\qquad\qquad 3x + 4y = 8 \tag{5}$

In this case equations (4) and (5) contradict one another, so there is no solution to the system of equations. By inspection, the planes represented by the three equations are not parallel, and must therefore form a prism.

Matrix form

In example 1 the system of equations can be written as

$$\begin{pmatrix} 1 & 2 & 1 \\ 3 & -1 & -1 \\ 1 & 5 & 2 \end{pmatrix} \begin{pmatrix} x \\ y \\ z \end{pmatrix} = \begin{pmatrix} 5 \\ 2 \\ 7 \end{pmatrix}$$

The inverse of the matrix $\begin{pmatrix} 1 & 2 & 1 \\ 3 & -1 & -1 \\ 1 & 5 & 2 \end{pmatrix}$ is $\dfrac{1}{5} \begin{pmatrix} 3 & 1 & -1 \\ -7 & 1 & 4 \\ 16 & -3 & -7 \end{pmatrix}$

So

$$\begin{pmatrix} x \\ y \\ z \end{pmatrix} = \frac{1}{5} \begin{pmatrix} 3 & 1 & -1 \\ -7 & 1 & 4 \\ 16 & -3 & -7 \end{pmatrix} \begin{pmatrix} 5 \\ 2 \\ 7 \end{pmatrix} = \frac{1}{5} \begin{pmatrix} 10 \\ -5 \\ 25 \end{pmatrix} = \begin{pmatrix} 2 \\ -1 \\ 5 \end{pmatrix}$$

Therefore the vector $\begin{pmatrix} 2 \\ -1 \\ 5 \end{pmatrix}$ is mapped on to $\begin{pmatrix} 5 \\ 2 \\ 7 \end{pmatrix}$ by the transformation

with matrix $\begin{pmatrix} 1 & 2 & 1 \\ 3 & -1 & -1 \\ 1 & 5 & 2 \end{pmatrix}$.

In examples 2 and 3, with matrix equations

$$\begin{pmatrix} 3 & 1 & 1 \\ 1 & 2 & 1 \\ 1 & -3 & -1 \end{pmatrix} \begin{pmatrix} x \\ y \\ z \end{pmatrix} = \begin{pmatrix} 5 \\ 1 \\ 3 \end{pmatrix} \quad \text{and} \quad \begin{pmatrix} 1 & 3 & 1 \\ 2 & 1 & -1 \\ 1 & -2 & -2 \end{pmatrix} \begin{pmatrix} x \\ y \\ z \end{pmatrix} = \begin{pmatrix} 2 \\ 1 \\ 4 \end{pmatrix}$$

respectively, the determinant of each matrix is zero. Therefore in neither case is there a unique solution.

In example 2

$$\begin{pmatrix} 3 & 1 & 1 \\ 1 & 2 & 1 \\ 1 & -3 & -1 \end{pmatrix}\begin{pmatrix} \lambda \\ -4+2\lambda \\ 9-5\lambda \end{pmatrix} = \begin{pmatrix} 3 & 1 & 1 \\ 1 & 2 & 1 \\ 1 & -3 & -1 \end{pmatrix}\begin{pmatrix} 0 \\ -4 \\ 9 \end{pmatrix} + \lambda\begin{pmatrix} 3 & 1 & 1 \\ 1 & 2 & 1 \\ 1 & -3 & -1 \end{pmatrix}\begin{pmatrix} 1 \\ 2 \\ -5 \end{pmatrix}$$

$$= \begin{pmatrix} 5 \\ 1 \\ 3 \end{pmatrix} + \begin{pmatrix} 0 \\ 0 \\ 0 \end{pmatrix} = \begin{pmatrix} 5 \\ 1 \\ 3 \end{pmatrix}$$

So all the points on the line $\begin{pmatrix} x \\ y \\ z \end{pmatrix} = \begin{pmatrix} 0 \\ -4 \\ 9 \end{pmatrix} + \lambda\begin{pmatrix} 1 \\ 2 \\ -5 \end{pmatrix}$ are mapped

on to $\begin{pmatrix} 5 \\ 1 \\ 3 \end{pmatrix}$. The transformation represented by the matrix is a 'squash' of

the 3-dimensional space on to the plane $-x + 2y + z = 0$, and the image

vector $\begin{pmatrix} 5 \\ 1 \\ 3 \end{pmatrix}$ lies in this plane.

In example 3 there is no vector which can be mapped on to $\begin{pmatrix} 2 \\ 1 \\ 4 \end{pmatrix}$. This is

because the matrix $\begin{pmatrix} 1 & 3 & 1 \\ 2 & 1 & -1 \\ 1 & -2 & -2 \end{pmatrix}$ represents a 'squash' of the 3-

dimensional space on to the plane $x - y + z = 0$. However, the image

vector $\begin{pmatrix} 2 \\ 1 \\ 4 \end{pmatrix}$ does not lie in this plane, so no vector can be mapped on to it.

Combination of vectors

Interpreting the equations of example 1 in vector form:

$$x\begin{pmatrix} 1 \\ 3 \\ 1 \end{pmatrix} + y\begin{pmatrix} 2 \\ -1 \\ 5 \end{pmatrix} + z\begin{pmatrix} 1 \\ -1 \\ 2 \end{pmatrix} = \begin{pmatrix} 5 \\ 2 \\ 7 \end{pmatrix}$$

it can be shown that the vectors $\begin{pmatrix} 1 \\ 3 \\ 1 \end{pmatrix}, \begin{pmatrix} 2 \\ -1 \\ 5 \end{pmatrix}, \begin{pmatrix} 1 \\ -1 \\ 2 \end{pmatrix}$ are independent.

This means geometrically that they do not all lie in the same plane, and that any 3-dimensional vector can be made up from a linear combination of them. In particular

$$\begin{pmatrix} 5 \\ 2 \\ 7 \end{pmatrix} = 2\begin{pmatrix} 1 \\ 3 \\ 1 \end{pmatrix} - 1\begin{pmatrix} 2 \\ -1 \\ 5 \end{pmatrix} + 5\begin{pmatrix} 1 \\ -1 \\ 2 \end{pmatrix}$$

In example 2, where

$$x\begin{pmatrix} 3 \\ 1 \\ 1 \end{pmatrix} + y\begin{pmatrix} 1 \\ 2 \\ -3 \end{pmatrix} + z\begin{pmatrix} 1 \\ 1 \\ -1 \end{pmatrix} = \begin{pmatrix} 5 \\ 1 \\ 3 \end{pmatrix}$$

the vectors $\begin{pmatrix} 3 \\ 1 \\ 1 \end{pmatrix}, \begin{pmatrix} 1 \\ 2 \\ -3 \end{pmatrix}, \begin{pmatrix} 1 \\ 1 \\ -1 \end{pmatrix}$ are linearly dependent since

$$\begin{pmatrix} 3 \\ 1 \\ 1 \end{pmatrix} + 2\begin{pmatrix} 1 \\ 2 \\ -3 \end{pmatrix} - 5\begin{pmatrix} 1 \\ 1 \\ -1 \end{pmatrix} = \begin{pmatrix} 0 \\ 0 \\ 0 \end{pmatrix}$$

Therefore the three vectors are coplanar. But since the vector $\begin{pmatrix} 5 \\ 1 \\ 3 \end{pmatrix}$ lies in the same plane, it can be made up as a combination of these three vectors in an infinite number of ways.

Similarly in example 3, the vectors $\begin{pmatrix} 1 \\ 2 \\ 1 \end{pmatrix}, \begin{pmatrix} 3 \\ 1 \\ -2 \end{pmatrix}, \begin{pmatrix} 1 \\ -1 \\ -2 \end{pmatrix}$ are dependent, and therefore coplanar. But in this case the vector $\begin{pmatrix} 2 \\ 1 \\ 4 \end{pmatrix}$ does not lie in the plane so defined. Therefore no possible combination of the vectors can produce $\begin{pmatrix} 2 \\ 1 \\ 4 \end{pmatrix}$, and there is no solution.

* Cramer's rule for solving equations

An alternative method for solving the equations

$$a_1 x + b_1 y + c_1 z = d_1$$
$$a_2 x + b_2 y + c_2 z = d_2$$
$$a_3 x + b_3 y + c_3 z = d_3$$

uses the properties of cofactors developed in sections 16.4 and 16.5.

Writing the equations in matrix form, we can multiply by the adjoint matrix **M***.

So $\quad$ **Mx** = **d**

$\Rightarrow \quad$ **M*Mx** = **M*d**

$\Rightarrow \quad\quad$ Δ**Ix** = **M*d**

$\Rightarrow \quad\quad$ Δ**x** = **M*d**

$\Rightarrow \quad \begin{pmatrix} \Delta x \\ \Delta y \\ \Delta z \end{pmatrix} = \begin{pmatrix} A_1 & A_2 & A_3 \\ B_1 & B_2 & B_3 \\ C_1 & C_2 & C_3 \end{pmatrix} \begin{pmatrix} d_1 \\ d_2 \\ d_3 \end{pmatrix} = \begin{pmatrix} A_1 d_1 + A_2 d_2 + A_3 d_3 \\ B_1 d_1 + B_2 d_2 + B_3 d_3 \\ C_1 d_1 + C_2 d_2 + C_3 d_3 \end{pmatrix}$

But $\quad A_1 d_1 + A_2 d_2 + A_3 d_3 = \begin{vmatrix} d_1 & b_1 & c_1 \\ d_2 & b_2 & c_2 \\ d_3 & b_3 & c_3 \end{vmatrix} = \det(\mathbf{d}, \mathbf{b}, \mathbf{c})$

where $\det(\mathbf{d}, \mathbf{b}, \mathbf{c})$, which we will call Δ_1, is simply the determinant Δ with column 1 (vector **a**) replaced by the column vector **d**.

Similarly $B_1 d_1 + B_2 d_2 + B_3 d_3 = \Delta_2$, which is Δ with its second column replaced by **d**, and $C_1 d_1 + C_2 d_2 + C_3 d_3 = \Delta_3$, which is Δ with its third column replaced by **d**.

So $\quad \Delta x = \Delta_1, \quad \Delta y = \Delta_2, \quad \Delta z = \Delta_3$

This solution in determinant form is known as *Cramer's rule*, and is clearly capable of generalisation to any number of equations. It gives the solutions of the equations in a very neat form, but it suffers the serious disadvantage that evaluation of determinants is usually extremely tedious.

Example 4

Solve the equations

$$x + y - z = 1$$
$$x - y + z = 2$$
$$2x \quad\quad - z = 4$$

Here $\quad \Delta = \begin{vmatrix} 1 & 1 & -1 \\ 1 & -1 & 1 \\ 2 & 0 & -1 \end{vmatrix} = 1(1) - 1(-3) - 1(2) = 2$

$\Delta_1 = \begin{vmatrix} 1 & 1 & -1 \\ 2 & -1 & 1 \\ 4 & 0 & -1 \end{vmatrix} = 1(1) - 1(-6) - 1(4) = 3$

$\Delta_2 = \begin{vmatrix} 1 & 1 & -1 \\ 1 & 2 & 1 \\ 2 & 4 & -1 \end{vmatrix} = 1(-6) - 1(-3) - 1(0) = -3$

$$\Delta_3 = \begin{vmatrix} 1 & 1 & 1 \\ 1 & -1 & 2 \\ 2 & 0 & 4 \end{vmatrix} = 1(-4) - 1(0) + 1(2) = -2$$

So $2x = 3, \quad 2y = -3, \quad 2z = -2$

$\Rightarrow \quad x = \frac{3}{2}, \quad y = -\frac{3}{2}, \quad z = -1$

(as we could have found more simply by elimination!)

Exercise 16.6

In nos **1–5** solve, where possible, the sets of simultaneous equations.

1 $\begin{aligned} x + y + z &= 7 \\ x - y + 2z &= 4 \\ 3x - 2y - 5z &= 3 \end{aligned}$

2 $\begin{aligned} x + y - z &= 6 \\ 4x - y - 2z &= 5 \\ 5x + y + 2z &= 4 \end{aligned}$

3 $\begin{aligned} x \quad\ \ - z &= 4 \\ 2x + y + 3z &= -2 \\ 4x + 2y + 5z &= -3 \end{aligned}$

4 $\begin{aligned} x - y \quad\ \ &= 3 \\ x + 2y + 4z &= 5 \\ 3x - 6y - 4z &= 7 \end{aligned}$

5 $\begin{aligned} x + 2y + 3z &= 5 \\ 2x + 5y + 2z &= 1 \\ x + 4y - 5z &= -3 \end{aligned}$

6 Given that

$$A = \begin{pmatrix} 1 & -2 & 0 \\ 2 & -5 & -4 \\ 3 & -6 & 1 \end{pmatrix}, \quad B = \begin{pmatrix} 29 & -2 & -8 \\ 14 & -1 & -4 \\ -3 & 0 & 1 \end{pmatrix}$$

find **AB** and **BA**.

Hence, or otherwise, solve the equations:

a) $\begin{aligned} x - 2y \quad\ \ &= 3 \\ 2x - 5y - 4z &= 5 \\ 3x - 6y + z &= 9 \end{aligned}$

b) $\begin{aligned} 29x - 2y - 8z &= -3 \\ 14x - y - 4z &= 1 \\ -3x \quad\ \ + z &= -1 \end{aligned}$ (L)

7 Find the solution sets of the equations

$$\begin{aligned} x - 2y + 2z &= 1 \\ 2x - y + z &= 2 \\ x - 5y + kz &= 1 \end{aligned}$$

in the cases   **a)** $k = -1$,   **b)** $k = 5$.

Give geometrical interpretations of your solutions. (SMP)

8 Solve the system of equations

$$x \quad + 2z = 2$$
$$x + y + z = \lambda$$
$$\lambda x + y \quad = 1$$

in the cases   **a)** $\lambda = 1$,   **b)** $\lambda = \frac{1}{2}$. Interpret your answers geometrically.

(SMP)

9 Find the solution set of the simultaneous equations

$$2x + 4y - 3z + \ \ 8 = 0$$
$$x - 5y + 3z - 10 = 0$$
$$13x + 5y + pz + \ \ q = 0$$

when
a) $p = -6$, $q = 10$
b) $p = -5$, $q = 3$
c) $p = -6$, $q = 9$
In each case interpret your solution geometrically, with particular reference to the line of intersection of the planes $2x + 4y - 3z + 8 = 0$ and $x - 5y + 3z - 10 = 0$. (AEB, 1983)

10 Given the system of equations

$$2\lambda x - \ y + \ z = -1$$
$$x + \lambda y + 3z = \ \ 1$$
$$3x - 3y - \lambda x = -\lambda$$

a) solve the equations in the case $\lambda = 0$ and give a brief geometric explanation why your answer is unique;
b) find the solution set of the system of equations for those values of the parameter λ for which there is no unique solution. (OC)

11 Discuss the solution of the system of equations

$$x + 2y + \lambda z = 0$$
$$x + 3y + \ z = 3$$
$$\lambda x + 8y + 5z = 6$$

for all possible values of the parameter λ, and interpret geometrically the three essentially different cases that arise. The solution-set in each case should be stated. (OC)

12 Find necessary and sufficient conditions on λ and on μ for the equations

$$x - y + 3z = 1$$
$$2x - y + \ z = 0$$
$$\lambda x + y - 5z = \mu$$

to have no solution.

Show that the planes given by

$$x - y + 3z = 1$$
$$2x - y + z = 0$$
$$y - 5z = -2$$

intersect in a line L and find a vector equation for L.

Find an equation of the plane which contains L and the point $(3, 2, 1)$.

(JMB)

13 Prove that if α, β are real numbers and $\alpha \neq 0$, then the equations

$$x + y + 2z = 5$$
$$2x + y + z = 2$$
$$3x + y + \alpha z = \beta$$

have a unique solution. Prove further, that if $\alpha = 0$, then these equations have a solution only if β takes one particular value, which should be found. Find all the solutions of the equations in this case.   (os)

14 Given that k is a real constant, solve completely, when possible, the simultaneous equations

$$kx + 2y + 8z = 0$$
$$x - y + 2kz = 0$$
$$x + y + 6z = 2k$$

discussing carefully any special cases.   (os)

15 If ω denotes one of the cube roots of 1 other than 1 itself, solve the equations

$$x + y + z = a$$
$$x + \omega y + \omega^2 z = b$$
$$x + \omega^2 y + \omega z = c$$

for x, y, z in terms of a, b, c. Give your answer in as simple terms as possible.   (SMP)

16 Find the condition that the equations

$$ax + by + cz = 0$$
$$bx + cy + az = 0$$
$$cx + ay + bz = 0$$

have a solution other than $(x, y, z) = (0, 0, 0)$.

17 Find the values of λ for which the equations

$$4x - 6y - z = \lambda x$$
$$x - 4y - z = \lambda y$$
$$2x + 3y + z = \lambda z$$

have a solution other than $x = y = z = 0$, and find the ratios $x:y:z$ for each of these values of λ.   (OC)

18 Consider the three homogeneous equations

$$\begin{pmatrix} a_1 & b_1 & c_1 \\ a_2 & b_2 & c_2 \\ a_3 & b_3 & c_3 \end{pmatrix} \begin{pmatrix} x \\ y \\ z \end{pmatrix} = \begin{pmatrix} 0 \\ 0 \\ 0 \end{pmatrix}$$

Prove that, if solutions other than $x = y = z = 0$ exist, then $\Delta = 0$ (the 3×3 determinant of the coefficients).

Prove conversely, that if $\Delta = 0$, then solutions other than the trivial ones may be found.

[Note that $\Delta x = \Delta_1 = 0$, etc., and that if $\Delta = 0$, then any arbitrary value of x will satisfy. Note also that

$$\begin{pmatrix} a_2 & b_2 & c_2 \\ a_3 & b_3 & c_3 \end{pmatrix} \begin{pmatrix} x \\ y \\ z \end{pmatrix} = \begin{pmatrix} 0 \\ 0 \end{pmatrix} \text{ are satisfied by } \begin{pmatrix} x \\ y \\ z \end{pmatrix} = \begin{pmatrix} A_1 \\ B_1 \\ C_1 \end{pmatrix}.]$$

19 Find a necessary and sufficient condition on α which ensures that the vectors

$$\begin{pmatrix} -1 \\ 4 \\ 7 \end{pmatrix}, \begin{pmatrix} 3 \\ -2 \\ 1 \end{pmatrix}, \begin{pmatrix} 4 \\ -1 \\ \alpha \end{pmatrix}$$

are linearly independent in $\mathbb{R}^3$.

Investigate the solution of the equations

$$-x + 4y + 7z = 2$$
$$3x - 2y + z = 4$$
$$4x - y + \alpha z = \alpha + 2$$

for all values of the constant α.   (JMB)

20 Show that the only values of λ for which the simultaneous equations

$$\begin{aligned} x + (\lambda - 4)y + 2z &= 0 \\ 2x - 6y + (\lambda + 3)z &= 0 \\ (\lambda - 5)x + 12y - 8z &= 0 \end{aligned}$$

have a solution in which x, y and z are not all zero are $\lambda = 1$ and $\lambda = 4$.

Show that, when $\lambda = 4$, the three planes represented by the above equations meet in a line L. Show also that L intersects the line whose equations are

$$\frac{x - 8}{4} = y + 1 = \frac{z - 2}{10}$$

Interpret the three equations geometrically in the case when $\lambda = 1$.   (JMB)

* 16.7 Systematic reduction: inverse matrices

In section 16.6 we solved a set of three simultaneous equations by combining the equations in such a way as to eliminate, or successively reduce, the number of variables. We now tackle this 'reduction' method in a still more systematic way, which can be programmed for a computer, and also extended to a larger number of equations and variables.

Consider the set of equations

$$x - y - z = 1$$
$$x + 3y + 2z = 8$$
$$2x + 4y + z = 5$$

which can for convenience be written in matrix form:

$$\begin{pmatrix} 1 & -1 & -1 \\ 1 & 3 & 2 \\ 2 & 4 & 1 \end{pmatrix} \begin{pmatrix} x \\ y \\ z \end{pmatrix} = \begin{pmatrix} 1 \\ 8 \\ 5 \end{pmatrix}$$

First we eliminate the x coefficients in rows 2 and 3. This is done by the row operations $\mathbf{r}'_2 = \mathbf{r}_2 - \mathbf{r}_1$ and $\mathbf{r}'_3 = \mathbf{r}_3 - 2\mathbf{r}_1$, which give:

$$\begin{pmatrix} 1 & -1 & -1 \\ 0 & 4 & 3 \\ 0 & 6 & 3 \end{pmatrix} \begin{pmatrix} x \\ y \\ z \end{pmatrix} = \begin{pmatrix} 1 \\ 7 \\ 3 \end{pmatrix}$$

Next we make the y coefficient in row 2 equal to 1 by $\mathbf{r}'_2 = \frac{1}{4}\mathbf{r}_2$:

$$\begin{pmatrix} 1 & -1 & -1 \\ 0 & 1 & \frac{3}{4} \\ 0 & 6 & 3 \end{pmatrix} \begin{pmatrix} x \\ y \\ z \end{pmatrix} = \begin{pmatrix} 1 \\ \frac{7}{4} \\ 3 \end{pmatrix}$$

Then we eliminate the y coefficient in row 3 by $\mathbf{r}'_3 = \mathbf{r}_3 - 6\mathbf{r}_2$:

$$\begin{pmatrix} 1 & -1 & -1 \\ 0 & 1 & \frac{3}{4} \\ 0 & 0 & -\frac{3}{2} \end{pmatrix} \begin{pmatrix} x \\ y \\ z \end{pmatrix} = \begin{pmatrix} 1 \\ \frac{7}{4} \\ -\frac{15}{2} \end{pmatrix}$$

The matrix is now said to be in *triangular* or *echelon* form, and it is possible at this stage to obtain the solutions of the equations by evaluating z from the third row, substituting in the second row to obtain y, and then substituting in the first row to obtain x. (This method of solution by reduction to echelon form is sometimes referred to as *pivotal condensation*.) However we shall continue the process to obtain the solutions explicitly.

Clearly the vector $\begin{pmatrix} x \\ y \\ z \end{pmatrix}$ does not need to be written out each time, and the present stage in the working can be summarised as:

$$\left(\begin{array}{ccc|c} 1 & -1 & -1 & 1 \\ 0 & 1 & \frac{3}{4} & \frac{7}{4} \\ 0 & 0 & -\frac{3}{2} & -\frac{15}{2} \end{array}\right)$$

$$\mathbf{r}_3' = -\tfrac{2}{3}\mathbf{r}_3 \qquad \left(\begin{array}{ccc|c} 1 & -1 & -1 & 1 \\ 0 & 1 & \frac{3}{4} & \frac{7}{4} \\ 0 & 0 & 1 & 5 \end{array}\right)$$

$$\begin{aligned} \mathbf{r}_1' &= \mathbf{r}_1 + \mathbf{r}_3 \\ \mathbf{r}_2' &= \mathbf{r}_2 - \tfrac{3}{4}\mathbf{r}_3 \end{aligned} \qquad \left(\begin{array}{ccc|c} 1 & -1 & 0 & 6 \\ 0 & 1 & 0 & -2 \\ 0 & 0 & 1 & 5 \end{array}\right)$$

$$\mathbf{r}_1' = \mathbf{r}_1 + \mathbf{r}_2 \qquad \left(\begin{array}{ccc|c} 1 & 0 & 0 & 4 \\ 0 & 1 & 0 & -2 \\ 0 & 0 & 1 & 5 \end{array}\right)$$

Writing this out in full,

$$\begin{pmatrix} 1 & 0 & 0 \\ 0 & 1 & 0 \\ 0 & 0 & 1 \end{pmatrix}\begin{pmatrix} x \\ y \\ z \end{pmatrix} = \begin{pmatrix} 4 \\ -2 \\ 5 \end{pmatrix} \quad \Rightarrow \quad \begin{pmatrix} x \\ y \\ z \end{pmatrix} = \begin{pmatrix} 4 \\ -2 \\ 5 \end{pmatrix}$$

Elementary matrices

The effect of each of the row operations is equivalent to pre-multiplying the matrix or vector by another matrix.

For instance, the operation $\mathbf{r}_2' = \mathbf{r}_2 - \mathbf{r}_1$ can be achieved by pre-multiplying

by the matrix $\begin{pmatrix} 1 & 0 & 0 \\ -1 & 1 & 0 \\ 0 & 0 & 1 \end{pmatrix}$, and the operation $\mathbf{r}_3' = -\tfrac{2}{3}\mathbf{r}_3$ by pre-

multiplying by the matrix $\begin{pmatrix} 1 & 0 & 0 \\ 0 & 1 & 0 \\ 0 & 0 & -\frac{2}{3} \end{pmatrix}$

One other type of row operation which is sometimes useful is that of interchanging rows. Pre-multiplying by the matrix $\begin{pmatrix} 0 & 1 & 0 \\ 1 & 0 & 0 \\ 0 & 0 & 1 \end{pmatrix}$ will interchange $\mathbf{r}_1$ and $\mathbf{r}_2$.

Matrices which perform such row operations are known as *elementary matrices*.

The elementary matrix associated with a given row operation can be obtained very easily by performing that row operation on the identity matrix.

For example

$$\mathbf{r}_3' = \mathbf{r}_3 + 4\mathbf{r}_1 : \begin{pmatrix} 1 & 0 & 0 \\ 0 & 1 & 0 \\ 0 & 0 & 1 \end{pmatrix} \rightarrow \begin{pmatrix} 1 & 0 & 0 \\ 0 & 1 & 0 \\ 4 & 0 & 1 \end{pmatrix} \quad \text{and} \quad \text{pre-multiplication by}$$

the elementary matrix $\begin{pmatrix} 1 & 0 & 0 \\ 0 & 1 & 0 \\ 4 & 0 & 1 \end{pmatrix}$ will effect this row operation on any other

3×3 matrix (or vector).

Exercise 16.7a

1–3 Use the method of row reduction to find solutions to nos **1–3** in exercise 16.6.

4 By using the method of row reduction on nos **4** and **5** in exercise 16.6, explain what happens in this method when there is not a unique solution to the set of equations. How is it possible to tell whether there is no solution to the equations, or an infinite number of solutions?

5 Use row operations to solve the following sets of equations:

a) $2x + 2y - 3z = 0$
$3x + 2y - 10z = 7$
$3x + 4y + z = 8$

b) $2x - 7y + 4z = 2$
$4x + 2z = -3$
$3x - 2y + 7z = 1$

c) $x + y + z = 1$
$x - 2y + z = -2$
$2x + 3y - 2z = -1$

d) $\begin{pmatrix} 1 & 1 & -1 \\ 1 & 2 & 3 \\ 1 & 3 & -1 \end{pmatrix} \begin{pmatrix} x \\ y \\ z \end{pmatrix} = \begin{pmatrix} 1 \\ -1 \\ 1 \end{pmatrix}$

6 Find numbers p, q, r such that:

$$p \begin{pmatrix} 1 \\ 1 \\ 2 \end{pmatrix} + q \begin{pmatrix} -1 \\ 1 \\ -1 \end{pmatrix} + r \begin{pmatrix} 1 \\ 2 \\ 3 \end{pmatrix} = \begin{pmatrix} 1 \\ 0 \\ 2 \end{pmatrix}$$

7 Solve where possible the equations

$$\begin{pmatrix} 1 & 0 & -5 \\ 2 & 1 & 2 \\ -4 & -3 & -16 \end{pmatrix} \begin{pmatrix} x \\ y \\ z \end{pmatrix} = \begin{pmatrix} a \\ b \\ c \end{pmatrix}$$

in the cases $\begin{pmatrix} a \\ b \\ c \end{pmatrix} =$ **a)** $\begin{pmatrix} 0 \\ 0 \\ 0 \end{pmatrix}$, **b)** $\begin{pmatrix} 1 \\ 1 \\ 1 \end{pmatrix}$, **c)** $\begin{pmatrix} 5 \\ 0 \\ 10 \end{pmatrix}$, **d)** $\begin{pmatrix} 5 \\ 1 \\ 10 \end{pmatrix}$.

8 Describe a method of finding the solution of three simultaneous linear equations and illustrate your answer by solving the system:

$$3.9x + 2.5y - 0.5z = 4.3$$
$$5.2x + 2.9y - 1.6z = 10.9$$
$$2.6x + 1.8y - 0.7z = 2.8$$

working to three significant figures.   (MEI)

9 Find the values of l, m, n such that the product **BA** of the matrices

$$\mathbf{A} = \begin{pmatrix} 1 & 2 & 1 \\ 4 & 1 & 2 \\ -10 & 3 & 4 \end{pmatrix}, \quad \mathbf{B} = \begin{pmatrix} 1 & 0 & 0 \\ l & 1 & 0 \\ m & n & 1 \end{pmatrix}$$

is of the form

$$\mathbf{BA} = \begin{pmatrix} p_1 & q_1 & r_1 \\ 0 & q_2 & r_2 \\ 0 & 0 & r_3 \end{pmatrix}$$

Hence, or otherwise, solve the set of equations

$$\mathbf{Ax} = \mathbf{y}$$

where $\mathbf{x} = \begin{pmatrix} x_1 \\ x_2 \\ x_3 \end{pmatrix}$ and $\mathbf{y} = \begin{pmatrix} 0 \\ 8 \\ -4 \end{pmatrix}$ (L)

10 Reduce the matrix equation

$$\begin{pmatrix} 1 & 2 & -1 \\ 2 & 1 & \alpha \\ \alpha & 1 & 2 \end{pmatrix} \begin{pmatrix} x \\ y \\ z \end{pmatrix} = \begin{pmatrix} \beta \\ 0 \\ 0 \end{pmatrix} \quad (\alpha, \beta \text{ real numbers})$$ (1)

to the form

$$\begin{pmatrix} * & * & * \\ 0 & * & * \\ 0 & 0 & * \end{pmatrix} \begin{pmatrix} x \\ y \\ z \end{pmatrix} = \begin{pmatrix} * \\ * \\ * \end{pmatrix} \quad \text{(where the asterisks denote real numbers).}$$

Hence show that equation (1) has a unique solution if $\alpha^2 \neq 4$, has a non-empty solution set if $\alpha = 2$ whatever the value of β, but has no solution when $\alpha = -2$ unless $\beta = 0$.

Find the solution set when

a) $\alpha = -1, \quad \beta = 1$
b) $\alpha = -2, \quad \beta = 0$ (SMP)

Inverse matrices

The process of row reduction now leads to an alternative method of obtaining the inverse of a 3×3 matrix.

The solution of the set of equations

$$
\begin{aligned}
x - \ y - \ z &= 1 \\
x + 3y + 2z &= 8 \\
2x + 4y + \ z &= 5
\end{aligned}
$$

was achieved by transforming the matrix

$$
\mathbf{M} = \begin{pmatrix} 1 & -1 & -1 \\ 1 & 3 & 2 \\ 2 & 4 & 1 \end{pmatrix}
$$

into the identity matrix

$$
\mathbf{I} = \begin{pmatrix} 1 & 0 & 0 \\ 0 & 1 & 0 \\ 0 & 0 & 1 \end{pmatrix}
$$

by a sequence of row operations. Each of these row operations was equivalent to pre-multiplication by an elementary matrix, e.g. for the first operation, $\mathbf{r}'_2 = \mathbf{r}_2 - \mathbf{r}_1$, the elementary matrix was

$$
\mathbf{E}_1 = \begin{pmatrix} 1 & 0 & 0 \\ -1 & 1 & 0 \\ 0 & 0 & 1 \end{pmatrix}
$$

and for the second row operation, $\mathbf{r}'_3 = \mathbf{r}_3 - 2\mathbf{r}_1$,

$$
\mathbf{E}_2 = \begin{pmatrix} 1 & 0 & 0 \\ 0 & 1 & 0 \\ -2 & 0 & 1 \end{pmatrix}
$$

So the combined effect of multiplication by these elementary matrices is summarised by the equation

$$
\mathbf{E}_8 \dots \mathbf{E}_2 \mathbf{E}_1 \mathbf{M} = \mathbf{I}
$$

Hence
$$
\mathbf{M}^{-1} = \mathbf{E}_8 \dots \mathbf{E}_2 \mathbf{E}_1 = \mathbf{E}_8 \dots \mathbf{E}_2 \mathbf{E}_1 \mathbf{I}
$$

This expression for $\mathbf{M}^{-1}$ appears very complicated, but the effect of successive multiplication by each elementary matrix is just to carry out its particular row operation.

So if, as each row operation is carried out on $\mathbf{M}$, the same operation is applied to $\mathbf{I}$, by the time $\mathbf{M}$ is transformed to the identity matrix, $\mathbf{I}$ itself will have been transformed to $\mathbf{M}^{-1}$. The process is illustrated below.

$$\mathbf{M} = \begin{pmatrix} 1 & -1 & -1 \\ 1 & 3 & 2 \\ 2 & 4 & 1 \end{pmatrix} \qquad\qquad \begin{pmatrix} 1 & 0 & 0 \\ 0 & 1 & 0 \\ 0 & 0 & 1 \end{pmatrix} = \mathbf{I}$$

$$\begin{pmatrix} 1 & -1 & -1 \\ 0 & 4 & 3 \\ 2 & 4 & 1 \end{pmatrix} \quad \mathbf{r}_2' = \mathbf{r}_2 - \mathbf{r}_1 \quad \begin{pmatrix} 1 & 0 & 0 \\ -1 & 1 & 0 \\ 0 & 0 & 1 \end{pmatrix}$$

$$\begin{pmatrix} 1 & -1 & -1 \\ 0 & 4 & 3 \\ 0 & 6 & 3 \end{pmatrix} \quad \mathbf{r}_3' = \mathbf{r}_3 - 2\mathbf{r}_1 \quad \begin{pmatrix} 1 & 0 & 0 \\ -1 & 1 & 0 \\ -2 & 0 & 1 \end{pmatrix}$$

$$\begin{pmatrix} 1 & -1 & -1 \\ 0 & 1 & \frac{3}{4} \\ 0 & 6 & 3 \end{pmatrix} \quad \mathbf{r}_2' = \tfrac{1}{4}\mathbf{r}_2 \quad \begin{pmatrix} 1 & 0 & 0 \\ -\frac{1}{4} & \frac{1}{4} & 0 \\ -2 & 0 & 1 \end{pmatrix}$$

$$\begin{pmatrix} 1 & -1 & -1 \\ 0 & 1 & \frac{3}{4} \\ 0 & 0 & -\frac{3}{2} \end{pmatrix} \quad \mathbf{r}_3' = \mathbf{r}_3 - 6\mathbf{r}_2 \quad \begin{pmatrix} 1 & 0 & 0 \\ -\frac{1}{4} & \frac{1}{4} & 0 \\ -\frac{1}{2} & -\frac{3}{2} & 1 \end{pmatrix}$$

$$\begin{pmatrix} 1 & -1 & -1 \\ 0 & 1 & \frac{3}{4} \\ 0 & 0 & 1 \end{pmatrix} \quad \mathbf{r}_3' = -\tfrac{2}{3}\mathbf{r}_3 \quad \begin{pmatrix} 1 & 0 & 0 \\ -\frac{1}{4} & \frac{1}{4} & 0 \\ \frac{1}{3} & 1 & -\frac{2}{3} \end{pmatrix}$$

$$\begin{pmatrix} 1 & -1 & -1 \\ 0 & 1 & 0 \\ 0 & 0 & 1 \end{pmatrix} \quad \mathbf{r}_2' = \mathbf{r}_2 - \tfrac{3}{4}\mathbf{r}_3 \quad \begin{pmatrix} 1 & 0 & 0 \\ -\frac{1}{2} & -\frac{1}{2} & \frac{1}{2} \\ \frac{1}{3} & 1 & -\frac{2}{3} \end{pmatrix}$$

$$\begin{pmatrix} 1 & -1 & 0 \\ 0 & 1 & 0 \\ 0 & 0 & 1 \end{pmatrix} \quad \mathbf{r}_1' = \mathbf{r}_1 + \mathbf{r}_3 \quad \begin{pmatrix} \frac{4}{3} & 1 & -\frac{2}{3} \\ -\frac{1}{2} & -\frac{1}{2} & \frac{1}{2} \\ \frac{1}{3} & 1 & -\frac{2}{3} \end{pmatrix}$$

$$\mathbf{I} = \begin{pmatrix} 1 & 0 & 0 \\ 0 & 1 & 0 \\ 0 & 0 & 1 \end{pmatrix} \quad \mathbf{r}_1' = \mathbf{r}_1 + \mathbf{r}_2 \quad \begin{pmatrix} \frac{5}{6} & \frac{1}{2} & -\frac{1}{6} \\ -\frac{1}{2} & -\frac{1}{2} & \frac{1}{2} \\ \frac{1}{3} & 1 & -\frac{2}{3} \end{pmatrix} = \mathbf{M}^{-1}$$

In practice of course a number of these row operations can be carried out at the same time.

In the process of forming the inverse, it is not essential to pass through the echelon stage, providing one contrives to produce zeros in the correct positions. Moreover the process involving row operations could be replaced by one using column operations, but row and column operations must never be mixed, since column operations are equivalent to *post*-multiplication by elementary matrices.

This method of inverting matrices by using *either* row operations *or* column operations can be used for matrices of any size, and is the means which is most widely used in practice.

Exercise 16.7b

1 Use row operations to find the inverses of the following matrices:

a) $\begin{pmatrix} 4 & 3 \\ 2 & 2 \end{pmatrix}$
 b) $\begin{pmatrix} 7 & -5 \\ 5 & -4 \end{pmatrix}$
 c) $\begin{pmatrix} a & b \\ c & d \end{pmatrix}$

2 Use row operations to find the inverses of the following matrices:

a) $\begin{pmatrix} 2 & 2 & -3 \\ 3 & 2 & -10 \\ 3 & 4 & 1 \end{pmatrix}$
 b) $\begin{pmatrix} 2 & -1 & 4 \\ 4 & 0 & 2 \\ 3 & -2 & 7 \end{pmatrix}$
 c) $\begin{pmatrix} 1 & 1 & 1 \\ 1 & -2 & 1 \\ 2 & 3 & -2 \end{pmatrix}$

d) $\begin{pmatrix} 1 & 1 & -1 \\ 1 & 2 & 3 \\ 1 & 3 & -1 \end{pmatrix}$
 e) $\begin{pmatrix} 4 & 6 & 5 \\ -3 & 8 & 14 \\ 9 & -4 & 2 \end{pmatrix}$

3 Find the inverse of the matrix $\mathbf{M} = \begin{pmatrix} 1 & -1 & 1 \\ 3 & -9 & 5 \\ 1 & -3 & 3 \end{pmatrix}$.

Rewrite the set of equations

$$\begin{aligned} x - y + z &= 3 \\ 3x - 9y + 5z &= 6 \\ x - 3y + 3z &= 13 \end{aligned}$$

in a form using the matrix **M**, and hence solve the equations.   (MEI)

4 You are given the matrix

$$\mathbf{M} = \begin{pmatrix} 0 & 6 \\ 1 & 3 \end{pmatrix}$$

Three matrices $\mathbf{E}_1, \mathbf{E}_2, \mathbf{E}_3$ are to be found such that the product $\mathbf{E}_3\mathbf{E}_2\mathbf{E}_1\mathbf{M}$ is the identity matrix. These three matrices are to be 'elementary matrices', i.e. matrices having one of the forms:

a) $\begin{pmatrix} 0 & 1 \\ 1 & 0 \end{pmatrix}$

b) $\begin{pmatrix} a & 0 \\ 0 & 1 \end{pmatrix}$ or $\begin{pmatrix} 1 & 0 \\ 0 & a \end{pmatrix}$ for some number a

c) $\begin{pmatrix} 1 & b \\ 0 & 1 \end{pmatrix}$ or $\begin{pmatrix} 1 & 0 \\ b & 1 \end{pmatrix}$ for some number b

Find suitable matrices E_1, E_2, E_3.

Describe in words three transformations of the plane which, applied in succession in the correct order, will map the origin onto itself, the point $(0, 1)$ onto $(1, 0)$, and the point $(6, 3)$ onto $(0, 1)$.   (SMP)

5   A is the matrix

$$\begin{pmatrix} 1 & 0 & -3 \\ 4 & 1 & -12 \\ 8 & 2 & -23 \end{pmatrix}$$

Find an elementary matrix E_1 such that E_1A has third row $(0\,0\,*)$, where $*$ stands for a non-zero number.

Hence find a set of three elementary matrices E_1, E_2, E_3 such that

$$E_3E_2E_1A = I$$

Use the elementary matrices to find A^{-1}; also find the value of the determinant of A.   (SMP)

6   What transformations are associated with these elementary matrices?

a) $\begin{pmatrix} 3 & 0 & 0 \\ 0 & 1 & 0 \\ 0 & 0 & 1 \end{pmatrix}$ **b)** $\begin{pmatrix} 0 & 1 & 0 \\ 1 & 0 & 0 \\ 0 & 0 & 1 \end{pmatrix}$ **c)** $\begin{pmatrix} 1 & 0 & 2 \\ 0 & 1 & 0 \\ 0 & 0 & 1 \end{pmatrix}$

7   Write down the 3×3 elementary matrices whose post-multiplication effects the following column operations:
a) $c_2' = 4c_2$ **b)** $c_1' = c_1 + 3c_2$ **c)** exchange c_1 and c_3

8   Find the inverse of the matrix $\begin{pmatrix} 1 & 0 & 2 & -1 \\ 2 & 1 & 3 & 0 \\ 0 & 0 & -2 & 1 \\ 3 & 1 & 1 & 0 \end{pmatrix}$.

Miscellaneous problems

1 $Oxyz$ make a set of mutually perpendicular right-handed axes in three-dimensional space. The transformation Q_1 is a quarter-turn about the ray Ox, its sense being clockwise if one looks out from the origin along the ray; and Q_2 is a quarter-turn about a ray from $(0, 0, 1)$ parallel to Oy, clockwise if one looks out from $(0, 0, 1)$ along the ray. Find the images of a general point (x, y, z) under $\mathbf{Q}_1$ and under $\mathbf{Q}_2$, and hence find the image of (x, y, z) under $\mathbf{Q}_2\mathbf{Q}_1$ (i.e., $\mathbf{Q}_1$ followed by $\mathbf{Q}_2$). Prove that the transformation $\mathbf{Q}_2\mathbf{Q}_1$ has no fixed point.

Give the equations of four planes with the property that $\mathbf{M}_4\mathbf{M}_3\mathbf{M}_2\mathbf{M}_1 = \mathbf{Q}_2\mathbf{Q}_1$, where $\mathbf{M}_i$ denotes the reflection in the ith plane. Explain why $\mathbf{Q}_2\mathbf{Q}_1$ cannot be expressed as the product of fewer than four reflections.   (SMP)

2   Square matrices $\mathbf{A}$, $\mathbf{B}$ of the same order have the property that $\mathbf{A}^T\mathbf{A} = \mathbf{I}$, $\mathbf{B}^T\mathbf{B} = \mathbf{I}$ (where $\mathbf{A}^T$, $\mathbf{B}^T$ denote the transposes of $\mathbf{A}$, $\mathbf{B}$, and $\mathbf{I}$ is the unit matrix). Prove that:
a) if $\mathbf{C} = \mathbf{A}^{-1}$, then $\mathbf{C}^T\mathbf{C} = \mathbf{I}$;
b) if $\mathbf{D} = \mathbf{AB}$, then $\mathbf{D}^T\mathbf{D} = \mathbf{I}$.
What can you deduce about the set of matrices of the given order such that $\mathbf{X}^T\mathbf{X} = \mathbf{I}$?   (SMP)

3   $\mathbf{A}$ is a non-singular 3×3 matrix; $\mathbf{A}^T$ denotes the transpose and $\mathbf{A}^{-1}$ is the inverse. Prove that

$$(\mathbf{AB})^T = \mathbf{B}^T\mathbf{A}^T \quad \text{and that} \quad (\mathbf{A}^T)^{-1} = (\mathbf{A}^{-1})^T$$

Denote by $\mathbf{A}^*$ the matrix $(\mathbf{A}^T)^{-1}$. Operations e, r, s, t are defined on the set of all non-singular 3×3 matrices by

$$e(\mathbf{A}) = \mathbf{A}, \quad r(\mathbf{A}) = \mathbf{A}^*, \quad s(\mathbf{A}) = \mathbf{A}^{-1}, \quad t(\mathbf{A}) = \mathbf{A}^T$$

If tr denotes the composition 'first r, then t', prove that $tr = s$, justify the other entries in the composition table, and complete the table.

1st 2nd	e	r	s	t
e				
r			s	
s				
t	t	s	r	e

                                                                        (MEI)

4   (A statistical application of the transpose of a matrix.)
Suppose we have two variables x, y, and a number of observed pairs of values have been recorded: (x_i, y_i), $i = 1, 2, \ldots, n$. It is supposed theoretically that the points ought to lie on a straight line, whose equations we will take to be $y = mx + c$. We cannot of course satisfy all the n equations,

$$y_i = mx_i + c, \quad \text{or} \quad \begin{pmatrix} x_1 & 1 \\ x_2 & 1 \\ \vdots & \vdots \\ x_n & 1 \end{pmatrix} \begin{pmatrix} m \\ c \end{pmatrix} = \begin{pmatrix} y_1 \\ y_2 \\ \vdots \\ y_n \end{pmatrix}$$

for the two unknowns m and c. However the best values of m and c according to the 'principle of least squares' may be found by *pre-multiplying by the transpose* of the $n \times 2$ matrix, thus:

$$\begin{pmatrix} x_1 & x_2 & \cdots & x_n \\ 1 & 1 & \cdots & 1 \end{pmatrix} \begin{pmatrix} x_1 & 1 \\ x_2 & 1 \\ \vdots & \vdots \\ x_n & 1 \end{pmatrix} \begin{pmatrix} m \\ c \end{pmatrix} = \begin{pmatrix} x_1 & x_2 & \cdots & x_n \\ 1 & 1 & \cdots & 1 \end{pmatrix} \begin{pmatrix} y_1 \\ y_2 \\ \vdots \\ y_n \end{pmatrix}$$

Proceed to show that the values of m and c determined by this equation are those which make the line pass through the mean centre of the n points, and have gradient m, where

$$m = \frac{s_{xy}}{s_x^2}$$

(see gradient of regression line of y on x, in section 15.7).

Find the equation of the regression line of y on x through the five points $(-3, -1), (-2, 3), (2, 2), (5, 6), (8, 5)$.

5   Given any 3×3 non-zero matrix $\mathbf{A}$, show that a necessary condition for the simultaneous equations $\mathbf{Ax} = \lambda\mathbf{x}$ to have a non-zero solution $\mathbf{x}$ is that the scalar λ must satisfy an algebraic cubic equation. Show further that if the cubic equation is given by $a\lambda^3 + b\lambda^2 + c\lambda + d = 0$, then the matrix $\mathbf{A}$ satisfies the corresponding equation

$$\det(a\mathbf{A}^3 + b\mathbf{A}^2 + c\mathbf{A} + d\mathbf{I}) = 0$$

where $\mathbf{I}$ is the 3×3 unit matrix.

If $\mathbf{A} = \begin{pmatrix} 1 & 0 & -1 \\ 0 & 2 & 1 \\ 0 & 0 & 3 \end{pmatrix}$, show that the above cubic in this case is

$$\lambda^3 - 6\lambda^2 + 11\lambda - 6 = 0$$

and verify that $\mathbf{A}^3 - 6\mathbf{A}^2 + 11\mathbf{A} - 6\mathbf{I}$ has zero determinant.   (MEI)

6   Express the following simultaneous linear equations in matrix form:

$$\begin{aligned} x + y + 2z &= 4 & X + Y + Z &= 3x \\ y - 3z &= -2 & X - 2Y + Z &= 3y \\ 2x + 5y - 5z &= k & Y - Z &= z \end{aligned}$$

Find the value of k for which they are consistent, and obtain the general solution for X, Y, Z in that case. Give a geometrical illustration.   (MEI)

7   What condition must be satisfied by the constants $a, b, c, l, m,$ and n in order for the system of equations

$$- ny + mz = a$$
$$nx \qquad - lz = b$$
$$-mx + ly \qquad = c$$

to have solutions?

If this condition is satisfied, show that any solution of the equations will then also satisfy the additional equation $ax + by + cz = 0$, and write down the most general solution in this case.

Do any of these solutions also satisfy the equation $lx + my + nz = 0$?

8 A 4×4 determinant contains 24 terms. When there is a single zero among its 16 elements, there will be $24 - 6 = 18$ non-zero terms. Find the greatest and the least possible number of non-zero terms when *three* zero elements are present, showing where they may be placed.

9 If n is a positive integer, show that $b - c$, $c - a$, $a - b$ are factors of the determinant

$$\begin{vmatrix} 1 & 1 & 1 \\ a & b & c \\ a^{n+2} & b^{n+2} & c^{n+2} \end{vmatrix}$$

and that the remaining factor is the sum of all products of a, b, c of degree n (with repetitions of a, b, c included).

Obtain the corresponding result when the last row of the determinant is replaced by a^{-n}, b^{-n}, c^{-n}.   (OC)

10 Given that $\begin{vmatrix} x^3 & a^2 + 1 & 1 \\ a^3 & x^2 + 1 & 1 \\ 1 & x^2 + a^2 & 1 \end{vmatrix} = 0$, show that $x = 1$, or $x = a$, or

$x = -a/(a + 1)$ will satisfy the equation. Are there any other roots?

11 Given that a, b, c, d, e are five consecutive integers in ascending order, show that the determinant

$$\begin{vmatrix} a^2 & b^2 & c^2 \\ b^2 & c^2 & d^2 \\ c^2 & d^2 & e^2 \end{vmatrix}$$

has a value which is independent of the value of a and find this value.

                                                                (JMB)

12 Expand the determinants $\begin{vmatrix} a & b & 0 \\ b & a & b \\ 0 & b & a \end{vmatrix}$, $\begin{vmatrix} a & b & 0 & 0 \\ b & a & b & 0 \\ 0 & b & a & b \\ 0 & 0 & b & a \end{vmatrix}$, and continue with similar determinants of order 5, 6, ..., etc.

Show that the determinant
$$\begin{vmatrix} 1 & 1 & 0 & 0 & 0 & . & . & . & . \\ 1 & 1 & 1 & 0 & 0 & . & . & . & . \\ 0 & 1 & 1 & 1 & 0 & . & . & . & . \\ 0 & 0 & 1 & 1 & 1 & . & . & . & . \\ . & . & . & . & . & & & & \\ . & . & . & . & . & & & & \\ . & . & . & . & . & & & & \end{vmatrix}$$

must have the value 0, 1 or -1.

Epilogue

How many numbers are there?

We have come a long way since first learning to count and now retrace our steps to this apparently childish question. By now, however, we are aware of a variety of sets of numbers. In arithmetic modulo 2, for instance, there are just two elements, 0 and 1, whilst in the system of complex numbers there is a far richer variety including such diverse numbers as

$$0, \quad 1, \quad -3, \quad \tfrac{2}{3}i, \quad \sqrt{2}, \quad \sqrt{3} - i, \quad \text{etc.}$$

In this epilogue we shall confine ourselves to the real numbers $\mathbb{R}$ and shall start by looking again at their subset of positive integers, $\mathbb{Z}^+$.

The positive integers, $\mathbb{Z}^+$

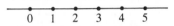

A child very soon realises that there is no end to the sequence of natural numbers, or positive integers. Even so, they are capable of being *counted*, or set in order, because this is precisely what is done when we say 'one, two, three, four, five, . . . : we name them in sequence, so that—given life and breath—any given number would ultimately be reached. We therefore say that the positive integers $\mathbb{Z}^+$ are *countable*.

The integers, $\mathbb{Z}$

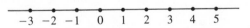

When, however, we consider *all* the integers (positive, negative and zero) an immediate difference is apparent, that there is no obvious starting-point for such an operation of counting. Even so, they can clearly be set down in the order

$$0, \quad +1, \quad -1, \quad +2, \quad -2, \quad +3, \quad -3, \quad \ldots$$

Here again, every integer—positive, zero or negative, and however large—has its place in this sequence, which therefore provides us with a means of threading our way through them. So, just as for natural numbers, *the set $\mathbb{Z}$ of all integers is countable.*

The rational numbers, $\mathbb{Q}$

There are, of course, many gaps between the integers and into these gaps we can place other quotients of integers like $\frac{1}{2}, \frac{5}{3}, -\frac{7}{6}$, so obtaining the set $\mathbb{Q}$ of rational numbers:

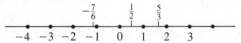

It is clear that there is no end to the number of such rationals that can be placed between two integers. The interval between 0 and 1, for instance, could be divided into tenths, hundredths, thousandths or millionths, and each of the points of subdivision would correspond to a rational number. So the number of rationals between 0 and 1 is clearly infinite.

This is, of course, just as true of the number of rationals that can be placed between any two given rationals, however close they are. For if p/q and r/s are the given rational numbers (p, q, r, s being integers), their mean is

$$\frac{1}{2}\left(\frac{p}{q} + \frac{r}{s}\right) = \frac{ps + qr}{2qs}$$

Now $ps + qr$ and $2qs$ are integers, so this mean is itself a rational number and we see that between any two rational numbers there is always another, and, therefore, by continued bisection, an infinity.

There is, therefore, certainly no shortage of rational numbers and the former question again arises. Is it still possible to count our way through this multitude of rationals, or are they too numerous? Can we construct a sequence in which every rational number has its place?

This question was first raised in 1873 by the German mathematician Georg Cantor and did not prove difficult for him to answer. Rather, as we shall see, his genius lay in *asking* the question.

Let us start by considering the positive rationals, together with the number 0. First we construct the pattern

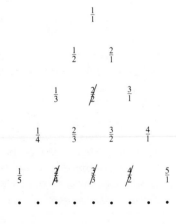

and delete any number which is not in its lowest terms. Now if this pattern is continued indefinitely, every positive rational number will have its place. We notice, for instance, that the fifth row consists of all positive rationals whose numerator and denominator add up to five and the sixth row consists of those whose numerator and denominator (when expressed in their lowest terms) add up to six. Similarly the rational number $\frac{9}{11}$ would occur in the 20th row; and, more generally, the number p/q (if expressed in its lowest terms) in the $(p + q)$th row.

So, although the positive (and zero) rationals are extremely numerous, we have invented a way of counting them:

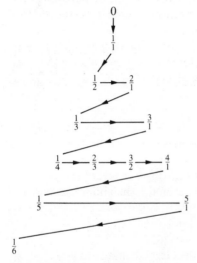

Now we can just as easily count *all* the rationals (positive, zero and negative) simply by placing each negative rational immediately after the corresponding positive rational:

$$0; \quad +1, -1; \quad +\tfrac{1}{2}, -\tfrac{1}{2}, +2, -2; \quad +\tfrac{1}{3}, -\tfrac{1}{3}, \ldots$$

Hence the set $\mathbb{Q}$ of all rationals is countable.

The continuum of real numbers, $\mathbb{R}$

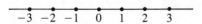

There is such a multitude of rational numbers, with an infinity of them inside any interval, that the question naturally arises whether there is room for any other kind of number. Now we have seen that Euclid had effectively answered this question about 300 BC when he showed that $\sqrt{2}$ is not a rational number and we naturally ask whether such numbers are rare or common. Are most numbers rational or irrational? Certainly we have seen that the rational numbers are countable. Can we similarly count all the numbers on the straight line, rational and irrational?

This question too was first asked by Cantor and at the beginning of December in 1873, he answered it by proving one of the most famous results of mathematics, that of the *non-enumerability of the continuum: the real numbers* $\mathbb{R}$, *rational and irrational, are too numerous to count.*

Let us consider the numbers between 0 and 1 and *suppose* that they *can* be arranged in a sequence

$$a_1, a_2, a_3, a_4, \ldots$$

which includes every such number.

Now each of these numbers can be expressed as a decimal. It is true that there is a slight difficulty about decimals which terminate, since they can be represented either with a recurring 0 or a recurring 9: $\frac{3}{20}$, for instance, can be written

either as   0.150000...
or as   0.149999...

But in such cases we simply stipulate that a recurring 9 is never to be used. So each number can be represented uniquely in decimal form, and we can write

$$a_1 = 0.\alpha_{11}\alpha_{12}\alpha_{13}\alpha_{14} \cdots$$
$$a_2 = 0.\alpha_{21}\alpha_{22}\alpha_{23}\alpha_{24} \cdots$$
$$a_3 = 0.\alpha_{31}\alpha_{32}\alpha_{33}\alpha_{34} \cdots$$
$$a_4 = 0.\alpha_{41}\alpha_{42}\alpha_{43}\alpha_{44} \cdots$$

as a sequence containing *every* number between 0 and 1.

Having done this, we now attempt to construct another number

$$b = 0.\beta_1\beta_2\beta_3\beta_4 \cdots$$

which is different from every *a*. Deciding to be as perverse as we possibly can, we see that this could be achieved if *b* were

different from a_1 in its first decimal place,
different from a_2 in its second decimal place,
different from a_3 in its third decimal place, and so on.

We therefore *choose* $\beta_1, \beta_2, \beta_3 \ldots$ by the following rule:

if $\alpha_{nn} \neq 5$, choose $\beta_n = 5$,
and   if $\alpha_{nn} = 5$, choose $\beta_n = 4$.

Then $\beta_n \neq \alpha_{nn}$, for all values of *n*. Hence $b = 0.\beta_1\beta_2\beta_3 \ldots$ is different from every single member of the sequence $a_1, a_2, a_3 \ldots$

By this means, therefore, we have succeeded in constructing a number between 0 and 1 which is not a member of this sequence. Hence our initial assumption is contradicted, and we see that it is *not* possible to count even the real numbers between 0 and 1. It is, therefore, certainly impossible to count all the real numbers $\mathbb{R}$, so that *the continuum is not enumerable*.

This, of course, provides us with another proof, over two thousand years later than that of Euclid, that there are other numbers besides the rationals. For the rational numbers were countable, but the continuum is certainly not countable, so must be far more numerous; and whereas Euclid's proof produces a single irrational number, Cantor produced an uncountable infinity all at once.

Transcendental numbers

It is now interesting to speculate about the numbers which constitute this uncountable multitude. Which are they? A first guess would be that they are irrational numbers like $\sqrt{2}$, $\sqrt[3]{5}$ and $\sqrt[4]{\frac{7}{11}}$. But this is mistaken. For the positive rational numbers $\mathbb{Q}^+$, as they are countable, can be written

$$r_1, \quad r_2, \quad r_3 \ldots$$

so their square roots are $\sqrt{r_1}, \sqrt{r_2}, \sqrt{r_3} \ldots$

their cube roots are $\sqrt[3]{r_1}, \sqrt[3]{r_2}, \sqrt[3]{r_3} \ldots$

their fourth roots are $\sqrt[4]{r_1}, \sqrt[4]{r_2}, \sqrt[4]{r_3} \ldots$ etc.

Now we can count through this set of numbers by this route:

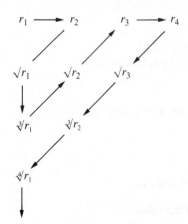

and so the set of *all positive rational numbers and all their roots is countable*.

It can similarly be shown that the set of all polynomial equations with rational coefficients have roots (known as the set of *algebraic* numbers) which are also countable. So it is clear that there must exist a vast uncountable multitude of numbers which are not the roots of any such equation. These numbers, which are not algebraic, are called *transcendental*.

Which numbers are transcendental? Certainly it is clear that no rational number or root of a rational number is transcendental, and we might well wonder what comprises this vast galaxy. Now by coincidence it was also proved in 1873 that e cannot be the root of any algebraic equation with rational coefficients, so must be just such a transcendental number; and in 1882 it was shown that π is another. Even today, however, we know very few kinds of naturally occurring transcendental numbers: ln 2, e^{π}, $2^{\sqrt{2}}$, sin 1 have been proved transcendental, but they are rarities.

What is known is simply that when all rationals, together with all roots of all rationals and all roots of all algebraic equations with rational coefficients, have been removed from the continuum, it has hardly been diminished, and that the infinitely vaster multitude of transcendental numbers has remained secretly in hiding. And in this, it might be said, these numbers are typical of the whole of mathematics, where much is known but much more is not.

Revision exercises

Chapter 11

1   The position vectors of the vertices of a triangle ABC referred to a given origin are $\mathbf{a}$, $\mathbf{b}$, $\mathbf{c}$. P is a point on AB such that $AP/PB = \frac{1}{2}$, Q is a point on AC such that $AQ/QC = 2$, and R is a point on PQ such that $PR/RQ = 2$. Prove that the position vector of R is $\frac{4}{9}\mathbf{a} + \frac{1}{9}\mathbf{b} + \frac{4}{9}\mathbf{c}$.

  Prove that R lies on the median BM of the triangle ABC, and state the value of BR/RM. (SMP)

2   The position vectors, relative to the origin O, of the points A and B are respectively $\mathbf{a}$ and $\mathbf{b}$. State, in terms of $\mathbf{a}$ and $\mathbf{b}$, the position vector of the point T which lies on AB and is such that $\mathbf{AT} = 2\mathbf{TB}$.

  Find the position vector of the point M on OT produced such that BM and OA are parallel.

  If AM is produced to meet OB produced in K, determine the ratio OB:BK. (OC)

3   In the triangle ABC, P is the point on BC such that $BP:PC = \lambda:\mu$. Show that

$$(\lambda + \mu)\mathbf{AP} = \lambda\mathbf{AC} + \mu\mathbf{AB}$$

The non-collinear points A, B and C have position vectors $\mathbf{a}$, $\mathbf{b}$ and $\mathbf{c}$ respectively with respect to an origin O. The point M on AC is such that $AM:MC = 2:1$ and the point N on AB is such that $AN:NB = 2:1$. Show that

$$\mathbf{BM} = \tfrac{1}{3}\mathbf{a} - \mathbf{b} + \tfrac{2}{3}\mathbf{c}$$

and find a similar expression for $\mathbf{CN}$.

  The lines BM and CN intersect at L. Given that

$$\mathbf{BL} = r\mathbf{BM} \quad \text{and} \quad \mathbf{CL} = s\mathbf{CN}$$

where r and s are scalars, express $\mathbf{BL}$ and $\mathbf{CL}$ in terms of r, s, $\mathbf{a}$, $\mathbf{b}$ and $\mathbf{c}$. Hence, by using triangle BLC, or otherwise, find r and s. (AEB, 1983)

4   The base of a solid hemisphere of radius a is firmly attached to one plane end, also of radius a, of a solid cylinder of length l, thus forming one solid of revolution. The material is the same and uniform throughout the composite body. Find the position of its centre of mass.

If the body can rest in equilibrium with any point of the surface of the hemi-spherical portion resting on a horizontal plane, show that $a = l\sqrt{2}$.

(MEI)

5   Prove, by integration, that the centre of mass of a uniform solid right circular cone, of vertical height h and base radius r, is at a distance $3h/4$ from the vertex of the cone.

Such a cone is joined to a uniform solid right circular cylinder, of height h, base radius r and made of the same material so that the plane base of the cone coincides with a plane face of the cylinder. Find the distance of the centre of mass of the composite solid from the centre of the base of the cone.

When the composite solid hangs in equilibrium from a point A on the circumference of the base of the cone, the line joining A to the vertex of the cone is horizontal. Show that $4r = h\sqrt{5}$.   (L)

6   Prove that the centre of mass of a uniform semi-circular lamina of radius r is at a distance $4r/3\pi$ from the centre of the semi-circle.

From a uniform lamina in the form of a semi-circle of radius $2a$ a concentric semi-circular portion of radius a is removed. Find the distance from the centre of the semi-circles to the centre of mass of the remaining material.   (JMB)

7   OABC is a square of side $2a$. $\mathbf{i}, \mathbf{j}$ are unit vectors along OA, OC. The mid-point of AB is L; the mid-point of BC is M; OL, AM meet at P; BP meets OA at N. Show that the segment $\mathbf{OP}$ can be measured by the vector $\lambda(2a\mathbf{i} + a\mathbf{j})$ and also by the vector $2a\mathbf{i} + \mu(2a\mathbf{j} - a\mathbf{i})$. Hence determine λ and μ. Prove that ON $= \frac{2}{3}$OA.   (OC)

8   The points A and B have coordinates $(2, 1, 1)$ and $(0, 5, 3)$ respectively. Find the equation of the line AB in terms of a parameter. If C is the point $(5, -4, 2)$ find the coordinates of the point D on AB such that CD is perpendicular to AB.

Find the equation of the plane containing AB and perpendicular to the line CD.   (MEI)

9   Points P, Q and R have position vectors $\mathbf{p}, \mathbf{q}$ and $\mathbf{r}$. If $\mathbf{p} = (1 - \alpha)\mathbf{q} + \alpha\mathbf{r}$, for some number α, describe the position of P relative to Q and R.

O, A, B, C are four non-coplanar points in space. A, B and C have position vectors $\mathbf{a}, \mathbf{b}, \mathbf{c}$ relative to O. The position vector of V is $2\mathbf{a} - \mathbf{c}$, and of W is $-2\mathbf{a} + 3\mathbf{b}$. If VW meets the plane OBC in U, find the position vector of U and show that U is on BC.

Use scalar products to show that if V is in the plane through O perpendicular to OB, and W is in the plane through O perpendicular to OC, then U is in the plane through O perpendicular to OA.   (OC)

10   In a tetrahedron PQRS, the edges PQ, RS are perpendicular to the faces

PRS, PQS respectively; L is the mid-point of PS and M is the mid-point of QR. Prove that

$$PQ^2 + RS^2 = QR^2 - PS^2$$
$$PM = SM = \tfrac{1}{2}QR$$
$$LM^2 = PQ^2 + RS^2 \tag{OC}$$

11 a) The plane π passes through the points with position vectors $p\mathbf{i}$, $q\mathbf{j}$, $r\mathbf{k}$, where $pqr \neq 0$. Show that c, the perpendicular distance of the origin from the plane π, satisfies

$$\frac{1}{p^2} + \frac{1}{q^2} + \frac{1}{r^2} = \frac{1}{c^2}$$

b) A straight line l through the point A with position vector $\mathbf{a}$ is parallel to a unit vector $\mathbf{e}$. The point R lies on the line l and has position vector $\mathbf{r}$. Show that

$$(\mathbf{r} - \mathbf{a}) \times \mathbf{e} = \mathbf{0}$$

The point P has position vector $\mathbf{p}$. Show that the perpendicular distance of P from the line l is

$$|(\mathbf{p} - \mathbf{a}) \times \mathbf{e}|$$

Obtain the perpendicular distance from the origin to the straight line through the points $(5, 1, -2)$ and $(2, -1, 4)$. (C)

12 Sketch the curve C which has polar equation

$$r = a(2 - \sin \theta), \quad 0 \leqslant \theta < 2\pi$$

A point P moves around C in such a way that OP has a constant angular velocity ω. Find the velocity of P; and find the acceleration of P when $\theta = \tfrac{1}{2}\pi$. (SMP)

Chapter 12

1 Sketch on one Argand diagram the loci corresponding to the sets A and B defined by

$$A = \{z : |z - (3 - i)| = 4\}$$
$$B = \{z : \arg z = \pi/4\}$$

Deduce, or show otherwise, that the set $A \cap B$ contains just one complex number and give the modulus of this number. (JMB)

2   In an Argand diagram, the origin and the point representing the complex number $(1 + i)$ form two vertices of an equilateral triangle. Find, in any form, the complex number represented by the third vertex, given that its real part is positive.   (L)

3   Find the modulus and argument of the complex number $\dfrac{1 - 3i}{1 + 3i}$.

Show that, as the real number t varies, the point representing $\dfrac{1 - it}{1 + it}$ in the Argand diagram moves round a circle, and write down the radius and centre of the circle.   (O)

4   Find the set of fixed points of the mapping

$$z \mapsto z^4 + iz$$

of the complex plane into itself. Plot these fixed points on an Argand diagram.

(A 'fixed point' of a mapping is a point that maps to itself.)   (SMP)

5   Define the modulus and argument of a complex number $z = x + iy$.

The modulus of the complex number a is r_1 and its argument is θ_1; those of b are r_2 and θ_2. Express the moduli and arguments of $a^2, a^{-1}, a^*, ab, a/b$ and a/b^* in terms of $r_1, r_2, \theta_1,$ and θ_2.

The complex numbers a, b, c, d, e, f are represented in the Argand diagram by the points A, B, C, D, E, F respectively. Show that $\dfrac{b - a}{c - a} = \dfrac{e - d}{f - d}$ is a necessary and sufficient condition for the triangles ABC and DEF to be similar and hence, or otherwise, show that this relation implies that $\dfrac{a - b}{c - b} = \dfrac{d - e}{f - e}$.   (MEI)

6   Prove that the non-real cube roots of unity are

$$-\frac{1}{2} \pm i\frac{\sqrt{3}}{2}$$

These roots are represented in an Argand diagram by the points A, B and the number $z = -2$ is represented by the point C. Show that the area of the sector of the circle with centre C through A and B which is bounded by CA, CB and the minor arc AB is $\frac{1}{2}\pi$.   (JMB)

7   Copy the diagram of the complex plane, and give clearly the geometrical significance of

$$\arg\left(\frac{z_3 - z_1}{z_3 - z_2}\right)$$

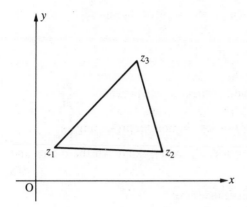

On separate diagrams, mark in the set of all points which satisfy

a) $\arg\left(\dfrac{z - i}{z + 1}\right) = \pi$

b) $\arg\left(\dfrac{z - i}{z + 1}\right) = 0$

c) $\arg\left(\dfrac{z - i}{z + 1}\right) = \tfrac{1}{2}\pi$ (SMP)

8   Given that $z_1 = 1 + i\sqrt{3}, z_2 = 1 - i\sqrt{3}$, plot the points corresponding to z_1 and z_2 on an Argand diagram, and obtain the modulus and argument of each.

Show that $z_1^3 = z_2^3$, and indicate the points corresponding to $z_1 + z_2$, $z_1 z_2$ and z_1^3 on your diagram.

Hence, or otherwise, obtain a quadratic equation with real coefficients which has z_1 as a root.

Find the values of the real constants a and b if the equation

$$z^3 + z^2 + az + b = 0$$

also has z_1 as a root.   (L)

9   The complex number

$$z_1 = 1 + ic$$

where c is a positive real number, is such that z_1^3 is real. Prove that the only possible value for c is $\sqrt{3}$.

Given that

$$z_2 = 1 - i\sqrt{3}$$

prove that the complex numbers z_1/z_2 and z_2/z_1 are the non-real cube roots of unity.   (JMB)

10   The variable non-zero complex number z satisfies the equation

$$|z - i| = 1$$

By writing z in the form $r(\cos\theta + i\sin\theta)$ with $r > 0$, prove that the modulus of z is $2\sin\theta$. Obtain the modulus and argument of $1/z$ in terms of θ, and show that the imaginary part of $1/z$ is constant. Show in an Argand diagram the loci of the points which represent **a)** z, **b)** $1/z$.   (JMB)

11   **a)** Show that the points in the Argand diagram corresponding to the values of z for which

$$(1 - z)^n = z^n$$

all lie on the line whose equation is $\operatorname{Re}(z) = \frac{1}{2}$.
b) Using de Moivre's theorem, or otherwise, expand $\cos 6\theta$ in powers of $\cos\theta$.   (L)

12   The complex numbers z_1, z_2, z_3 are represented on an Argand diagram by the points A, B, C and

$$z_1 = 4i, \quad z_2 = 2\sqrt{3} - 2i, \quad z_3 = -2\sqrt{3} - 2i$$

a) Show that the triangle ABC is equilateral.
b) Show that z_1^2 and $z_2 z_3$ are represented by the same point in the Argand diagram.
c) Find the equation of the circle OBC, where O is the origin, in the $|z - a| = b$ form and also in the $f(x, y) = 0$ form.
d) Describe the locus of the point P, given that P represents the complex number z in the Argand diagram and

$$\arg\left(\frac{z - z_1}{z - z_2}\right) = \frac{\pi}{2}$$ (AEB, 1981)

13   Given that α is the complex number $1 - \sqrt{3}i$, write $(z - \alpha)(z - \alpha^*)$ as a quadratic expression in z with real coefficients (where α^* denotes the complex conjugate of α).

Express α in modulus–argument form. Deduce the values of α^2 and α^3, and show that α is a root of

$$z^3 - z^2 + 2z + 4 = 0$$

Find all three roots of this equation.

Plot these roots on the complex plane as triangle ABC (where vertex A corresponds to the root in the first quadrant, vertex B to the real root).

What complex number represents the displacement **BA**? What complex number can be used by multiplication to rotate **BA** through 30° anti-clockwise?

If, in a 30° anti-clockwise rotation about B, $A \rightarrow A'$, find the complex number that corresponds to the point A'.   (SMP)

14 Write down, in polar form, the five roots of the equation $z^5 = 1$. Show that, when these five roots are plotted on an Argand diagram, they form the vertices of a regular pentagon of area $\frac{5}{2} \sin \frac{2}{5}\pi$.

By combining appropriate pairs of these roots, prove that for $z \neq 1$,

$$\frac{z^5 - 1}{z - 1} = (z^2 - 2z \cos \tfrac{2}{5}\pi + 1)(z^2 - 2z \cos \tfrac{4}{5}\pi + 1)$$

Use this result to deduce that $\cos \frac{2}{5}\pi$ and $\cos \frac{4}{5}\pi$ are the roots of the equation

$$4x^2 + 2x - 1 = 0 \qquad \qquad \text{(AEB, 1983)}$$

15 a) Points P and Q represent the complex numbers $z (= x + iy)$ and $w (= u + iv)$ in the z-plane and the w-plane respectively. Given that z and w are connected by the relation

$$w = \frac{z - i}{z + i}$$

and that the locus of P is the x-axis, find the Cartesian equation of the locus of Q and sketch the locus of Q on an Argand diagram.
b) Find the cube roots of -1 in the form $a + ib$, where a and b are real. Hence, or otherwise, solve the equation

$$(z - 1)^3 + z^3 = 0 \qquad \qquad \text{(L)}$$

16 Prove that, when n is a positive integer,

$$(\cos \theta + i \sin \theta)^n = \cos n\theta + i \sin n\theta$$

Given that

$$z = \cos \theta + i \sin \theta$$

and assuming that the result above is also true for negative integers, show that

$$z^n - z^{-n} = 2i \sin n\theta$$

Hence, or otherwise, prove that

$$16 \sin^5 \theta = \sin 5\theta - 5 \sin 3\theta + 10 \sin \theta$$

Find all the solutions of the equation

$$4 \sin^5 \theta + \sin 5\theta = 0$$

which lie in the interval $0 \leqslant \theta \leqslant 2\pi$. \qquad \qquad (JMB)

17 If $\quad z_1 = r_1(\cos \theta_1 + i \sin \theta_1) \quad$ and $\quad z_2 = r_2(\cos \theta_2 + i \sin \theta_2)$, prove that

$$z_1 z_2 = r_1 r_2 [\cos(\theta_1 + \theta_2) + i \sin(\theta_1 + \theta_2)]$$

The complex number z is equal to $\cos\theta + i\sin\theta$. Express each of the following complex numbers in modulus–argument form, giving the argument as a function of θ in each case:

$$1/z, \quad z^2, \quad -iz \quad \text{and} \quad (2+2i)z$$

If the point in the complex plane representing z moves anti-clockwise around the part of the circle $|z| = 1$ from $z = 1$ to $z = -1$, describe carefully (with the aid of a diagram) the motion of the points represented by

a) $z + 2 - 3i$

b) $1/z$

c) z^2

d) $-iz$

e) $(2+2i)z$

In each case give the complex numbers representing the starting and the finishing positions.   (SMP)

18 a) Show that if $z = \cos\theta + i\sin\theta$, then

$$\cos n\theta = \tfrac{1}{2}(z^n + z^{-n})$$

Using this formula verify that $2\cos 2\theta = (2\cos\theta)^2 - 2$.
   If $z^5 = 1$ and $z \neq 1$, show that $\cos\theta$ satisfies the equation

$$4c^2 + 2c - 1 = 0$$

b) Show the roots of $z^5 = 1$ on an Argand diagram and, by using the triangle inequality or otherwise, show that

$$\sqrt{\{(x - \cos 2\pi/5)^2 + (y - \sin 2\pi/5)^2\}} \\ + \sqrt{\{(x - \cos 4\pi/5)^2 + (y - \sin 4\pi/5)^2\}} \geqslant 2\sin\tfrac{1}{5}\pi$$

for all real x and y.
   Show clearly on your diagram the points (x, y) for which the *equality* holds.   (MEI)

Chapter 13

1 a) Find the solution of the differential equation

$$\frac{dy}{dx} = \frac{\sin^2 x}{y^2}$$

which also satisfies $y = 1$ when $x = 0$.

b) Find the solution of the differential equation

$$\frac{d^2y}{dx^2} = 1 - 4y$$

which also satisfies $y = 1$ when $x = 0$ and $y = \frac{1}{4}$ when $x = \frac{1}{8}\pi$.   (O)

2   Find the general solution of the differential equation

$$\frac{d^2x}{dt^2} + 2\frac{dx}{dt} - 3x = -12e^{-t}$$

Hence find the particular solution such that $x = 1$ when $t = 0$ and $x \to 0$ as $t \to +\infty$. In this particular case, show that $x \leqslant \sqrt{2}$ for all positive values of t.   (C)

3   **a)** Solve the differential equation

$$\frac{d^2x}{dt^2} + 5\frac{dx}{dt} + 4x = 5 + 4t$$

given that, when $t = 0$, $x = 0$ and $\frac{dx}{dt} = 1$.

b) Find the general solution of the differential equation

$$(x^2 + 1)\frac{dy}{dx} + y = \tan^{-1}x$$ (C)

4   **a)** Find the general solution of the differential equation

$$t\frac{dx}{dt} - 3x = t^5$$

giving x explicitly in terms of t.

b) Find the solution of the differential equation

$$\frac{d^2y}{dx^2} + 4\frac{dy}{dx} + 4y = 0$$

given that the solution curve passes through the point $(0, 1)$ and has a minimum point at $x = 1$. Sketch this solution curve.   (C)

5   Using the substitution $t = \ln x$, change the independent variable from x to t in the differential equation

$$x^2\frac{d^2y}{dx^2} + x\frac{dy}{dx} - y = \ln x$$

and hence, or otherwise, find the general solution of the equation, expressing y in terms of x.

Find the particular solution for which $y = 0$ when $x = 1$, and $\frac{dy}{dx} = 3$ when $x = 1$.   (L)

6 a) Find the solution of the differential equation

$$\frac{d^2y}{dx^2} + \frac{dy}{dx} = 6(x + y)$$

for which $y = 0, \dfrac{dy}{dx} = -1$ at $x = 0$.

b) By first differentiating with respect to x, or otherwise, show that the differential equation

$$\left(\frac{dy}{dx}\right)^2 - x\left(\frac{dy}{dx}\right) + y = 0$$

is satisfied by a family of straight lines, the equations of which are of the form $y = a(x - a)$, and by a single parabola. Find where a line of the family intersects the parabola. Give a rough sketch of the solution-curves, showing in particular how the lines are related to the parabola.

(oc)

7 A particle is moving vertically downwards in a medium which exerts a resistance to the motion proportional to the speed of the particle. It is released from rest at O at zero time, and at time t its speed is v and its position is at a distance z below O. If the terminal velocity is U, prove that

a) $\dfrac{dv}{dt} = g\left(1 - \dfrac{v}{U}\right)$

b) $gz + Uv = Ugt$

(o)

8 A car of mass m is moving along a straight horizontal road with its engine working at a constant rate P. The car starts from rest and after a time interval t its speed is v. The resistance to motion is kv, where k is a positive constant. Prove that

$$v^2 = \frac{P}{k}\left(1 - e^{-2kt/m}\right)$$

Obtain the maximum speed of the car under these conditions, and find, in terms of m and k, the time taken to reach half this maximum speed.   (c)

9 If left undisturbed, the population P of mice on a small island would increase at 2% of its current value every day. However, it is estimated that the number of mice killed each day by predators is $0.8\sqrt{P}$. Express these facts in the form of an equation for dP/dt in terms of P, where t is the time in days. (In this model, P and t are to be treated as continuous variables.)

a) If when $t = 0$ the population is 1000 mice, use a step-by-step method to estimate how it will change over the next 20 days. Use a step-length of $\delta t = 10$, and work to the nearest whole number of mice at each step.

b) Use your differential equation to show that if P exceeds a certain critical value then the population will continually increase, but that if P is below that value it will continually decrease. (SMP)

10 At each point (x, y) on a curve, the square of the sum of the x and y coordinates is equal to the gradient of the curve at that point. Express this property in the form of a differential equation.

Using the substitution $z = x + y$, show that

$$\frac{\mathrm{d}z}{\mathrm{d}x} = 1 + z^2$$

and find the general solution of this equation. Given that the curve passes through the origin, find the equation of the curve.

Show that the equation of the tangent to the curve at the point where $x = \pi/4$ is

$$y = x + 1 - \frac{\pi}{2} \qquad\qquad \text{(JMB)}$$

11 Whilst accelerating in third gear, the speed ($v\,\mathrm{m\,s}^{-1}$) of a car increases from $10\,\mathrm{m\,s}^{-1}$ to $14\,\mathrm{m\,s}^{-1}$. Over this range of speeds the accelerating force on the car is given by $F\,\mathrm{N}$, where $F = 12000/v$, and resistances to motion may be neglected compared with this force. If the mass of the car is $1000\,\mathrm{kg}$, write down a differential equation connecting v with t, the time in seconds.

Solve this equation, and hence find the time taken by the car to accelerate through this range of speeds. (SMP)

12 During a chemical reaction two substances A and B decompose. The number of grams, x, of substance A present at time t is given by

$$x = \frac{10}{(1 + t)^3}$$

There are y grams of B present at time t and $\dfrac{\mathrm{d}y}{\mathrm{d}t}$ is directly proportional to the product of x and y. Given that $y = 20$ and $\dfrac{\mathrm{d}y}{\mathrm{d}t} = -40$ when $t = 0$, show that

$$\frac{\mathrm{d}y}{\mathrm{d}t} = \frac{-2y}{(1 + t)^3}$$

Hence determine y as a function of t.

Determine the amount of substance B remaining when the reaction is essentially complete. (AEB, 1982)

Chapter 14

1   A particle of mass 4 kilograms executes simple harmonic motion with amplitude 2 metres and period 10 seconds. The particle starts from rest at time $t = 0$. Find its maximum speed and the time at which half the maximum speed is first attained. Find also the maximum value of the magnitude of the force required to maintain the motion. (You may leave your answers in terms of π.)   (JMB)

2   A rope is made of nylon which stretches under tension. Its unstretched length is 40 m. When it is stretched by x m, the force in the rope is $1200x$ N. Find the work done in stretching the rope to a length $(40 + l)$ m, stating a unit.

   A mountaineer of mass 80 kg, standing on a ledge, attaches one end of this rope to a metal ring anchored to the ledge, and fastens the other end to himself. He then falls from the ledge in a free vertical drop. If, when he *first* comes to rest, the rope is stretched by an amount l m, form an equation for l. Hence find how far below the ledge he is first brought to rest. (Take g to be $10 \,\mathrm{m\,s^{-2}}$.)   (SMP)

3   Show that the potential energy of a light elastic string of modulus λ and the natural length l, extended to a length $l + y$, is

$$\frac{\lambda y^2}{2l}$$

   One end of a light elastic string of natural length l and modulus $4mg$ is attached to a fixed point O. The other end is attached to a particle P of mass m. The particle is projected vertically downwards from O with speed $\sqrt{(4gl)}$. By using the principle of conservation of energy, or otherwise, find the speed of P when P is at a depth x below O, where $x > l$, and show that the greatest depth below O attained by P is $5l/2$. Find also the maximum value of the speed.

   Show that the particle subsequently rises to a maximum height $3l/2$ above O.   (JMB)

4   Three small beads A, B, C of mass $2m$, m, $4m$ respectively, are threaded in that order on a smooth, horizontal, straight wire. Initially the beads are at rest. The bead A is projected towards B and after the collision the two beads coalesce to form a single bead moving with speed V. Find the intial speed of projection of A and the loss in kinetic energy in the collision.

   The combined bead strikes C, the coefficient of restitution for this collision being $\frac{1}{6}$. Find the velocity of C after this collision and find also the magnitude of the impulse on C.   (C)

5   A child, standing on horizontal ground, throws a ball from a height of 1 m towards a smooth vertical high wall, the bottom of which is 6 m from

the child's feet. The initial velocity of the ball has a magnitude of $9\,\text{m s}^{-1}$, and is inclined at $45°$ above the horizontal.

Show that the ball strikes the wall at a height of about $2.64\,\text{m}$ above the ground, and that just before striking the wall its vertical component of velocity is about $2.88\,\text{m s}^{-1}$ downwards. Take g as $9.8\,\text{m s}^{-2}$.

Given that the coefficient of restitution between the ball and the wall is $\frac{1}{2}$, find the horizontal distance from the bottom of the wall at which the ball strikes the ground.   (MEI)

6 Three identical particles, not in contact, lie in a straight line on a frictionless horizontal plane. The coefficient of restitution between the particles is e. One of the end particles is projected towards the centre particle with speed U.

Show that the speed of the middle particle after three collisions is $\frac{1}{8}(1 - e^2)(3 - e)U$ and find the speeds of the other particles at this stage. Determine the condition on e for there to be no further collisions.   (MEI)

7 A particle is released from rest at a height of $20\,\text{m}$ above a horizontal plane. The particle falls freely under gravity to strike the plane. The coefficient of restitution between the particle and the plane is e. Find
   a) the height reached after the third bounce;
   b) the time between the first and second bounce.
   Given that the total time from the instance of release until the particle finally comes to rest is 4 seconds, find
   c) the value of e;
   d) the total distance covered by the particle.
   (Take g as $10\,\text{m s}^{-2}$.)   (L)

8 Forces $(2P, 0)$, $(0, P)$, $(2P, 0)$, $(0, P)$, act through the respective points $(0, 0)$, $(a, 0)$, (a, a) and $(0, a)$ on a lamina in the (x, y) plane. Prove that they will be in equilibrium with a force $(-4P, -2P)$, and find the equation of the line of action of this force.   (O)

9 In this question **i** and **j** are unit vectors in the directions of rectangular axes Ox, Oy respectively.

A system of forces consists of a force $11\mathbf{i} + 25\mathbf{j}$ acting at the point with coordinates $(3, 0)$, a force $-20\mathbf{i} + 15\mathbf{j}$ acting at $(1, 2)$ and a force $-30\mathbf{i} + 40\mathbf{j}$ acting at $(2, -1)$, the forces being measured in newtons and the coordinates in metres. The system is equivalent to a force through O and a couple. Show that this force has magnitude $89\,\text{N}$, and find its direction. Find also the moment of the couple.

Find the equation of the line of action of the resultant of the system, and verify that it passes through the point $(\frac{9}{4}, 0)$.

A further force $a\mathbf{i} + b\mathbf{j}$ acting at the point $(2, 1)$ is introduced. The system is now equivalent to a force of magnitude $F\,\text{N}$ acting along the line

$3y = 4x$ and a couple with anti-clockwise moment $41\,\text{N}\,\text{m}$. Find the values of a, b and F. (c)

10 A rectangular lamina ABCD, with $AB = 4a$, $BC = 3a$, is acted upon by forces of magnitude P along AB, $2P$ along BC, $3P$ along CD, $4P$ along DA and $5P$ along DB, the direction of each force being indicated by the order of the letters. Find the magnitude of the resultant of the forces on the lamina; find also the tangent of the angle between the direction of the resultant and AB.

The point M, on BA produced, is such that the sum of the moments, about M, of the forces on the lamina is zero. Find the length of AM. (c)

11 A particle of mass $3m$ is attached to the end A, and a particle of mass m is attached to the end B, of a light rod AB of length $8a$. The rod hangs from a fixed point O by means of light inelastic strings OA, OB, each of length $5a$. Prove that in equilibrium the inclination of the rod to the horizontal is θ where $\tan\theta = \frac{2}{3}$, and find the tension in the string OB. (o)

12 Prove that the centre of mass of a uniform right circular cone of semi-vertical angle α and height h is on the axis of the cone at a distance $\frac{3}{4}h$ from the vertex.

Such a cone is supported with its base area in contact with a smooth vertical wall by means of a light inelastic string joining the vertex of the cone to a point on the wall vertically above the centre of the base. Find the maximum possible length of the string. (o)

13 Explain what is meant by **a)** the coefficient of friction, **b)** limiting friction.

A uniform rod AB, of mass $4M$ and length l, is smoothly jointed at the end A to a fixed straight horizontal wire AC. The end B is connected by means of a light inextensible string of length l to a bead of mass M which can slide on the wire, the coefficient of friction between the bead and the wire being μ. Show that, when the friction is limiting, the angle of inclination, θ, of the rod to the horizontal wire is given by

$$\cot\theta = 2\mu \qquad\qquad\text{(AEB, 1981)}$$

14 Two uniform rods AB, BC of equal lengths are freely jointed together at B. The mass of AB is m and that of BC is $2m$. The rods stand in equilibrium in a vertical plane on a rough horizontal plane, and the angle $ABC = 2\alpha$. The coefficient of friction at A is μ_1, and the coefficient of friction at C is μ_2. Find the horizontal and vertical components of the reaction between the rods at B, and prove that $\mu_1 \geqslant \frac{3}{5}\tan\alpha$ and $\mu_2 \geqslant \frac{3}{7}\tan\alpha$. (o)

15 A uniform rod of weight $4W$ and length $2a$ is maintained in a horizontal position by two light inextensible strings, each of length a, which are attached to the ends of the rod. To the other ends of the strings are

attached small rings each of weight W. The rings can slide on a rough horizontal bar, the coefficient of friction between each ring and the bar being $\frac{1}{2}$. Show that the greatest possible distance between the rings is $16a/5$, and find the tension in the strings when the rings are in this position. (C)

16 A particle, of mass m, moves such that its velocity at any instant is given by $\dot{\mathbf{r}} = \mathbf{\Omega} \times \mathbf{r}$, where $\mathbf{\Omega}$ is a constant vector and $\mathbf{r}$ is its position vector. Given that $\mathbf{\Omega} = \omega\mathbf{k}$ (ω constant) and $\mathbf{r} = x\mathbf{i} + y\mathbf{j} + z\mathbf{k}$, show that

a) $\ddot{\mathbf{r}} = -\omega^2(x\mathbf{i} + y\mathbf{j})$
b) $\mathbf{\Omega}$ is perpendicular to $\dot{\mathbf{r}}$

Write down the resultant force acting on the particle and find the moment of this force about the origin.

Verify that $\mathbf{r} = a\cos\omega t\mathbf{i} + a\sin\omega t\mathbf{j} + b\mathbf{k}$, where a and b are constants, gives the position vector for a possible motion of the particle and show that, in this case, the speed of the particle is constant. (AEB, 1982)

17 A uniform solid sphere of mass $5m$ and radius a is fixed to the end of a thin uniform rod of mass $3m$ and length $4a$ so that the centre of the sphere lies on the extended axis of the rod. The other end of the rod is freely pivoted at a point in such a way that the system can swing in a vertical plane under the influence of a vertical gravitational field g.

Find the moment of inertia of the system about its pivot.

If the system is balanced vertically above the pivot and given a slight displacement, find its angular speed at its lowest point. (MEI)

18 a) A particle of mass $0.05\,\text{kg}$ is performing oscillations in simple harmonic motion between two fixed points. Given that the maximum speed of the particle is $\pi/5\,\text{m s}^{-1}$ and that the maximum force exerted on it is $\pi^2/100\,\text{N}$, determine the amplitude and the periodic time of oscillation.

b) A flywheel is rotating about its axis with an angular velocity of 20 radians per second; its moment of inertia about its axis is $0.15\,\text{kg m}^2$. Determine the constant torque which must be applied in order to reduce the angular velocity to 5 radians per second in a time of $2.5\,\text{s}$.

c) A body of mass $0.4\,\text{kg}$ is suspended from a fixed point by a light string of length $0.9\,\text{m}$. The body moves in a horizontal circle, making 2 revolutions every $3\,\text{s}$. Calculate the tension in the string, and find, to the nearest tenth of a degree, the inclination of the string to the vertical. (AEB, 1983)

19 A satellite is in steady circular orbit at height h above the surface of the earth. At this height the gravitational pull on the satellite has magnitude $GmM/(R + h)^2$, where m is the mass of the satellite, M is the mass of the earth, R is the radius of the earth, and G is constant. Find the speed of the satellite, and show that it will take a time

$$2\pi(R + h)^{3/2}(GM)^{-1/2}$$

to travel once round the orbit. (SMP)

20 Mars has a mass of 6.46×10^{23} kg and a radius of 3380 km. Given that G (Newton's gravitational constant) is 6.67×10^{-11} m^3 kg^{-1} s^{-2}, estimate the acceleration due to gravity at the Martian surface.

Assuming atmospheric resistance can be neglected, estimate the minimum speed at which a Martian space rocket should be launched vertically so as to escape completely from the influence of the planet's gravity. In this case, determine the speed of the rocket when it is 8000 km from the centre of the planet. (MEI)

Chapter 15

1 The discrete random variables, X and Y, are independent and have the following probability distributions:

r	1	2	3	4	5
P$(X = r)$	0	$\frac{1}{4}$	$\frac{1}{3}$	$\frac{1}{3}$	$\frac{1}{12}$
P$(Y = r)$	$\frac{1}{5}$	0	$\frac{2}{5}$	0	$\frac{2}{5}$

Calculate the mean and variance of each of X and Y. Give the probability distribution of $Z = X - Y$. What are the mean and variance of Z? (O)

2 The continuous random variable X has p.d.f.

$$\text{f}(x) = \begin{cases} 30x^2(1 - x)^2 & (0 \leqslant x \leqslant 1) \\ 0 & \text{(elsewhere)} \end{cases}$$

Sketch the graph of f(x), and find the mean of X. Obtain also the variance of X.

Find the probability that X takes a value within 0.1 of its mean. Find also the probability that the mean of 20 independent observations from this distribution takes a value within 0.1 of its mean. (MEI)

3 A ferry across a river can carry up to four cars. Cars arrive at the ferry on one side of the river in such a way that the time T minutes for four cars to arrive, measured from 20 minutes before the ferry is due to depart, has the probability density function

$$ate^{-t/10}, \quad t \geqslant 0$$

where a is a constant. Show that $a = 1/100$, and find the mean and the variance of T.

[You may use the result that, for any positive integer n,

$$\int_0^\infty t^n e^{-t/10} \, dt = n! \, 10^{n+1}]$$

The ferry service adopts the policy that, if there are not four cars on the ferry when they are due to depart, they will wait for up to another five minutes in the hope of completing their load. Find the probabilities that
a) the ferry is not delayed in departing;
b) the ferry is delayed for five minutes.   (JMB)

4   Brown and White Ltd, the sugar refining company, pack their product into nominally 1 kilogram bags. The actual masses may be assumed to be a random sample from a normal distribution with mean equal to the machine preset value and standard deviation 20 grams. Current quality control regulations insist that at least 95% of the bags have mass 975 grams or greater. What should the machine preset value be?

With the advent of a new government, the quality control regulations are rationalised. Now, whatever the nominal mass, the product must be packed such that at least 99% of the bags have mass equal to 95% of the nominal mass or greater. What should the new machine preset value be?

Brown and White decide to market new 5 kilogram bumper bags made up of five of the original 1 kilogram bags. Assuming that these new bumper packs with nominal mass 5 kilograms must satisfy the revised quality control regulations in a similar way to the 1 kilogram bags, what should the new machine preset value be? What percentage of their product would be saved in the long run if only bumper packs were marketed as opposed to only 1 kilogram bags?   (AEB, 1982)

5   The following table gives the number f_r of each of 519 equal time intervals in which r radioactive atoms decayed:

number of decays (r)	0	1	2	3	4	5	6	7	8	$\geqslant 9$
observed number of intervals (f_r)	11	41	73	105	107	82	55	28	9	8

Estimate the mean and variance of r.

Suggest, with justification, a theoretical distribution from which the data could be a random sample. Hence calculate expected values of f_r and comment briefly on the agreement between these and the observed values.

In the experiment each time interval was of length 7.5 s. In a further experiment, 1000 time intervals each of length 5 s are to be examined. Estimate the number of these intervals within which no atoms will decay.   (MEI)

6   The random variable X has probability density function f given by

$$f(x) = \begin{cases} 1, & 0 \leqslant x \leqslant 1 \\ 0, & \text{otherwise} \end{cases}$$

Write down the mean of X, and determine its standard deviation.

Suppose X_1 and X_2 are two independent random variables, each having the same distribution as X. Determine the mean and standard deviation of $(X_1 - X_2)$, and verify that these are the same as those given by the probability density function g, where

$$g(x) = \begin{cases} 1 + x, & -1 \leqslant x < 0 \\ 1 - x, & 0 \leqslant x \leqslant 1 \\ 0, & \text{otherwise} \end{cases}$$

(MEI)

7   Three cards labelled 1, 2, 3 are thoroughly shuffled and are then examined to determine the cards whose labels correspond with their positions in the arrangement. The total number of cards in their correct positions is $y = x_1 + x_2 + x_3$, where $x_i = 1$ if card i is in the ith position and $x_i = 0$ otherwise. Give the values of $P(x_i = 1)$ and $P(x_i = 0)$ and use them to find $E(x_i)$. Hence determine $E(y)$.

Show that $E(x_i^2) = \frac{1}{3}$. Explain why $P(x_i x_j = 1) = \frac{1}{6}$ when $i \neq j$ and determine $E(x_i x_j)$. *Hence* find $E(y^2)$ and show that $Var(y) = 1$.   (SMP)

8   At one stage in the manufacture of an article, a cylindrical rod with a circular cross-section has to fit into a circular socket. Quality control measurements show that the distribution of rod diameters is Normal with mean 5.01 cm and standard deviation 0.03 cm, while that of socket diameters is independently Normal with mean 5.11 cm and standard deviation 0.04 cm. If components are selected at random for assembly, what proportion of rods will not fit?

Rods and sockets are randomly paired for delivery to customers. Batches for delivery are made up of n such pairs. What is the largest value of n for which the probability that all the rods in the batch fit into their respective sockets is greater than 0.9? Given that $n = 30$, find the probability that not more than one rod will fail to fit into its socket. (MEI)

9   **a)** Two players, A and B, play a game in which A throws two fair dice simultaneously. For each throw of the dice B charges A a fee of 2 pence and pays A the same number of pence as the difference between the scores of the two dice. Determine whether A or B would expect to win in the long run and calculate A's expectation over a series of 90 throws.

b) A company produces a type of electrical component and the lifetimes of the components have a standard deviation of 50 hours. The manufacturer claims that, on average, the lifetime of a component is 2000 hours.

A randomly selected sample of 100 components had a mean lifetime of 1986 hours. Show that, on this evidence, the manufacturer's claim is not justified at the 5% level.

If the mean lifetime of 1986 hours had been obtained from a sample of n components, determine the greatest value of n for which the manufacturer's claim would not have been rejected at the 5% level.

(AEB, 1982)

10 When a standard lubricant is used on the bearings of a wheel, the length of time for which the wheel will spin in an experiment has mean 200 s and standard deviation 10 s. A 'new improved' lubricant is tried out and five spins take times

209, 224, 196, 219, 212

seconds. Test, assuming that the standard deviation is unchanged and stating clearly all the other assumptions you need to make, whether the new lubricant has significantly increased the spinning time. (SMP)

11 A market researcher has found that only 192 out of the 800 people he has interviewed are in favour of a new hypermarket being built near their town, which has a population of 15000. How many inhabitants should he expect from this information to want the development? The researcher claims that it is unlikely that as many as 4000 of the town's residents actually want the new development. Test his claim (stating carefully any assumptions you need to make). (SMP)

12 A random variable X has a rectangular distribution between a and b ($b > a$), so that its probability density function is given by

$$f(x) = (b - a)^{-1}, \quad a \leqslant x \leqslant b,$$
$$f(x) = 0, \qquad \text{otherwise}$$

Prove that the mean and variance of X are $(a + b)/2$ and $(b - a)^2/12$, respectively.

Ten metal rods are measured to the nearest millimetre and the lengths (in millimetres) obtained are

93, 87, 91, 89, 86, 95, 87, 90, 90, 92

The measured length of a rod of true length c mm is a random variable which has a rectangular distribution between $(c - 0.5)$ mm and $(c + 0.5)$ mm. Assuming that the sum of the measured lengths of the ten rods may be approximated by a Normal distribution, with variance equal to the sum of the ten individual variances, determine 95 per cent symmetric confidence intervals for

a) the mean of the true lengths of the ten rods;
b) the sum of the true lengths of the ten rods. (JMB)

13 **a)** Samples of 10 oranges are taken from a crate of 100 oranges whose mean weight is 152 g with a standard deviation of 7 g. Calculate the expected value of the mean of such samples and the variance of the distribution of means.

b) A basket contained ten oranges which were then picked out one by one. Their weights in order of being picked out were:

162, 160, 156, 153, 148, 139, 157, 152, 143, 150 grams

Obtain Spearman's rank correlation coefficient as a measure of the tendency to pick the biggest oranges first.   (AEB, 1981)

14 In an investigation into prediction using the stars and planets, a celebrated astrologer Horace Cope predicted the ages at which thirteen young people would first marry. The complete data, of predicted and actual ages at first marriage, are now available and are summarised in the following table:

person	A	B	C	D	E	F	G	H	I	J	K	L	M
predicted age, x (years)	24	30	28	36	20	22	31	28	21	29	40	25	27
actual age, y (years)	23	31	28	35	20	25	45	30	22	27	40	27	26

a) Draw a scatter diagram of these data.

b) Calculate the equation of the regression line of y on x and draw this line on the scatter diagram.

c) Comment upon the results obtained, particularly in view of the data for person G. What further action would you suggest?   (AEB, 1981)

15 A toy company specialises in high precision scale models of vintage cars. Before being allowed on to the assembly line employees must complete a training course and receive an assessment, in the range 1–20, of their aptitude for this kind of work. The better suited to this work would expect a high course assessment. The table below gives the number of complete cars assembled in one day by 80 employees, together with their training course assessment figure.

		number of complete cars assembled in one day				
		1–3	4–6	7–9	10–12	13–15
training course assessment	1–5	5	7	4	0	0
	6–10	1	7	9	6	0
	11–15	0	2	8	11	5
	16–20	0	1	1	4	9

Calculate the product–moment correlation coefficient for these data.

(AEB, 1981)

Chapter 16

1 **a)** Given $\mathbf{M} = \begin{pmatrix} k & 1 \\ 0 & k \end{pmatrix}$, calculate $\mathbf{M}^2$ and $\mathbf{M}^3$. Suggest a form for $\mathbf{M}^n$ and confirm your suggestion, using the method of proof by induction.

b) Prove that, for any $n \times n$ matrices $\mathbf{A}$ and $\mathbf{B}$,

$$\mathbf{AB} = \mathbf{BA}$$

if and only if $(\mathbf{A} - k\mathbf{I})(\mathbf{B} - k\mathbf{I}) = (\mathbf{B} - k\mathbf{I})(\mathbf{A} - k\mathbf{I})$ for all values of the real number k.

c) Prove that, for any $n \times n$ matrices $\mathbf{A}$ and $\mathbf{B}$, $(\mathbf{AB})^{\mathrm{T}} = \mathbf{B}^{\mathrm{T}}\mathbf{A}^{\mathrm{T}}$, where $\mathbf{A}^{\mathrm{T}}$ is the transpose of $\mathbf{A}$. (MEI)

2 Draw a diagram with two coordinate axes, and on it show the unit vectors

$$\mathbf{u} = \begin{pmatrix} \cos \alpha \\ \sin \alpha \end{pmatrix} \quad \text{and} \quad \mathbf{v} = \begin{pmatrix} -\sin \alpha \\ \cos \alpha \end{pmatrix}$$

taking α to be an acute angle (to be clearly marked). What is the relationship between between the directions of $\mathbf{u}$ and $\mathbf{v}$?

A transformation of the plane has a matrix denoted by $\mathbf{P}_\alpha$, where

$$\mathbf{P}_\alpha = \begin{pmatrix} \cos^2 \alpha & \cos \alpha \sin \alpha \\ \cos \alpha \sin \alpha & \sin^2 \alpha \end{pmatrix}$$

Prove that $\mathbf{P}_\alpha \mathbf{u} = \mathbf{u}$ and that $\mathbf{P}_\alpha \mathbf{v} = \mathbf{0}$, the zero vector. Use these facts to describe the transformation in geometrical terms.

If $\beta = \alpha + \frac{1}{2}\pi$, write down the corresponding transformation matrix $\mathbf{P}_\beta$ in terms of α in its simplest form, and show that $\mathbf{P}_\beta \mathbf{P}_\alpha = \mathbf{O}$, the zero matrix. Give a geometrical explanation for this result. (SMP)

3 Linear transformations L_k of three dimensional space into itself are given by the matrices

$$\mathbf{A}_k = \begin{pmatrix} 1 & 1 & 1 \\ 1 & k & 0 \\ 1 & 0 & 0 \end{pmatrix}$$

a) Find the image of the plane $x + y + z = 0$ under L_1.

b) Find the point which maps to $(1, 2, 3)$ under L_2.

c) Prove that L_0 maps the whole space onto a plane and give the equation of this plane. (OC)

4 The matrix $\mathbf{M}$ is defined by

$$\mathbf{M} = \begin{pmatrix} 2 & -1 & 1 \\ 1 & 1 & 2 \\ -1 & 1 & 0 \end{pmatrix}$$

Show that there exists a non-zero vector $\mathbf{r}$, where $\mathbf{r} = \begin{pmatrix} x \\ y \\ z \end{pmatrix}$, and a real number k with $\mathbf{Mr} = k\mathbf{r}$, if and only if the determinant of

$$\begin{pmatrix} 2-k & -1 & 1 \\ 1 & 1-k & 2 \\ -1 & 1 & -k \end{pmatrix}$$

is zero. Find the values of k for which this determinant is zero. For each of these values find a non-zero vector $\mathbf{r}$ for which $\mathbf{Mr} = k\mathbf{r}$.

The transformation T maps $\mathbf{r}$ to $\mathbf{r}'$, where $\mathbf{r}' = \mathbf{Mr}$. Find

a) a vector equation of the line L_1, every point of which is mapped to the origin by T;

b) a vector equation of the line L_2 for which $T(L_2) = L_2$;

c) a vector equation of the plane π for which $T(\pi) = \pi$.   (JMB)

5 Show that the matrix

$$\mathbf{A} = \begin{pmatrix} 1 & 0 & 0 \\ 0 & \cos\alpha & -\sin\alpha \\ 0 & \sin\alpha & \cos\alpha \end{pmatrix}$$

represents a rotation about the x-axis of angle α. Write down a similar matrix $\mathbf{B}$ representing a rotation about the y-axis of angle β.

Verify that the effect of $\mathbf{A}$ followed by $\mathbf{B}$ is different from the effect of $\mathbf{B}$ followed by $\mathbf{A}$. If α and β are both very small, however, show that the operations are approximately commutative.

Suggest expressions for the elements of $\mathbf{A}^n$, where n is a positive integer, and use induction to test the correctness of your conjecture.

Prove that it is permissible to drop the restriction that n is positive.

(OC)

6 Explain what is meant by the statement that the m column vectors $\mathbf{x}_1$, $\mathbf{x}_2, \ldots, \mathbf{x}_m$, where each vector consists of n elements, are linearly independent.

Find the matrix $\mathbf{A}$, where $\mathbf{y} = \mathbf{Ax}$ represents a linear transformation for which

$$\begin{pmatrix} 3 \\ 1 \\ 2 \end{pmatrix} = \mathbf{A}\begin{pmatrix} 1 \\ 0 \\ 0 \end{pmatrix}, \quad \begin{pmatrix} 2 \\ 3 \\ 1 \end{pmatrix} = \mathbf{A}\begin{pmatrix} 0 \\ 1 \\ 0 \end{pmatrix}, \quad \begin{pmatrix} 3 \\ 2 \\ 1 \end{pmatrix} = \mathbf{A}\begin{pmatrix} 0 \\ 0 \\ 1 \end{pmatrix}$$

Find the images of $\mathbf{x}_1 = \begin{pmatrix} 1 \\ 1 \\ 1 \end{pmatrix}$, $\mathbf{x}_2 = \begin{pmatrix} 4 \\ 3 \\ -1 \end{pmatrix}$, and $\mathbf{x}_3 = \begin{pmatrix} 5 \\ 4 \\ 0 \end{pmatrix}$ under this transformation.

Show that
a) x_1 and x_2 are linearly independent;
b) Ax_1 and Ax_2 are linearly independent;
c) x_1, x_2 and x_3 are linearly dependent. (L)

7 Given that

$$A = \begin{pmatrix} 1 & 2 & 1 \\ -3 & 2 & -4 \\ 2 & -1 & 3 \end{pmatrix}$$

find det A and A^{-1}. (MEI)

8 Factorise completely the determinant

$$\begin{vmatrix} 1 & a + bc & a^2 + b^2c^2 \\ 1 & b + ca & b^2 + c^2a^2 \\ 1 & c + ab & c^2 + a^2b^2 \end{vmatrix}$$ (JMB)

9 a) Given that

$$D = \begin{vmatrix} a^2 & a & 1 \\ b^2 & b & 1 \\ c^2 & c & 1 \end{vmatrix}$$

express D in a factorised form.
b) Given the matrix A, where

$$A = \begin{pmatrix} 0 & 0 & 0 \\ 1 & 0 & 0 \\ 0 & 1 & 0 \end{pmatrix}$$

find A^2 and A^3. Hence show that

$$I + At + A^2\frac{t^2}{2!} + \ldots + A^n\frac{t^n}{n!} + \ldots = \begin{pmatrix} 1 & 0 & 0 \\ t & 1 & 0 \\ \frac{1}{2}t^2 & t & 1 \end{pmatrix}$$

where I is the (3×3) unit matrix and t is a real number. (L)

10 State the condition necessary in order that a square matrix A shall have an inverse A^{-1}.
Prove that, if A and B are two $(n \times n)$ non-singular matrices, then

$$(AB)^{-1} = B^{-1}A^{-1}$$

Find A^{-1} given that

$$A = \begin{pmatrix} 3 & 2 & -1 \\ 1 & -1 & 2 \\ 2 & 4 & -1 \end{pmatrix}$$

Solve the system of equations

$$
\begin{aligned}
3x + 2y - z &= -5 \\
x - y + 2z &= 11 \\
2x + 4y - z &= -10
\end{aligned}
$$

(L)

11   The transformation matrix **A**, where

$$
\mathbf{A} = \tfrac{1}{9}\begin{pmatrix} 1 & -4 & 8 \\ 4 & -7 & -4 \\ 8 & 4 & 1 \end{pmatrix}
$$

has transpose $\mathbf{A}^{\mathrm{T}}$. Show that

$$
\mathbf{A}\,\mathbf{A}^{\mathrm{T}} = \mathbf{A}^{\mathrm{T}}\mathbf{A} = \mathbf{I}
$$

where **I** is the unit matrix.
   Prove that the planes

$$
\begin{aligned}
x - 4y + 8z &= 0 \\
4x - 7y - 4z &= 0 \\
8x + 4y + z &= 0
\end{aligned}
$$

meet in only one point, and find the coordinates of the point P of intersection of the planes

$$
\begin{aligned}
x - 4y + 8z &= 30 \\
4x - 7y - 4z &= 3 \\
8x + 4y + z &= 15
\end{aligned}
$$

Find also the coordinates of the point P′ into which P is transformed by **A**.

(L)

12   Given that the matrix

$$
\mathbf{M} = \begin{pmatrix} 1 & 1 & -1 \\ 2 & 3 & -5 \\ 3 & 1 & a \end{pmatrix}
$$

where a is a constant, find
a) the value of a for which the simultaneous equations

$$
\mathbf{M}\mathbf{x} = \mathbf{O}
$$

have a solution $\mathbf{x} \neq \mathbf{O}$.
b) the value of a for which the simultaneous equations

$$
\mathbf{M}\mathbf{x} = \mathbf{x}
$$

have a solution $\mathbf{x} \neq \mathbf{O}$.
   Find the solution **x** in each case and give brief geometrical interpretations of them in terms of transformations.

$$\text{In this question } \mathbf{x} = \begin{pmatrix} x_1 \\ x_2 \\ x_3 \end{pmatrix} \text{ and } \mathbf{O} = \begin{pmatrix} 0 \\ 0 \\ 0 \end{pmatrix}.$$

(MEI)

13   Show that the simultaneous equations

$$kx + y + z = 1$$
$$2x + ky - 2z = -1$$
$$x - 2y + kz = -2$$

have a unique solution except for three values of k which are to be found.
   Show that, when $k = 1$, the planes represented by the equations meet at a point P which lies in the plane $y = 1$, and that, when $k = -1$, the planes meet in a line L which also lies in the plane $y = 1$. Find the perpendicular distance of P from L. (JMB)

14   Find the solution set of the system of equations

$$x + y + z = -1$$
$$x + 2y + 3z = -2$$
$$3x - 2y - 7z = 2$$

If x, y, z represent coordinates of points in three-dimensional space, state
a)   what is represented geometrically by the solution set;
b)   what is represented geometrically by each of the three equations;
c)   how the geometrical figures represented by the equations are related to each other. (SMP)

Answers to exercises

Chapter 11

Exercise 11.1a

1. a) $\frac{1}{2}(a+b)$ b) $\frac{3}{4}a+\frac{1}{4}b$ c) $\frac{1}{4}a+\frac{3}{4}b$ d) $\frac{3}{2}a-\frac{1}{2}b$ e) $-\frac{1}{2}a+\frac{3}{2}b$

Exercise 11.1b

1. a) $\frac{1}{2}(b+c), \frac{1}{2}(c+a), \frac{1}{2}(a+b)$ b) $\frac{1}{3}(a+b+c)$
2. a) $\frac{2}{5}b+\frac{3}{5}c$ b) $\frac{1}{4}a+\frac{3}{4}c$ c) $\frac{1}{3}a+\frac{2}{3}b$ d)e)f) $\frac{1}{6}a+\frac{1}{3}b+\frac{1}{2}c$
3. a) $-2b+3c$ b) $\frac{3}{2}c-\frac{1}{2}a$ c) $-a+2b$
4. $r=\frac{1}{2}(1-t)a+\frac{1}{2}tb+\frac{1}{2}tc, \frac{1}{4}(a+b+c)$
5. $\lambda=\frac{4}{5}, \mu=\frac{2}{5}$

Exercise 11.2a

1. a) $\frac{1}{3}(b+c+d)$ b) $\frac{1}{4}(a+b+c+d)$ c) $\frac{1}{4}(a+b+c+d)$
3. a) $\frac{1}{6}a+\frac{1}{3}b+\frac{1}{2}c$ b) $(\frac{1}{2},\frac{3}{2},\frac{7}{5})$ c) $-0.4i+2.1j-0.3k$
4. $12\,\text{cm}$ 5. $\frac{4}{7}, \frac{26}{7}$

Exercise 11.2b

1. a) $(1\frac{3}{5}, 2\frac{2}{7})$ b) $(1\frac{17}{28}, 1\frac{23}{70})$ c) $(5\frac{2}{5}, 0)$ d) $(\frac{2}{3}h, \frac{1}{3}a)$
2. a) $(1\frac{3}{4}, 0)$ b) $(1\frac{43}{62}, 0)$ c) $(6, 0)$ d) $(\frac{3}{4}h, 0)$
3. $\bar{x}=\frac{3}{8}a$ 5. $\frac{5}{24}\pi a^3, \frac{27}{40}a$ 6. $x=2.71$

Exercise 11.3

1. $i-j, i+j, -i-j+2k$
2. a) $i-j, i+j, i-j+2k$ b) $90°$ c) $54.7°, 90°$ d) $39.2°, 39.2°$

Exercise 11.4

1. a) $r=e^\theta$ b) $2e^{2t}, 2e^{2t}; 0, 4e^{2t}$
3. $-20t^2, 10; -20t^2, 2k, 10$

Miscellaneous problems

7 $2a/\sqrt{3}$

9 $\dfrac{\cos^2 \beta + \cos^2 \gamma - 2\cos \alpha \cos \beta \cos \gamma}{\sin^2 \alpha}$

Chapter 12

Exercise 12.1

1 a) $3 + 7i, 1 - i$ **b)** $10 + 2i, 2 + 8i$ **c)** $10, 2i$ **d)** $-3 - 6i, 1 + 4i$
  e) $4 + i, 2 + 3i$ **f)** $7 + 7i, 1 - 3i$
2 a) $10 + 5i$ **b)** $11 + 2i$ **c)** $-13 - 19i$ **d)** $-3 - 4i$
  e) $-20 - 20i$ **f)** $-11 - 2i$
3 a) $0.4 - 0.2i$ **b)** $0.6 + 0.2i$ **c)** $0.02 - 0.14i$ **d)** $1 + 2i$ **e)** $-i$
  f) $-0.088 + 0.016i$ **g)** $1 + 2i$ **h)** $\cos \theta + i \sin \theta$
4 $\frac{2}{13} + \frac{3}{13}i$
5 $p = 7k, q = 3k$ $(k \in \mathbb{Z})$
6 a) $(x^3 - 3xy^2) + i(3x^2 y - y^3)$ **b)** $(x^2 - y^2 - 2ixy)(x^2 + y^2)^{-2}$

  c) $\left(x + \dfrac{x}{x^2 + y^2}\right) + i\left(y - \dfrac{y}{x^2 + y^2}\right)$

7 $(4 + 5i)(2 - 3i), 83^2 + 1^2 = 71^2 + 43^2$
8 $-3 - i$
9 $-1 + 2i$
10 $-\frac{4}{11}, \frac{2}{11}$
11 a) $\pm(2 + i)$ **b)** $\pm(3 - 2i)$ **c)** $\pm(1 + 3i)$ **d)** $\pm\left(\dfrac{1}{\sqrt{2}} - \dfrac{1}{\sqrt{2}}i\right)$

12 $2 - i$
13 $1, 2$
14 $1, -2; 0.3 + 2.1i$
16 a) $-2 \pm 5i$ **b)** $1\frac{1}{2} \pm 2i$ **c)** $i(-1 \pm \sqrt{2})$ **d)** $\frac{1}{2}(\pm\sqrt{7} - i)$
17 a) $z^2 - 6z + 4 = 0$ **b)** $z^2 + 6z + 10 = 0$ **c)** $z^2 - 4z + 7 = 0$
  d) $z^2 - (1 + i)z + (2 - i) = 0$
18 a) $5, 7; 11$
  b) $-1, 7; -13; z_1^2 + z_2^2 < 0 \Rightarrow z_1, z_2$ complex, but not conversely
19 a) $\frac{1}{2}(1 - i^{n+1})(1 + i)$ **b)** $0.2(1 - 2i)\{1 - (-2i)^n\}$
  c) $\frac{1}{2}i\{1 - (n + 1)i^n + ni^{n+1}\}$
20 $A = \frac{1}{2}i, B = -\frac{1}{2}i$

Exercise 12.2a

4 a) $5 + 5i$ **b)** $-6 + 12i$ **c)** 5 **d)** $5 + 14i$ **e)** $-13 + 21i$
8 $-2 + i, 6 - 3i$ **9** $-4 + 4i, -2i, -5 - i$

10 $6 + 2i, 7 - i$; or $0, 1 - 3i$
11 $7\frac{1}{2} + 1\frac{1}{2}i; -5^{\underline{1}}\ 4\frac{1}{2}i$
12   **a)** $1 + 5i$ **b)** $\overline{\frac{1}{3}} + 6i$ **c)** $2 + 3\frac{1}{2}i$
13 $-2 - 2i$

Exercise 12.2b

1   **a)** $\sqrt{2}, -\pi/4$ **b)** $2\sqrt{2}, 3\pi/4$ **c)** $\sqrt{2}, -3\pi/4$ **d)** $3, \pi$ **e)** $4, \pi/2$
  f) $\sqrt{10}, 0.32$ **g)** $\sqrt{10}, -1.25$ **h)** $5, 0.93$ **i)** $13, 1.97$ **j)** $\sqrt{29}, 1.19$
2   **a)** $-4i$ **b)** -5 **c)** $5\sqrt{2} - 5\sqrt{2}i$ **d)** $2\sqrt{3} + 2i$ **e)** $2\sqrt{3} - 2i$
  f) $-\dfrac{1}{\sqrt{2}} - \dfrac{1}{\sqrt{2}}i$
3   **a)** $5\sqrt{2}, 1.71$ **b)** $\dfrac{1}{\sqrt{2}}, -\dfrac{\pi}{4}$ **c)** $\sqrt{2}, \dfrac{\pi}{4}$ **d)** $5, 0.93$ **e)** $10, 2.50$
  f) $5\sqrt{5}, 2.03$
4   **a)** $a^n = \cos\frac{2}{3}n\pi + i\sin\frac{2}{3}n\pi$ **b)** $b^n = 2^n[\cos\frac{1}{4}n\pi - i\sin\frac{1}{4}n\pi]$
  c) $ab = 2[\cos\frac{5}{12}\pi + i\sin\frac{5}{12}\pi]; a/b = \frac{1}{2}[\cos\frac{11}{12}\pi + i\sin\frac{11}{12}\pi]$
5   **a)** $-i$ **b)** $-i$ **c)** $-\frac{1}{2} - \frac{1}{2}\sqrt{3}i$ **d)** $-i$
6   **a)** $5, 5\pi/6$ **b)** $125, \pi$
7 $2\cos 2\theta, 5\theta$
8 $x^2 - 2rx\cos\theta + r^2 = 0; \cos\theta = \dfrac{-p}{\sqrt{q}}$
9   **a)** $4, \pi/3$ **b)** $\frac{1}{2}, -\pi/6$
10 $\sqrt{2}, -\pi/4; 2\sqrt{2}, -3\pi/4; 1, -\pi/3$
11 $2\sqrt{2}, \pi/4; 2\sqrt{2}, -\pi/4; \mathrm{Re}(z_1^8) = 2^{12}; \mathrm{Im}(z_2^6) = \pm 2^9$
12   **a)** $2\cos\dfrac{\theta}{2}\left(\cos\dfrac{\theta}{2} + i\sin\dfrac{\theta}{2}\right)$ **b)** $i\tan\dfrac{\theta}{2}$ **c)** $\frac{1}{2}\sec\theta$ **d)** $i\tan\left(\dfrac{\theta}{2} + \dfrac{\pi}{4}\right)$
13 $(3 - 2i) \pm \sqrt{3}(1 + i)$
14 $\operatorname{cosec}\theta, \pi/2 - \theta$
15 $1, \pi/2; 1, \pi/4$
16 $1, \pi/3; 1, \pi/2$

Exercise 12.3

5 $1 + i, 1\frac{1}{2}$
6 $2 - i, \frac{1}{2}$
7 $1 + i, 2$
8 $1 + 3i, -3; \lambda = 30$
9 $(z + 2)(z + 2 + 2i)(z + 2 - 2i)$
10 $1 - 2i; (x + 2)(x^2 - 2x + 5)$
11 $(z - 1)(z + 2)(z^2 - 4z + 13); 1, -2, 2 - 3i$
12 1 (repeated), $\frac{1}{2} \pm \frac{1}{2}\sqrt{3}i$
13 $2 \pm i, -2 \pm i$
14 $1 \pm i, 2 \pm 3i$

15 $4 - 3i$; $5, 0$ and $5, \pm \frac{2}{3}\pi$

16 $z^2 + z - 6$; $2 + i, -3 + i$

Exercise 12.4a

1 **a)** $\cos \frac{4}{5}\pi + i \sin \frac{4}{5}\pi$ **b)** $\cos(-140°) + i \sin(-140°)$
   c) $\cos \frac{2}{3}\pi + i \sin \frac{2}{3}\pi$ **d)** $\cos 80° + i \sin 80°$
   e) $10^8\{\cos(-1.14) + i \sin(-1.14)\}$ **f)** $32\{\cos \frac{1}{2}\pi + i \sin \frac{1}{2}\pi$
   g) $2^{7/4}\{\cos(-\frac{7}{24}\pi) + i \sin(-\frac{7}{24}\pi)\}$ **h)** $2^{20}\{\cos \frac{2}{3}\pi + i \sin \frac{2}{3}\pi\}$
   i) $2^{1/6}\{\cos \frac{1}{12}\pi + i \sin \frac{1}{12}\pi\}$ **j)** $\sqrt{2}\{\cos(-\frac{1}{6}\pi) + i \sin(-\frac{1}{6}\pi)\}$

2 **a)** $64\cos^6 \frac{1}{2}\theta (\cos 3\theta + i \sin 3\theta)$ **b)** $\cos(5\theta - \frac{1}{2}\pi) + i \sin(5\theta - \frac{1}{2}\pi)$
   c) $\sec^n \theta\{\cos n(\frac{1}{2}\pi - \theta) + i \sin n(\frac{1}{2}\pi - \theta)\}$ **d)** $\cos \alpha + i \sin \alpha$

4 -64

5 **a)** $\pm(\cos \frac{1}{8}\pi + i \sin \frac{1}{8}\pi), \pm(\sin \frac{1}{8}\pi - i \cos \frac{1}{8}\pi)$
   b) $-i, \cos \frac{3}{10}\pi + i \sin \frac{3}{10}\pi, \cos \frac{7}{10}\pi + i \sin \frac{7}{10}\pi, \cos \frac{1}{10}\pi - i \sin \frac{1}{10}\pi,$
   $\cos \frac{9}{10}\pi - i \sin \frac{9}{10}\pi$

6 $2^{1/12}\{\cos(\frac{1}{24}\pi + \frac{1}{3}k\pi) + i \sin(\frac{1}{24}\pi + \frac{1}{3}k\pi)\}$, $k = 0, 1, 2, 3, 4, 5$

7 **a)** $1, -\frac{1}{2} \pm \frac{1}{2}\sqrt{3}i$ **b)** $-1, \frac{1}{2} \pm \frac{1}{2}\sqrt{3}i$

8 $3 \sin \theta - 4 \sin^3 \theta, 4 \cos^3 \theta - 3 \cos \theta$

9 **a)** $32\cos^6 \theta - 48\cos^4 \theta + 18\cos^2 \theta - 1$
   b) $64\cos^6 \theta - 80\cos^4 \theta + 24\cos^2 \theta - 1$
   c) $\dfrac{5 \tan \theta - 10 \tan^3 \theta + \tan^5 \theta}{1 - 10 \tan^2 \theta + 5 \tan^4 \theta}$
   d) $32\cos^5\theta - 32\cos^3 \theta + 6\cos \theta$

10 **a)** $\frac{1}{4}\cos 3\theta + \frac{3}{4}\cos \theta$
   b) $\frac{1}{8}\cos 4\theta - \frac{1}{2}\cos 2\theta + \frac{3}{8}$
   c) $\frac{1}{16}\sin 5\theta - \frac{5}{16}\sin 3\theta + \frac{5}{8}\sin \theta$

11 $\cos(\pm\frac{1}{4}\pi + \frac{2}{3}n\pi) + i \sin(\pm\frac{1}{4}\pi + \frac{2}{3}n\pi)$, $n = 0, 1, 2$;
   $\frac{1}{32}\cos 6\theta + \frac{3}{16}\cos 4\theta + \frac{15}{32}\cos 2\theta + \frac{5}{16}$

13 $2^n \cos^n \frac{1}{2}\theta(\cos \frac{1}{2}n\theta + i \sin \frac{1}{2}n\theta)$; sum $= 2^{n+1}\cos^n \frac{1}{2}\theta \cos \frac{1}{2}n\theta$

14 $\cos n\theta + i \sin n\theta$

Exercise 12.4b

1 **a)** $\frac{1}{2}\{\cos(20° + k120°) + i \sin(20° + k120°)\}$, $k = 0, 1, 2$
   b) $\sqrt{2}\{\cos(-16° + k72°) + i \sin(-16° + k72°)\}$, $k = 0, 1, 2, 3, 4$
   c) $\cos(-20° + k45°) + i \sin(-20° + k45°)$, $k = 0, 1, 2, \ldots, 7$
   d) $\cos(\frac{1}{18}\pi + \frac{2}{9}k\pi) + i \sin(\frac{1}{18}\pi + \frac{2}{9}k\pi)$, $k = 0, 1, 2, \ldots, 8$
   e) $2^{3/4}\{\cos(-\frac{1}{4}\pi + \frac{1}{2}k\pi) + i \sin(-\frac{1}{4}\pi + \frac{1}{2}k\pi)\}$, $k = 0, 1, 2, 3$
   f) $2^{1/6}\{\cos(\frac{5}{36}\pi + \frac{1}{3}k\pi) + i \sin(\frac{5}{36}\pi + \frac{1}{3}k\pi)\}$, $k = 0, 1, 2, 3, 4, 5$
   g) $\cos(\frac{1}{8}\pi + \frac{1}{2}k\pi) + i \sin(\frac{1}{8}\pi + \frac{1}{2}k\pi)$, $k = 0, 1, 2, 3$
   h) $2^{1/8}\{\cos(\frac{3}{16}\pi + \frac{1}{2}k\pi) + i \sin(\frac{3}{16}\pi + \frac{1}{2}k\pi)\}$, $k = 0, 1, 2, 3$

2 $-1 - 3i, 3 - i, -3 + i$

3   **a)** $-2, 1 \pm \sqrt{3}i$ **b)** $\cos\frac{2}{7}k\pi + i\sin\frac{2}{7}k\pi, \quad k = 1, 2, 3, 4, 5, 6$

   c) $\cos\frac{1}{5}\pi \pm i\sin\frac{1}{5}\pi, \cos\frac{3}{5}\pi \pm i\sin\frac{3}{5}\pi$

   d) $\cos\frac{2}{5}k\pi + i\sin\frac{2}{5}k\pi, \quad k = 0, 1, 2, 3, 4$

4 $\left(-5 + \dfrac{7}{\sqrt{3}}\right)\omega - \left(5 + \dfrac{7}{\sqrt{3}}\right)\omega^2$

5 $\cos(\frac{1}{6}\alpha + \frac{1}{3}k\pi) + i\sin(\frac{1}{6}\alpha + \frac{1}{3}k\pi), \quad k = 0, 1, 2, 3, 4, 5$

6   **a)** $\pm 1, \pm i$ **b)** $-\frac{1}{2}, -\frac{1}{2} \pm \frac{1}{2}i$

7 $\cos(\frac{1}{10}\pi + \frac{2}{5}k\pi) + i\sin(\frac{1}{10}\pi + \frac{2}{5}k\pi), \quad k = 0, 1, 2, 3, 4; 1, -\sin\frac{3}{10}\pi$

11 $z^2 - 5z + 7 = 0; z^2 + z + 7 = 0$

Exercise 12.5

1   **a)** circle, centre the origin, radius 2
   b) circle, centre 2i, radius 3
   c) circle, centre $-2 + i$, radius $\sqrt{5}$
   d) half-line from the origin, at $2\pi/3$ to the real axis
   e) half-line from -2, at $\pi/2$ to the real axis
   f) half-line from $3 - 2i$, at $-3\pi/4$ to the real axis

2   **a)** imaginary axis
   b) mediator of -1 and i
   c) real axis
   d) mediator of the origin and $-2 - i$
   e) circle, centre $1\frac{1}{8}$, radius $\frac{3}{8}$
   f) circle, centre $2 - \frac{1}{3}i$, radius $\sqrt{40/3}$

3   **a)** semi-circle above the real axis, centre the origin, radius 1
   b) major arc of the circle, centre $2 + 2i$, radius 2, from 2 to 2i
   c) line segment joining 1 and $-i$

4   **a)** ellipse, foci the origin and $1 - i$

   b) ellipse $\dfrac{x^2}{4} + \dfrac{y^2}{3} = 1$

   c) hyperbola $4x^2 - \dfrac{4y^2}{3} = 1$

5   **a)** $x = 0$
   b) $x + y = 0$
   c) $y = 0$
   d) $4x + 2y + 5 = 0$
   e) $8x^2 + 8y^2 - 18x + 9 = 0$
   f) $3x^2 + 3y^2 - 12x + 2y = 1$

6   **a)** interior of circle, centre the origin, radius 3
   b) exterior of circle, centre -2, radius 2
   c) circumference and exterior of circle, centre $-4i$, radius 3
   d) exterior of circle, centre $3 - i$, radius 2

e) interior of ellipse in no. **4b)**
f) all points for which $y > 0$

7   $\sqrt{3} + i$

8   a) exterior of circle, centre the origin, radius 3
b) $x < 3$
c) region above the real axis, and below the half-line from -3, at $\pi/6$ to the real axis

9   a) circumference and interior of circle, centre $-4 + 2i$, radius $2\sqrt{5}$
b) region between half-lines from $-4 + 2i$, at angles $\pi/3$ and $\pi/2$ to the real axis

10   a) line $x = 1$
b) semi-circle, centre the origin, radius 2, above the real axis; $1 + \sqrt{3}i$

12   centre the origin, radius 1

14   a) 3   b) $2 + 2i$

15   a) parabola $y^2 = 4x$ b) points $\pm 1, \pm i$
c) semi-circle, centre the origin, radius 1, to the right of the imaginary axis

16   a) $4\sqrt{10/3}$ b) $2\frac{2}{3} - 2i; 4 + 2i$

Exercise 12.6

1   a) i   b) $\frac{1}{2} - \frac{1}{2}\sqrt{3}i$ c) $-\dfrac{1}{\sqrt{2}} + \dfrac{1}{\sqrt{2}}i$ d) $-\frac{1}{2} - \frac{1}{2}\sqrt{3}i$ e) $0.37i$
f) $0.54 + 0.84i$ g) $-0.42 - 0.91i$ h) $1.47 + 2.29i$

2   a) $\sqrt{2}e^{i\pi/4}$ b) $3e^{i\pi/2}$ c) $2e^{i\pi}$ d) $e^{-i\pi/4}$
e) $2e^{i\pi/6}$ f) $2e^{2i\pi/3}$ g) $\sqrt{5}e^{0.46i}$ h) $5e^{0.93i}$

3   $x^2 - 2\sqrt{3}x + 4 = 0$

4   a) $1, e^{\pm 2i\pi/3}$ b) $2e^{i\pi}, 2e^{\pm i\pi/3}$
c) $\sqrt{2}e^{i\pi/20}, \sqrt{2}e^{i\pi/4}, \sqrt{2}e^{9i\pi/20}, \sqrt{2}e^{-3i\pi/20}, \sqrt{2}e^{-7i\pi/20}$

5   $\frac{1}{13}e^{-2x}(-2\cos 3x + 3\sin 3x); \frac{1}{13}e^{-2x}(-3\cos 3x - 2\sin 3x)$

8   $\sqrt{3} + i, 1 + i; \dfrac{\sqrt{3} + 1}{2\sqrt{2}}$

10   $-(1 + i) \pm 8^{1/4}e^{3i\pi/8}$

Exercise 12.7a

2   $\dfrac{b}{a}\coth\theta$

4   a) $2\cosh 2x$ b) $\frac{1}{2}\sinh\frac{1}{2}x$ c) $x\sinh x + \cosh x$ d) $\coth x$
e) $\tanh x$ f) $-\operatorname{sech} x\tanh x$ g) $-\operatorname{cosech} x\coth x$ h) $\operatorname{cosech} x$

5   a) $\cosh x$ b) $\frac{1}{2}\sinh 2x$ c) $\tanh x$ d) $\ln\cosh x$
e) $x\sinh x - \cosh x$ f) $x\tanh x - \ln\cosh x$

6   a) $2\sinh 1 = 2.35$ b) $\frac{1}{2}\pi(\sinh 2 + 2) = 8.837$

7 a) $x + \dfrac{x^3}{3!} + \dfrac{x^5}{5!} + \cdots$ **b)** $1 + \dfrac{x^2}{2!} + \dfrac{x^4}{4!} + \cdots$

8 a) $\ln 2 \approx 0.69$ **b)** $0, \ln 3 \approx 1.10$

11 a) $\sin x \cosh y + i \cos x \sinh y, \quad r^2 = \tfrac{1}{2}(\cosh 2y - \cos 2x)$

   b) $\cos x \cosh y - i \sin x \sinh y, \quad r^2 = \tfrac{1}{2}(\cosh 2y + \cos 2x)$

   c) $\dfrac{\sin 2x + i \sinh 2y}{\cos 2x + \cosh 2y}, \quad r^2 = \dfrac{\cosh 2y - \cos 2x}{\cosh 2y + \cos 2x}$

12 $x = \tfrac{1}{2}\ln(2 + \sqrt 3), y = n\pi + \pi/3$

Exercise 12.7b

1 a) $\ln 2$ **b)** $\ln(2 + \sqrt 3)$ **c)** $\tfrac{1}{2}\ln\dfrac{1+x}{1-x}$

2 a) $\sinh^{-1}\tfrac{1}{2}x$ **b)** $\cosh^{-1}\tfrac{1}{3}x$ **c)** $\tfrac{1}{2}\sinh^{-1}\tfrac{2}{3}x$ **d)** $\tfrac{1}{3}\cosh^{-1}\tfrac{3}{2}x$

   e) $\cosh^{-1}(x+1)$ **f)** $\sinh^{-1}(x+1)$

3 a) $x\sinh^{-1}x - \sqrt{(1+x^2)}$ **b)** $x\cosh^{-1}x - \sqrt{(x^2-1)}$

   c) $\tfrac{1}{2}\{x\sqrt{(x^2+1)} + \sinh^{-1}x\}$ **d)** $\tfrac{1}{2}x\sqrt{(x^2-4)} - 2\cosh^{-1}\tfrac{1}{2}x$

4 a) $\sinh^{-1}2 = \ln(2+\sqrt 5) = 1.44$

   b) $\cosh^{-1}\tfrac{3}{2} - \cosh^{-1}1 = \ln\left(\dfrac{3}{2} + \dfrac{\sqrt 5}{2}\right) = 0.96$

   c) $\sqrt 3 - \tfrac{1}{2}\cosh^{-1}2 = \sqrt 3 - \tfrac{1}{2}\ln(2+\sqrt 3) = 1.07$

   d) $\tfrac{1}{2}\sqrt 5 + \tfrac{1}{4}\sinh^{-1}2 = \tfrac{1}{2}\sqrt 5 + \tfrac{1}{4}\ln(2+\sqrt 5) = 1.48$

5 $\dfrac{1}{a}\tanh^{-1}\dfrac{x}{a}$

6 $\ln 2 - \sinh^{-1}1$

Exercise 12.8a

1 a) translation

   b) half-turn about $1 - 2i$

   c) enlargement $\times 5$, centre $\tfrac{1}{2}i$

   d) enlargement $\times 2$ and rotation $+\tfrac{1}{2}\pi$ about O

   e) spiral similarity about O, $\times 5$ through $(-\tan^{-1}\tfrac{4}{3})$

   f) spiral similarity about O, $\times 5$ through an angle α

   g) enlargement $\times\tfrac{1}{2}$, centre -2

   h) spiral similarity about O, $\times\sqrt 2$ through $-\tfrac{1}{4}\pi$

   i) rotation through $+\tfrac{1}{2}\pi$ about $-\tfrac{1}{2} - \tfrac{1}{2}i$

2 a) $w = z + 3i$ **b)** $w = 3z \operatorname{cis} 135°$

   c) $w - i = \dfrac{3}{\sqrt 2}(-1 + i)(z - i)$ **d)** $w = -iz + 2$

   e) $w = -z + 8 + 2i$ **f)** $w = 4z - 6 - 3i$ **g)** $w = 3z - 2i$

   h) $w = 3z - 6i$ **i)** $w = -2iz + 1 + 2i$ **j)** $w = -2iz + 4 + 3i$

3 a) rotation $+\tfrac{1}{2}\pi$ about point 1 **b)** reflection in line $y = x - 1$

5 **a)** reflection in $x + y = 1$ **b)** reflection in $y = x \tan \alpha$
   c) glide reflection in line $y = x + 1$ **d)** glide reflection in $y = x - 1$
6 **a)** semi-circle $|z| = 3$ **b)** semi-circle $|z| = 2$ through $-2i, -2, 2i$
   c) semi-circle $|z| = 1$ through $-1, i, 1$ **d)** the same as **c)** moved $+2$ in
   direction of real axis
   e) semi-circle $|z| = 2$ through $-2i, 2, 2i$

Exercise 12.8b

1 **a)** whole circle $|z| = 1$, beginning and ending at -1
   b) whole circle $|z + 1| = 1$ **c)** quadrant of circle $|z| = 1$
2 **a)** the same translated $+1$ in direction of imaginary axis
   b) the same enlarged $\times 2$ in origin   **c)** the same rotated $+\frac{1}{2}\pi$ about
   origin
   d) the circle $|z| = 1$ anti-clockwise from 1 to 1, thence along real axis to
   O
   e) the semi-circle $|z| = 1$ clockwise through $1, -i, -1$, thence along real
   axis to $-\infty$, and from $+\infty$ to $+1$
3 semi-circle $|z| = 1$ through $1, i, -1$
4 **a)** $2, 2\theta; 1, -\theta$
   b)   i) twice anti-clockwise round the circle, centre the origin, radius 2,
        starting at $(1, 0)$
      ii) clockwise round the circle, centre the origin, radius 1, starting at
        $(1, 0)$
      iii) anti-clockwise round the circle, centre $(-2, 0)$, radius 1, starting
        at $(-1, 0)$
      iv) down the line $x = -\frac{1}{2}$
5 imaginary axis with $y > 0$
6 anti-clockwise along $|w| = 1$
7 $|z + 4 - 3i| = 5; 3x - 4y = -24$
8 coaxal system of Apollonius circles based on the points 1 and $-2i$; the
   origin side of the line $2x + 4y + 3 = 0$
10 $x = \text{constant} \mapsto \text{circles centre O, radius } e^x$
   $y = \text{constant} \mapsto \text{half-lines, arg } w = y$
11 three quarters of the circle $x^2 + y^2 - x + y = 0$, starting at O, passing
   through $1 - i$, and finishing at 1
12 circle, centre $\frac{1}{2}$, radius $\frac{1}{2}$
13 circle $|z| = a^2$; line $x = -\frac{1}{2}; x < -\frac{1}{2}; x < -\frac{1}{2}$
14 **a)** locus of P is a circle, centre O, radius $2a$; locus of Q is an ellipse, centre O, semi-major axis $\frac{5}{2}a$, semi-minor axis $\frac{3}{2}a$.
   b) minimum value is $2a$, when P is at a
15 circle, centre $1 + \frac{1}{4}i$, radius $\sqrt{17}/4$
16 **a)** $\dfrac{26}{25} + \dfrac{7}{25}i$ **b)** $-1 - i$ **c)** $\pm\left(\dfrac{1}{\sqrt{2}} + \dfrac{1}{\sqrt{2}}i\right)$ **d)** $y > x$

e) $x + y = -1$

17 **a)** $x + y = 0$ **b)** circle, centre $1\frac{1}{3} + \frac{1}{3}i$, radius $\dfrac{2\sqrt{2}}{3}$

  c) semi-circle through $1, 1 - i$ and $-i$

Miscellaneous problems

1 $a = \pm 2, b = \pm 3$; $(z^2 - 4z + 13)(z^2 + 4z + 13)$
3 rotation through $+\theta$ about c;   half-turn about $\sqrt{3}i$
4 **a)** $2 + 3i, \pm 1$
  b) the points of $|z| = 2$ in the third and fourth quadrants
5 **a)** $\cos(\pm\frac{1}{9}\pi + \frac{2}{3}k\pi) + i\sin(\pm\frac{1}{9}\pi + \frac{2}{3}k\pi)$, $k = 0, 1, 2$
  b) $\frac{1}{2} + iy$
  c) $\frac{1}{2}$
  d) $2^{-1/3}\{\cos\frac{2}{3}k\pi + i\sin\frac{2}{3}k\pi\}$, $k = 0, 1, 2$
  e) $2^{-1/2}\{\cos(\frac{2}{3}k\pi - \frac{1}{4}\pi) + i\sin(\frac{2}{3}k\pi - \frac{1}{4}\pi)\}$, $k = 0, 1, 2$
7 **a)** true   **b)** false, e.g. $a = 0, b = 1, c = i$
  c) true
8 **a)** $r = 1$; f(z) describes line segment $[-1, 1]$
  c) $\theta = 0$, real axis, $x \geqslant 1$
    $\theta = \pi$, real axis, $x \leqslant -1$
    $\theta = \pm\frac{1}{2}\pi$, imaginary axis
9 if U, P, Q represent $1, z, z^2$, and if $\angle$ UOP $= \theta$, then UP and PQ are inclined to real axis at angles 2θ, 3θ ($\triangle$s OUP, OPQ are isosceles)
10 $2, 1/\sqrt{5}$
12 $e^{-\pi/2} \times e^{2k\pi}$
13 $x = \text{constant} \mapsto \text{hyperbolae}$, $y = \text{constant} \mapsto \text{ellipses}$
14 $\begin{pmatrix} \cos 2\alpha & \sin 2\alpha \\ \sin 2\alpha & -\cos 2\alpha \end{pmatrix}$

Chapter 13

Exercise 13.1

2 **a)** $x\dfrac{dy}{dx} + y = 0$ **b)** $\dfrac{dy}{dx} = y$ **c)** $2x\dfrac{dy}{dx} = y$

  d) $y\dfrac{dy}{dx} = x$ **e)** $x^2\dfrac{d^2y}{dx^2} - 2x\dfrac{dy}{dx} + 2y = 0$ **f)** $\dfrac{d^2x}{dt^2} = x$

Exercise 13.2

1 6.3
2 0.49

3 -3.9

4 0.7

5 **a)** 3.7 **b)** 5.0

6 **a)** 0.82 **b)** 0.724; $\frac{2}{3}$; 23%, 9%

7 **a)** 1.35 **b)** 1.22; 1.10; 23%, 11%

8 **a)** 85°C **b)** 32 minutes

9 **a)** 80 m s^{-1} **b)** 5.3 seconds

11 **a)** 1.37 **b)** 1.86 **c)** 1.37

12 1.24; 1.55, 2.00, 2.59, 3.36

Exercise 13.3

1 **a)** $y = A\,e^x$ **b)** $y = A\,e^{-x^2/2}$ **c)** $y = x^2 + A$ **d)** $y = \dfrac{1}{x+c}$

    e) $y = Ax$ **f)** $2x^2 + y^2 = A$

2 **a)** $\dfrac{dy}{dx} = \dfrac{y}{x}, \quad \dfrac{dy}{dx} = -\dfrac{x}{y}, \quad x^2 + y^2 = A$

    b) $\dfrac{dy}{dx} = 1, \quad \dfrac{dy}{dx} = -1, \quad y = -x + A$

    c) $\dfrac{dy}{dx} = \dfrac{2y}{x}, \quad \dfrac{dy}{dx} = -\dfrac{x}{2y}, \quad x^2 + 2y^2 = A$

    d) $\dfrac{dy}{dx} = \dfrac{1}{2y}, \quad \dfrac{dy}{dx} = -2y, \quad y = A\,e^{-2x}$

    e) $\dfrac{dy}{dx} = -y, \quad \dfrac{dy}{dx} = \dfrac{1}{y}, \quad y^2 = 2x + A$

    f) $\dfrac{dy}{dx} = -\dfrac{2y}{x}, \quad \dfrac{dy}{dx} = \dfrac{x}{2y}, \quad x^2 - 2y^2 = A$

3 $\sin x = 2\sin y \;\Rightarrow\; y = \sin^{-1}(\tfrac{1}{2}\sin x)$

4 $y = 2\tan^{-1}\tfrac{1}{2}x$

5 $S = A\,e^{kt}$, 57000

7 100 years

8 160 000 km

9 $x = \dfrac{1}{1 + 9(\frac{2}{3})^{2t/3}}, \quad 26\%$

10 171 years, 141 years

11 $\dfrac{d\theta}{dt} = a - b\theta; \quad 3.76\,\text{min}\,(a = 21.3, b = 0.170)$

12 $a\dfrac{dM}{dt} + bM = f(t),\; 21.2\,\text{days}$

13 **a)** $v\dfrac{dv}{dx} = -g - kv^2$ **b)** $x = \dfrac{1}{2k}\ln\dfrac{g + ku^2}{g + kv^2}$ **c)** $\dfrac{1}{2k}\ln\left(1 + \dfrac{ku^2}{g}\right)$

    d) $\sqrt{\left(\dfrac{g}{k}\right)}$

14 $x = \left(\dfrac{10 - t}{10}\right)^3$, $\frac{1}{8}\,\text{kg}$

15 $132\,\text{min}$

Exercise 13.4

1 **a)** $y = A\,\mathrm{e}^{-x} + x - 1$ **b)** $y = x(\ln x + A)$ **c)** $y = \mathrm{e}^x + Ax^2$
 d) $y = A\,\mathrm{e}^{-x} + x^2 - 2x + 2$

2 **a)** $y = A(1 - x^2)^{-1/2} - 1$ **b)** $(1 + x)y = A\,\mathrm{e}^x - x + 1$

3 $A\,\mathrm{e}^{-t} + \frac{1}{2}(\sin t - \cos t)$, $\frac{1}{2}(\sin t - \cos t)$

4 $2\,\mathrm{e}^{-t}(1 - \mathrm{e}^{-t})$

5 **a)** $\mathrm{e}^{Rt/L}$; $A\,\mathrm{e}^{-(Rt/L)} + \dfrac{E_0}{R^2 + \omega^2 L^2}(R\sin\omega t - \omega L\cos\omega t)$;

 $\dfrac{LE_0}{R^2 + \omega^2 L^2}(R\sin\omega t + \omega L\cos\omega t)$

 b) $\mathrm{e}^{t/RC}$; $A\,\mathrm{e}^{-t/RC} + \dfrac{\omega RCE_0}{1 + \omega^2 R^2 C^2}(\omega RC\sin\omega t - \cos\omega t)$;

 $\dfrac{\omega RCE_0}{1 + \omega^2 R^2 C^2}(\omega RC\sin\omega t - \cos\omega t)$

Exercise 13.5

1 **a)** $y = A\,\mathrm{e}^{3x} + B\,\mathrm{e}^x$ **b)** $y = A\,\mathrm{e}^{5x} + B\,\mathrm{e}^{-x}$ **c)** $y = A\,\mathrm{e}^{2x} + B\,\mathrm{e}^{-2x}$
 d) $y = A + B\,\mathrm{e}^{4x}$ **e)** $x = A\cos 2t + B\sin 2t$
 f) $x = \mathrm{e}^t(A\cos t + B\sin t)$ **g)** $\mathrm{e}^{-t}(A\cos 2t + B\sin 2t)$
 h) $(A + Bt)\mathrm{e}^{-2t}$

2 **a)** $y = A\,\mathrm{e}^x + B\,\mathrm{e}^{2x} + 3$ **b)** $y = A\,\mathrm{e}^x + B\,\mathrm{e}^{2x} + \frac{1}{4}(2x^2 + 6x + 7)$
 c) $x = A\,\mathrm{e}^{-t} + B\,\mathrm{e}^{-2t} + \frac{1}{6}\mathrm{e}^t$
 d) $x = \mathrm{e}^{-t}(A\cos 2t + B\sin 2t) + \frac{1}{10}(2\sin t - \cos t)$

3 **a)** $y = A\,\mathrm{e}^{3x} + B\,\mathrm{e}^{-x} - \frac{1}{3}x + \frac{2}{9}$ **b)** $y = A\,\mathrm{e}^{-3x} + B\,\mathrm{e}^x - \frac{1}{4}\mathrm{e}^{-x}$
 c) $x = A\,\mathrm{e}^t + B\,\mathrm{e}^{2t} + \frac{1}{10}(\sin t + 3\cos t)$
 d) $\theta = \mathrm{e}^{-t}(A\cos t + B\sin t) - \frac{1}{5}(\sin t + 2\cos t)$

4 **a)** $y = \mathrm{e}^{2x} + 2\,\mathrm{e}^x$ **b)** $x = \mathrm{e}^{2t} - \mathrm{e}^{-2t}$ **c)** $\theta = \mathrm{e}^t + \mathrm{e}^{-2t} - 2$

5 $2(\mathrm{e}^{-2t} - \mathrm{e}^{-3t})$, $\frac{8}{27}\text{rad}$ ($\approx 17°$) when $t = \ln 1.5\,(\approx 0.4\,\text{s})$

6 $q = A\,\mathrm{e}^{-200t} + B\,\mathrm{e}^{-1000t} + \frac{1}{10}$

Miscellaneous problems

1 $805\,\text{s}$

2 $8\,mu^2/3P$

3 $x = \dfrac{e\,Et^2}{2m}$, $y = \dfrac{W}{\Omega}(1 - \cos\Omega t)$, $z = \dfrac{W}{\Omega}\sin\Omega t$

4 $\dfrac{a}{b}\sqrt{\dfrac{2h}{g}}$

5 $\dfrac{d\mathbf{r}}{dt} = -c\mathbf{v} + \mathbf{g}; \quad \mathbf{r} = \dfrac{1}{c}(ct + e^{-ct} - 1)\mathbf{g} + \dfrac{1}{c}(1 - e^{-ct})\mathbf{u}$

6 $\mathbf{A} = 18\mathbf{i} + 3\mathbf{j}, \quad \mathbf{B} = -2\mathbf{i} - 3\mathbf{j}$

7 $y - \dfrac{V\sin\alpha}{k} + \dfrac{g}{k^2}(1 - e^{-kt}) - \dfrac{gt}{k}$

10 $200\,°C, \quad T = 200 - 20\,e^{-t}(2\sin 2t + 9\cos 2t), \quad 0.895\,\text{min},$
409 °C, 1.448 min

Chapter 14

Exercise 14.1

1   **a)** $8\,\text{N}$ **b)** $1\,\text{N}$ **c)** $0.2\,\text{m}$ **d)** $0.05\,\text{m}$
2   **a)** $0.16\,\text{J}$ **b)** $2.5 \times 10^{-3}\,\text{J}$ **c)** $4\,\text{J}$ **d)** $0.25\,\text{J}$
3   **a)** $2 \times 10^3\,\text{N}\,\text{m}^{-1}$ **b)** $0.004\,\text{J}$
4   **a)** $5 \times 10^4\,\text{N}\,\text{m}^{-1}$ **b)** $2.5\,\text{J}$
5   **a)** $4\,\text{m}$ **b)** $400\,\text{J}$ **c)** $800\,\text{J}$
6   **a)** $200(2 + x)$ **b)** $25x^2$ **c)** $400 + 200x - 25x^2; 9.7\,\text{m}$
7 $4.06 \times 10^4\,\text{J}, 15\,\text{m}, 32.8\,\text{m}\,\text{s}^{-1}, 53.9\,\text{m}\,\text{s}^{-2}$

Exercise 14.2a

1   **a)** $4, 2\pi = 6.28\,\text{s}, 1/2\pi = 0.159\,\text{s}^{-1}, -1.66$
   b) $2, \pi/5 = 0.628\,\text{s}, 5/\pi = 1.59\,\text{s}^{-1}, 1.82$
   c) $5, \pi/50 = 0.0628\,\text{s}, 50/\pi = 15.9\,\text{s}^{-1}, -4.4$
   d) $a, 1/n; n, a\sin 4n\pi\,(= 0, \text{if } n \in \mathbb{Z})$
2   **a)** $10\sin t$ **b)** $3\sin\pi t$ **c)** $\frac{1}{2}\sin 200\pi t$

Exercise 14.2b

1   **a)** $\ddot{x} = -4x$ **b)** $3.14\,\text{s}$ **c)** $x = 3\cos 2t$ **d)** $6\,\text{m}\,\text{s}^{-1}$
3   **a)** $\frac{3}{2}$ **b)** $6.28\,\text{s}$ **c)** $\frac{1}{3}$ **d)** 0.13
4   **a)** $0.63\,\text{s}$ **b)** $0.68\,\text{s}$

7 $(\frac{1}{2}\pi + \frac{1}{2})\sqrt{\dfrac{a}{2g}}$

8 $mg(1 - \cos nt)$

9 $\dfrac{v}{\sqrt{(gl)}}, \quad \frac{1}{4}\pi\sqrt{\dfrac{l}{g}}$

10 $\ddot{\theta} = -\dfrac{4T}{ml}\theta, \quad \pi\sqrt{\dfrac{ml}{T}}$

Exercise 14.3

1   $1\frac{5}{6}, 3\frac{1}{3}\,\mathrm{m\,s}^{-1}$
2   $3.6\,\mathrm{m\,s}^{-1}, 0.4$
3   $-\frac{1}{2}, +2\,\mathrm{m\,s}^{-1}$
4   $1\,\mathrm{m\,s}^{-1}, \frac{1}{8}$
5   $10, 7.5\,\mathrm{m\,s}^{-1}$
8   $0.952, 40.7\,\mathrm{m}$

Exercise 14.4a

1   a) $\mathbf{i}+\mathbf{k}$ b) $-5\mathbf{i}+3\mathbf{j}-\mathbf{k}$ c) $2\mathbf{i}-\mathbf{j}$ d) $-11\mathbf{i}+9\mathbf{j}-5\mathbf{k}$
    e) $-4\mathbf{i}-3\mathbf{j}+\mathbf{k}$ f) $-12\mathbf{i}-6\mathbf{j}-12\mathbf{k}$
3   a) force $6\mathbf{i}+3\mathbf{j}$ along $3x-6y+4=0$ b) couple $4\mathbf{k}$
    c) equilibrium   d) force $2\mathbf{i}+15\mathbf{j}$ along $15x-2y-2=0$
4   $(30P, 30P)$ along $x-y=\frac{1}{3}a$;   force $(30P, 30P)$, couple $10aP$

Exercise 14.4b

1   $700\,\mathrm{N}, 800\,\mathrm{N}$; $75\,\mathrm{kg}$
2   a) $0.213\,\mathrm{N}$ b) $0.106\,\mathrm{N}$
3   $\frac{1}{3}W(4x-y+6), \frac{1}{3}W(-x+4y+6)$
4   $11\frac{1}{4}\,\mathrm{kg}, 8\frac{1}{4}\,\mathrm{kg}$ 5 $\frac{2}{3}W\cot\beta$ 6 $40°$
8   a) $W\tan 15°, \tan 15° \approx 0.27$ b) $\frac{1}{3}W, \tan 30° \approx 0.58$

    c) $\frac{1}{4}W, \dfrac{1}{2\sqrt{3}} \approx 0.29$

10   $\frac{5}{9}l$ from A, $\dfrac{5mgl^2}{6a}$, $\dfrac{mgl(18a-5l)}{12a}$
11   $\dfrac{2d-l\cos\theta}{l\sin\theta}$
12   a) $2\sqrt{13}P, 6x-8y=5a$ b) Pa c) $6Pa$

Exercise 14.5a

2   $\frac{5}{3}\mathbf{i}+\mathbf{j}$, towards O

Exercise 14.5b

1   $\frac{4}{3}M(a^2+b^2)$
2   a) $\frac{1}{2}Ma^2$ b) $\frac{3}{2}Ma^2$ c) $2Ma^2$
3   a) $\frac{5}{4}Ma^2$ b) $\frac{3}{2}Ma^2$ 4   a) $\frac{2}{5}Ma^2$ b) $\frac{7}{5}Ma^2$
5   a) $\frac{2}{3}Ma^2$ b) $\frac{5}{3}Ma^2$

6   a) $\frac{3}{10}Ma^2$ b) $\frac{3}{20}M(a^2+4h^2)$ c) $\frac{1}{20}M(3a^2+2h^2)$

Exercise 14.5c

1. **a)** $12 \, \mathrm{kg \, m^2 \, s^{-1}}$, $36 \, \mathrm{J}$
   b) $0.48 \, \mathrm{kg \, m^2 \, s^{-1}}$, $0.96 \, \mathrm{J}$
   c) $0.24 \, \mathrm{kg \, m^2 \, s^{-1}}$, $0.48 \, \mathrm{J}$
   d) $96 \, \mathrm{kg \, m^2 \, s^{-1}}$, $288 \, \mathrm{J}$
   e) $168 \, \mathrm{kg \, m^2 \, s^{-1}}$, $877 \, \mathrm{J}$
   f) $7.15 \times 10^{33} \, \mathrm{kg \, m^2 \, s^{-1}}$, $2.60 \times 10^{29} \, \mathrm{J}$

2. **a)** $2F/ma$; **b)** $ma\omega/2F$; **c)** $ma\omega^2/4F$

3. $5.82 \, \mathrm{N}$, $1.8 \, \mathrm{s}$, $58.2 \, \mathrm{N}$

4. **a)** $62.8 \, \mathrm{rad \, s^{-2}}$ **b)** $126 \, \mathrm{rad \, s^{-1}}$ **c)** $78.5 \, \mathrm{N}$

5. $3.65 \, \mathrm{kg \, m^2}$

6. **a)** mga, $\sqrt{(3g/2a)}$, $3g/4a$; **b)** $2mga$, $\sqrt{(3g/a)}$, 0

7. $38.7 \, \mathrm{rad \, s^{-1}}$

8. **a)** $\frac{1}{2}mgl\sin\theta$ **b)** $\ddot{\theta} = -\dfrac{3g}{2l}\sin\theta$ **c)** $\ddot{\theta} = -\dfrac{3g\theta}{2l}$ **d)** $2\pi\sqrt{(2l/3g)}$

9. **a)** $2\sqrt{\left(\dfrac{Mgx}{2M+m}\right)}$ **b)** $\dfrac{2Mg}{2M+m}$

10. $2\pi\sqrt{5a/3g}$

Exercise 14.6

1. **a)** $\mathbf{ML^{-1}T^{-2}}$ **b)** $\mathbf{ML^{-1}}$ **c)** $\mathbf{ML^{-2}}$ **d)** $\mathbf{T^{-1}}$
   e) $\mathbf{T^{-1}}$ **f)** $\mathbf{T^{-2}}$ **g)** 1 **h)** $\mathbf{ML^2}$
   i) $\mathbf{MT^{-1}}$ **j)** $\mathbf{M^{-1}L^3T^{-2}}$ **k)** $\mathbf{MT^{-2}}$ **l)** $\mathbf{MT^{-2}}$

2. **a)** $T \propto \dfrac{mv^2}{r}$ **b)** $h \propto \dfrac{E}{mg}$ **c)** $v \propto \sqrt{\dfrac{p}{\rho}}$ **d)** $f \propto \sqrt{\dfrac{k}{m}}$ **e)** $v \propto \sqrt{g\lambda}$

3. $T = kv^2\rho A$, $\mathbf{M^{-1}LT^2}$

Miscellaneous problems

1. $A\rho(1-k)u^2$, $4.5 \, \mathrm{m \, s^{-1}}$

3. $3 \, \mathrm{h} \, 35 \, \mathrm{min}$, $6.19 \times 10^6 \, \mathrm{N \, s}$, $1.03 \times 10^3 \, \mathrm{N}$

4. $\pi\sqrt{\dfrac{a}{g}}$

5. $l + \dfrac{mgl}{\lambda}\left[1 + \sqrt{\left(1 + \dfrac{2\lambda}{mg}\right)}\right]$,

   $\sqrt{\dfrac{2l}{g}} + \sqrt{\dfrac{ml}{\lambda}}\left[\tfrac{1}{2}\pi + \sin^{-1}\dfrac{1}{\sqrt{(1 + 2\lambda/mg)}}\right]$

6. C moves with constant velocity $\dfrac{P}{3m}$, $\dfrac{P^2}{3m}$

7 $P = \frac{1}{2}Av(v^2 + 2gh) \times 10^{-4}$, $\quad \ddot{\theta} = \dfrac{Av^2c}{10I}$, $\quad 0$

8 $M^{-1}L^0T$, $\quad \dfrac{dV}{dt} = -\lambda A\rho g V$

9 $\dfrac{I}{2\sqrt{(kC)}}\ln\dfrac{\sqrt{C} + \omega\sqrt{k}}{\sqrt{C} - \omega\sqrt{k}}$

10 MT^{-2}; $\alpha = -\frac{1}{2}, \beta = \frac{1}{2}, \gamma = \frac{3}{2}$

Chapter 15

Exercise 15.1a

1 $1.5, 0.75$ **2 a)** 2.92 **b)** 5.83 **c)** 1.46

3 $4.5, 8.25$ **4** $5.96, 1.52$

5 $4\frac{17}{36}$ **6** $1.94, 1.44$

7 $1.73, 3.13$ **8** $5.58, 0.60$

9 $3.61, 6.69$ **10** £2.78

11 $0.16, 0.04$; 1.24

12 a) $\frac{2}{11}$ **b)** $\frac{5}{11}$; $\quad 1.636, 0.595$; $1.636, 0.744$

13 $29.75, 230$; 0.5

14 $P(T = 30) = p$, $\quad P(T = 45) = 1 - p$; $\quad 45 - 15p, 225p(1 - p)$;
$P(T - 30) = \frac{1}{2}p$, $\quad P(T = 35) = \frac{1}{2}$, $\quad P(T = 45) = \frac{1}{2}(1 - p)$; $40 - \frac{15}{2}p$

Exercise 15.1b

1 $\dfrac{t}{2 - t}$; $\quad 2, 2$ **2 a)** $2, 1$ **b)** $18, 3$ **c)** $50, 5$

3 36, $\dfrac{48}{\sqrt{n}}$ **4** $\dfrac{3}{2}$, $\dfrac{3\sqrt{2}}{4}$

5 0.59; $\quad 7, 6.3$; $\quad 8.26$ **6** $4.49, 2.47$

7 a) $0.96, 0.88$ **b)** 1.58

9 $8, 7.92$; $\quad$ approximation np; $\quad 1.01\%$

10 a) $\frac{8}{27}$ **b)** $\frac{11}{27}$; $\quad \frac{20}{3}, \frac{40}{9}$

11 $(\frac{1}{6}t + \frac{1}{3}t^2 + \frac{1}{2}t^3)^5$; $\quad -\frac{1}{243}, \frac{121}{243}$

12 $\frac{2}{3} + \frac{1}{3}t$; $\quad \frac{1}{2} + \frac{1}{2}t$; $\quad 1\frac{5}{6}, \frac{17}{36}$

13 $\dfrac{t(1 - t^{10})}{10(1 - t)}$; $\dfrac{t^4(1 - t^{10})^4}{10^4(1 - t)^4}$; $\quad 0.012$

14 $\dfrac{t^2}{(6 - 5t)^2}$; $\quad 12, 60$

15 a) $(\frac{1}{5}t^4 + \frac{4}{5}t^{-1})^{25}, 0, 10$ **b)** $16.67, 11.78$

Exercise 15.2a

1 $a = 2$; $\dfrac{x^2}{8} + \dfrac{x}{4}$ **a)** $\frac{5}{32}$ **b)** $\frac{5}{8}$

2 $X = \sqrt[3]{2}$ **a)** 0.001 **b)** $\frac{7}{64}$

3 **a)** $\frac{1}{4}$ **b)** $\frac{2}{3}$

$$F(x) = \begin{cases} 0 & (x < 0) \\ \dfrac{x^2}{12} & (0 \leqslant x \leqslant 3) \\ 2x - \dfrac{x^2}{4} - 3 & (3 \leqslant x \leqslant 4) \\ 1 & (4 < x) \end{cases}$$

$\sqrt{6}$

4 $k = \frac{3}{4}$ **a)** $\frac{11}{16}$ **b)** $\frac{7}{27}$; $F(x) = \begin{cases} 0 & (x \leqslant -1) \\ \frac{1}{4}(2 + 3x - x^3) & (-1 \leqslant x \leqslant 1) \\ 1 & (1 \leqslant x) \end{cases}$

5 $a = \dfrac{\pi}{2}$; $F(x) = \begin{cases} 0 & (x < 0) \\ \frac{1}{2}(1 - \cos \pi x) & (0 \leqslant x \leqslant 1) \\ 1 & (1 < x) \end{cases}$ **a)** $\frac{1}{4}$ **b)** $\frac{1}{4}$

6 $f(t) = \dfrac{1}{2000} e^{-t/2000}$ **a)** 0.135 **b)** 0.145 **c)** 0.018

7 $r^2/16, r/8; \frac{5}{16}$

8 median $= -1$; $X_0 = 2.08$ **a)** $\frac{1}{4}$ **b)** $\frac{3}{4}$; $\frac{37}{64}$

Exercise 15.2b

1 $\frac{7}{6}, \frac{11}{36}$

2 $\frac{3}{2}, \frac{3}{4}$ **3** $\frac{7}{3}, \frac{13}{18}$ **4** $0, \frac{1}{5}$

5 $\frac{1}{2}, \frac{1}{4} - 2/\pi^2$

6 $f(x) = \dfrac{1}{b-a}$ for $a \leqslant x \leqslant b$, and zero elsewhere; $\dfrac{a+b}{2}$, $\dfrac{(b-a)^2}{12}$

7 $\alpha = \frac{3}{32}$; $2, \frac{4}{5}$; $\frac{5}{32}$

8 $\frac{3}{2}, \sqrt[3]{4}, \frac{3}{20}$

9 $2 + \dfrac{1}{\lambda}$, $2 + \dfrac{\ln 2}{\lambda}$, $\dfrac{1}{\lambda}$

10 $f(t) = \dfrac{1}{\mu} e^{-t/\mu}$; 0.135; 5000

11 $C = \dfrac{1}{2a}$, $\sigma = 1.53a$ **a)** 0.53 **b)** 1; $k = 1.28$

12 $\frac{3}{4}R, \frac{3}{80}R^2$

13 **a)** $2\frac{7}{8}$ kg **b)** £4.75, $\frac{3}{16}$

14 $\lambda = \frac{1}{2}$; $\frac{1}{2}\pi, \frac{1}{4}\pi^2 - 2, \frac{1}{2}\pi, \frac{1}{3}\pi$ and $\frac{2}{3}\pi$; 0.775

15 **a)** $k = 1/\pi$ **b)** $\frac{1}{2}$ **c)** $\frac{1}{3}$

16 λ, λ

Exercise 15.3a

1 **a)** 0.018   **b)** 0.729   **c)** 0.067
2 **a)** effectively zero   **b)** 0.999   **c)** 0.89
3 **a)** 0.052   **b)** 0.198
4 **a)** 0.131   **b)** 0.063
5 **a)** 0.222   **b)** 0.018   **c)** effectively zero
6 **a)** 0.827   **b)** 0.062   **c)** 0.021
7 **a)** 0.228   **b)** 0.051
8 **a)** 7.8%   **b)** 95.8%   **c)** 0.8%
9 $\frac{15}{37}$;   0.157
10 $n = 610$;   0.09
11 $n = 30, y = 24$
12 **a)** $\frac{1}{5} + \frac{4}{5}p$ **b)** $\dfrac{1-p}{1+4p}$ **c)** $8 + 32p$
   d) 0.13   **e)** 4

Exercise 15.3b

2 **a)** $0.5, 0.1$ **b)** $0.5, 0.05$
3 $1, 1;$ $1, \dfrac{1}{\sqrt{10}}$
4 $2.5, 0.112;$ 0.037
6 $0.157 \text{ kg}, 0.0012 \text{ kg};$ $0.038 \text{ kg};$ 0.6%;   154.53 kg

Exercise 15.4

1 **a)** 0.368   **b)** 0.368   **c)** 0.080
2 **a)** 0.330   **b)** 0.865
3 **a)** 0.018   **b)** 0.195   **c)** 0.156   **d)** 0.0003
4 **a)** 0.165   **b)** 0.269   **c)** 0.373
5 **a)** 0.003   **b)** 0.037
6 52
7 228, 211, 98, 30, 7, 1, 0 (to the nearest integers)
8 theoretical frequencies: 3, 9, 16, 17, 14, 9, 5, 2.5, 1, 0.4, 0—close correspondence with observed frequencies, suggesting lucky and unlucky tenants
9 $2, 1.97;$ so $\mu \approx \sigma^2$, as in case of Poisson distribution
10 50
11 1.05; 0.350, 0.367, 0.193, 0.0675, 0.0177;   105;   27
12 Poisson with mean $\frac{1}{4}n$;   113 matches
14 **a)** 0.9975   **b)** 0.849
15 **a)** 0.60   **b)** 0.91;   0.77;   0.07
16 $1 - e^{-2\mu};$ $\dfrac{1}{1 + e^{\mu}}$

17 $20, 60, 90, 90, 67, 40, 33$ **a)** £157.50   **b)** £167.50   **c)** £160.75
18   **a)** 0.406   **b)** 0.184   **c)** 0.37

Exercise 15.5a

2   **a)** $a\mathrm{E}(X) + b\mathrm{E}(Y)$, $a^2\,\mathrm{Var}(X) + b^2\,\mathrm{Var}(Y)$
 b) $a\mathrm{E}(X) + b\mathrm{E}(Y)$, $a^2\,\mathrm{Var}(X) + 2ab\,\mathrm{Cov}(X, Y) + b^2\,\mathrm{Var}(Y)$
3   **a)** $a\mathrm{E}(X) + b\mathrm{E}(Y) + c\mathrm{E}(Z)$ **b)** $a^2\,\mathrm{Var}(X) + b^2\,\mathrm{Var}(Y) + c^2\,\mathrm{Var}(Z)$
4   10.5, 8.75
5   10.5, 14.58   **6**   0, 2.42
7   Bob has the advantage; Alan's variance = 11.67, Bob's
 variance = 11.25
8   **a)** $6\frac{2}{3}$ **b)** $13\frac{1}{3}$ **c)** $26\frac{2}{3}$ **d)** $33\frac{1}{3}$
9   60 minutes, 3.16 minutes   **a)** 0.0057   **b)** 0.057   **c)** 0.565
10 $0, \frac{1}{2}$; $0, \frac{1}{4}$;   same mean and variance
11   **a)** 0.159   **b)** 0.182;   1, 3;   3, 5;
 c) 0.274   **d)** 0.310
12   **a)** 0.168   **b)** 0.027

13 $\sum_i \lambda_i \mu_i$, $\sqrt{\left(\sum_i \lambda_i^2 \sigma_i^2\right)}$;   150 s, 8.124 s   **a)** 0.7%   **b)** 11%

14   0.638
15 $\mathrm{Var}(Z) = \{(\lambda_1 a_1 + \lambda_2 b_1)^2 + (\lambda_1 a_2 + \lambda_2 b_2)^2\}\sigma^2$

Exercise 15.5b

1   **a)** 0.057   **b)** 0.57
2   **a)** 0.072   **b)** 0.165
3   0.048
4   **a)** 0.115   **b)** 0.008
5   **a)** effectively zero   **b)** 0.9987
6   0.001, suspicions justified
7   136
8   96

Exercise 15.5c

1   58.5, 31.7
2   5.08, 0.55
3   **a)** 1.79   **b)** 1.64   **c)** 1.61

4   **b)** $\sum_1^n \lambda_i = 1$

6 $np, np(1 - p)$; $\dfrac{\theta(1 - \theta)}{20}$, $\dfrac{\theta(1 - 2\theta)}{20}$; P_2 is better since it has a smaller
 variance.

7 $\dfrac{1}{2n}$

8 $\mathrm{Var}(\bar{X}) = \dfrac{\sigma^2}{n};\quad E(s^2) = \sigma^2;\quad \lambda = \dfrac{n-1}{m+n-2}$

Exercise 15.6a

1 significant at the 1% level
2 significant at the 5% level
3 **a)** significant **b)** not significant **c)** significant
   d) not significant evidence against the hypothesis
4 significant at 5% level that techniques are effective
5 significant at 1% level that support has decreased
6 **a)** not a significant result
   b) significant at 1% level that claim is not justified
7 **a)** significant **b)** not significant evidence against the hypothesis
8 significant at 5% level that cards were not drawn at random
9 consistent with hypothesis; significant at 0.1% level that hypothesis is incorrect
10 **a)** 0.0004 **b)** 0.043; not significant at 5% level that germination rate is less than 75%
11 significant at 1% level
12 highly significant, almost conclusive evidence
13 $(1 - X)^n$ **a)** 0.107 **b)** 0.028 **c)** 0.001; significant at 0.1% level that hypothesis is incorrect
14 0.109 **a)** 527 **b)** 537
15 475
16 not significant—consistent with a mean of 100; 0.484, 0.021 (using a two-tailed test)
17 $\frac{11}{32}$; 15
18 **a)** $\alpha = 0.159, \beta = 0.067$ **b)** $k = 1.65, \beta = 0.74$
   c) $n = 87, k = 0.5$

Exercise 15.6b

1 **a)** 58.8, 67.2 **b)** 58.0, 68.0 **c)** 56.5, 69.5
2 **a)** 1150.1, 1153.9 **b)** 1149.6, 1154.4
3 **a)** 154.6, 163.6 **b)** 153.2, 165.0
4 **a)** 7.35, 7.49 **b)** 7.32, 7.52
5 **a)** 54.1, 60.7 **b)** 53.05, 61.75
6 highly significant that there is bias; 0.2–0.6
7 1.9965, 0.0645; 0.0065; 1.98–2.01
8 50.6%, 64.4%

9   0.139, 0.205
10   binomial; $\mu = 100p$, $\sigma = \sqrt{(100p(1 - p))}$;   significant at the 1% level that the insurance does not appeal equally to men and women;   0.23, 0.41
11   no;   yes
12 x_1, $x_1 \pm 0.776$; $\bar{x}$, $\bar{x} \pm 0.163$

Exercise 15.7a

2 $y = 0.58x + 8.4$
3 $W = 0.36T + 40.4$; $W = 60$
4 $y = 1.4x - 26.3$
5   **a)** $y = 0.28x + 69.8$ **b)** $y = 3.6x - 36.4$
6 $y = 33.6x + 8.1$
7 $y = 1.24x - 24.4$
8 $F = 0.9I - 6.3$;   9
9 $y = -1.28x + 20$;   16.2
10   **a)** $y = 0.8x + 1.5$ **b)** $y = 1.9x - 3.1$
11   19.14, 15.03; $y = 0.33x + 8.8$
12 $\log_{10} Y = 0.021X - 37.2$
13   **a)** 15.2, 1.72   **b)** 149   **c)** 2.3   **d)** $a = 1.5, b = 0.015$

Exercise 15.7b

2   0.88, just significant at the 0.1% level
3   0.99, highly significant
4   0.92, highly significant
5   0.28, not significant
6   0.72, just significant at the 1% level
7   0.23, no significant correlation
8   0.35
9   correlation coefficient of -0.84, significant at the 1% level
10   0.65   **11**   0.38   **12**   0.65
13   **a)** 0.61, significant at 5% level   **b)** 0.45, not significant
   c) 0.93, significant at 0.1% level

Exercise 15.7c

1 $\rho = 0.66$, $\tau = 0.47$, not significant
2 $\rho = -0.08$, $\tau = -0.02$, tastes quite dissimilar
3   **a)** 0.79   **b)** 0.61;   both significant at the 1% level
4   0.90   **5** 0.22
6 $\rho = 0.79$, significant at the 1% level
7   **a)** $\rho = 0.92$ **b)** $\tau = 0.78$

8 **a)** $\tau = -0.71$ **b)** not quite significant
9 ACBDE, ABDCE, ABCED; 120; $\frac{1}{24}$
10 $\rho = 0.74$, $\tau = 0.62$, significant at the 1% level
11 $\tau = 0.67$, significant at the 1% level

Miscellaneous problems

1 $\binom{9}{3}\left(\frac{1}{6}\right)^4\left(\frac{5}{6}\right)^6$; $\binom{r-1}{k-1}p^k q^{r-k}$

2 $\dfrac{4(4t)^9\,e^{-4t}}{9!}$; $\frac{5}{8}$; $\dfrac{\lambda(\lambda t)^{k-1}\,e^{-\lambda t}}{(k-1)!}$

3 **a)** $\dfrac{(1-k)(1+t)}{1+(1-k)t}$ **b)** $\dfrac{(1-k)(1+2t)}{1+(2-k)t}$

4 **a)** $\frac{1}{4}$ **b)** $3\frac{3}{4}\,\text{min}$

5 **a)** 68 **b)** 404 **c)** 471

6 $\binom{n}{r}p^r(1-p)^{n-r}\dfrac{\lambda^n}{n!}e^{-\lambda}$; $\dfrac{(\lambda p)^r\,e^{-\lambda p}}{r!}$

7 36, 180

8 $50m + 16.3\sigma$

9 **a)** $\frac{1}{3}$ **b)** $\frac{19}{27}$

10 **a)** $-np(1-p)$ **b)** $\lambda = 1.96\sqrt{n}$ **c)** 0.077

11 **a)** $P(A_1) + P(A_2) + P(A_3) - P(A_1 \cap A_2) - P(A_2 \cap A_3) - P(A_3 \cap A_1)$
$\qquad + P(A_1 \cap A_2 \cap A_3)$;

$$\sum_i P(A_i) - \sum_{i \neq j}\sum P(A_i \cap A_j) + \cdots + (-1)^{n-1}P(A_1 \cap A_2 \ldots \cap A_n)$$

   b) $1 - \dfrac{1}{1!} + \dfrac{1}{2!} - \dfrac{1}{3!} + \cdots + (-1)^n\dfrac{1}{n!}$, which tends to e^{-1}

12 $P(t + \delta t) = (1 - \beta\delta t)P(t) + \alpha\delta t(1 - P(t))$

13 **a)** $p_n = \frac{1}{2}$ **b)** $p_n = \dfrac{1}{2}\left[1 + \left(\dfrac{a-b}{a+b}\right)^{n-1}\right]$

14 $A, 0.621$

15 $p_r = \dbinom{2N-r-1}{N-1}\dfrac{1}{2^{2N-r}}$; $\dbinom{2N-2}{N-1}\dfrac{2N-1}{2^{2N-1}}$

Chapter 16

Exercise 16.1

1 $\begin{pmatrix} 5 & 4 \\ -4 & 3 \\ 1 & -1 \end{pmatrix}$, $\begin{pmatrix} 6 & 2 \\ 0 & 4 \\ -2 & 8 \end{pmatrix}$, $\begin{pmatrix} 1 & -2 \\ 4 & 1 \\ -3 & 9 \end{pmatrix}$, $\begin{pmatrix} 9 & 10 \\ -12 & 5 \\ 5 & -11 \end{pmatrix}$, $\begin{pmatrix} 5 & -3 \\ 8 & 4 \\ -7 & 22 \end{pmatrix}$

2 $AB = \begin{pmatrix} 6 & 1 \\ 7 & 2 \end{pmatrix}$, $BA = \begin{pmatrix} 4 & 1 \\ 11 & 4 \end{pmatrix}$ $CD = \begin{pmatrix} 1 & -3 & 3 \\ -2 & 0 & 4 \\ 8 & -12 & 4 \end{pmatrix}$

$DC = \begin{pmatrix} -1 & 7 \\ -4 & 6 \end{pmatrix}$ $BD = \begin{pmatrix} -1 & 0 & 2 \\ 0 & -3 & 5 \end{pmatrix}$ $AD = \begin{pmatrix} -2 & 3 & 9 \\ 1 & -6 & 8 \end{pmatrix}$

$CA = \begin{pmatrix} 7 & 3 \\ 8 & 2 \\ 12 & 8 \end{pmatrix}$ $CB = \begin{pmatrix} 3 & 1 \\ 2 & 0 \\ 8 & 4 \end{pmatrix}$ **DB, DA, AC, BC**

impossible

3 $\begin{pmatrix} 1 & -3 \\ 1 & 0 \\ 1 & -1 \\ 0 & 1 \end{pmatrix} \begin{pmatrix} -1 & 2 & 7 & 3 \\ 8 & 1 & 2 & 0 \end{pmatrix} = \begin{pmatrix} -25 & -1 & 1 & 3 \\ -1 & 2 & 7 & 3 \\ -9 & 1 & 5 & 3 \\ 8 & 1 & 2 & 0 \end{pmatrix}$

4 a) $\begin{pmatrix} 1 & 0 & 0 \\ 0 & 1 & 0 \\ 0 & 0 & 1 \end{pmatrix}$ **b)** $\begin{pmatrix} -14 - 10\sqrt{2} & 27 - 11\sqrt{2} \\ -27 - 11\sqrt{2} & -14 + 10\sqrt{2} \end{pmatrix}$;

5 $\begin{pmatrix} 4 & 0 & 1 \\ 2 & 6 & 3 \\ -1 & 2 & 3 \end{pmatrix}$, $\begin{pmatrix} 2 & 3 & 0 \\ 1 & 4 & 2 \\ 3 & 5 & 6 \end{pmatrix}$

pre-multiplication exchanges the first and second rows of the matrix, post-multiplication exchanges the first and second columns

6 $\begin{pmatrix} 2 & 6 & 3 \\ 12 & 0 & 3 \\ -1 & 2 & 3 \end{pmatrix}$, $\begin{pmatrix} 1 & 4 & 2 \\ 3 & 9 & 0 \\ 3 & 5 & 6 \end{pmatrix}$

pre-multiplication multiplies the second row of a matrix by 3, post-multiplication multiplies the second column by 3

7 $\begin{pmatrix} 0 & 0 \\ 0 & 0 \end{pmatrix}$, $\begin{pmatrix} 12 & -6 \\ 24 & -12 \end{pmatrix}$

8 $\begin{pmatrix} 2 & 1 \\ 0 & 0 \end{pmatrix}$, $\begin{pmatrix} 2 & 1 \\ 0 & 0 \end{pmatrix}$

10 a) $\begin{pmatrix} 10 & 2 \\ 2 & 6 \end{pmatrix}$ **b)** $\begin{pmatrix} 10 & 2 & -2 \\ 2 & 4 & 2 \\ -2 & 2 & 2 \end{pmatrix}$

12 $\begin{pmatrix} a & ka \\ -\dfrac{a}{k} & -a \end{pmatrix}$; $\begin{pmatrix} a & b \\ c & -a \end{pmatrix}$ where $bc = 1 - a^2$

13 Only scalar matrices: $\begin{pmatrix} k & 0 & 0 \\ 0 & k & 0 \\ 0 & 0 & k \end{pmatrix}$ commute with all 3×3 square matrices

14 3D, $\begin{pmatrix} 0 & 0 \\ 0 & 0 \end{pmatrix}$

15 $\begin{pmatrix} a_1 b_1 - a_2 b_2 & a_1 b_2 + a_2 b_1 \\ -a_2 b_1 - a_1 b_2 & a_1 b_1 - a_2 b_2 \end{pmatrix}$, $\begin{pmatrix} a_1 b_1 - a_2 b_2 & a_2 b_1 + a_1 b_2 \\ -a_1 b_2 - a_2 b_1 & a_2 b_1 - a_2 b_2 \end{pmatrix}$,

multiplication of complex numbers: $(a_1 + ia_2)(b_1 + ib_2)$

16 $\begin{pmatrix} 13 & -21 \\ -21 & 34 \end{pmatrix}$, $\begin{pmatrix} 90 & -144 \\ -144 & 233 \end{pmatrix}$, etc. (Fibonacci)

17 $\begin{pmatrix} 1 & 0 & 0 \\ 10 & 1 & 0 \\ 10 & 0 & 1 \end{pmatrix}$

19 $\mathbf{C}^2 = \begin{pmatrix} 1 & -1 \\ 1 & 0 \end{pmatrix}$, $\mathbf{C}^6 = \begin{pmatrix} -1 & 0 \\ 0 & -1 \end{pmatrix}$

20 $a = 1, b = 1, c = -1, d = 4, e = 0, f = 3, g = 2, h = -1, k = -1$

Exercise 16.2a

1 a) $\begin{pmatrix} -1 & 0 \\ 0 & -1 \end{pmatrix}$, $\begin{pmatrix} 0 & -1 \\ 1 & 0 \end{pmatrix}$, $\begin{pmatrix} 0 & 1 \\ -1 & 0 \end{pmatrix}$,

$\begin{pmatrix} \dfrac{1}{2} & -\dfrac{\sqrt{3}}{2} \\ \dfrac{\sqrt{3}}{2} & \dfrac{1}{2} \end{pmatrix}$, $\begin{pmatrix} \dfrac{1}{2} & \dfrac{\sqrt{3}}{2} \\ -\dfrac{\sqrt{3}}{2} & \dfrac{1}{2} \end{pmatrix}$, $\begin{pmatrix} -\dfrac{1}{\sqrt{2}} & -\dfrac{1}{\sqrt{2}} \\ \dfrac{1}{\sqrt{2}} & -\dfrac{1}{\sqrt{2}} \end{pmatrix}$

b) $\begin{pmatrix} -1 & 0 \\ 0 & 1 \end{pmatrix}$, $\begin{pmatrix} 0 & 1 \\ 1 & 0 \end{pmatrix}$, $\begin{pmatrix} 0 & -1 \\ -1 & 0 \end{pmatrix}$, $\begin{pmatrix} 0.6 & 0.8 \\ 0.8 & -0.6 \end{pmatrix}$

c) $\begin{pmatrix} 3 & 0 \\ 0 & 3 \end{pmatrix}$, $\begin{pmatrix} 2 & 0 \\ 0 & 1 \end{pmatrix}$, $\begin{pmatrix} 1 & 0 \\ -1 & 1 \end{pmatrix}$

2 a) x and y stretches
b) enlargement and half-turn
c) enlargement $\times 5$ and rotation
d) x-shear
e) y-shear
f) reflection in $y = -x$ and enlargement

g) $\begin{pmatrix} 2 & -1 \\ 1 & 2 \end{pmatrix} = \begin{pmatrix} 1 & -\frac{1}{2} \\ 0 & 1 \end{pmatrix}\begin{pmatrix} 2 & 0 \\ 0 & 2 \end{pmatrix}$ (x-shear and enlargement)

h) reflection in $y = x \tan \frac{1}{2}\alpha$

i) reflection in line $y = x \tan (45° + \frac{1}{2}\alpha)$

j) rotation $+40°$

k) all points map into line $y = \frac{1}{2}x$

l) $(x, y) \mapsto (x, x)$ lying on $y = x$

3 reflections in $y = 0$, $x = 0$, $y = \pm x$; rotation through $\frac{1}{2}n\pi$, $(n \in \mathbb{Z})$

5 rotation of $\pi/4$ about the origin, reflection in $y = x/\sqrt{3}$

6 $\begin{pmatrix} \cos 2\theta & \sin 2\theta \\ \sin 2\theta & -\cos 2\theta \end{pmatrix}$

7 a) $\begin{pmatrix} 2 & 3 \\ -1 & -2 \end{pmatrix}\begin{pmatrix} 2 & -1 & 0 \\ 1 & 2 & 3 \end{pmatrix} = \begin{pmatrix} 7 & 4 & 9 \\ -4 & -3 & -6 \end{pmatrix}$

b) $\begin{pmatrix} 3 & 1 \\ 2 & 0 \end{pmatrix}\begin{pmatrix} 2 & -1 & 0 \\ 1 & 2 & 3 \end{pmatrix} = \begin{pmatrix} 7 & -1 & 3 \\ 4 & -2 & 0 \end{pmatrix}$

c) $\begin{pmatrix} 4 & -6 \\ -2 & 3 \end{pmatrix}\begin{pmatrix} 2 & -1 & 0 \\ 1 & 2 & 3 \end{pmatrix} = \begin{pmatrix} 2 & -16 & -18 \\ -1 & 8 & 9 \end{pmatrix}$;   all lying on line $2y + x = 0$

8 $\begin{pmatrix} 1 & 0 \\ 1 & 1 \end{pmatrix}\begin{pmatrix} 0 & 0 & 1 & 1 \\ -1 & 1 & 1 & -1 \end{pmatrix} = \begin{pmatrix} 0 & 0 & 1 & 1 \\ -1 & 1 & 2 & 0 \end{pmatrix}$ (parallelogram)

9 $\begin{pmatrix} -2 & 2 \\ 2 & 1 \end{pmatrix}\begin{pmatrix} 3 \\ -2 \end{pmatrix} = \begin{pmatrix} -10 \\ 4 \end{pmatrix}$

10 $-1, -\frac{11}{6}$

11 $M^2 = \begin{pmatrix} 7 & -6 \\ -18 & 19 \end{pmatrix}$; vectors in direction $(1, 1)$ unchanged; vectors in direction $(1, -3) \times 25$

12 a) $\pm 1; \begin{pmatrix} 3 \\ 1 \end{pmatrix}$, $\begin{pmatrix} 1 \\ -3 \end{pmatrix}$ **b)** $1, 6; \begin{pmatrix} 4 \\ -1 \end{pmatrix}$, $\begin{pmatrix} 1 \\ 1 \end{pmatrix}$ **c)** $4, -2; \begin{pmatrix} 1 \\ 1 \end{pmatrix}$, $\begin{pmatrix} 1 \\ -1 \end{pmatrix}$

13 $5, -1; y = \pm\frac{1}{3}x$

14 $q = \pm\sqrt{(1 - p^2)}$, $r = \mp\sqrt{(1 - p^2)}$, $s = p$; $\begin{pmatrix} \frac{3}{5} & \frac{4}{5} \\ -\frac{4}{5} & \frac{3}{5} \end{pmatrix}$, $\begin{pmatrix} \frac{3}{5} & -\frac{4}{5} \\ \frac{4}{5} & \frac{3}{5} \end{pmatrix}$; rotations through equal angles but in opposite senses.

Exercise 16.2b

1 rotation of $90°$ about the origin; reflection in $x = 0$; enlargement centre the origin, scale factor 2; shear parallel to the x-axis with $(0, 1) \rightarrow (1, 1)$

$$\mathbf{AB} = \begin{pmatrix} 0 & -1 \\ -1 & 0 \end{pmatrix} \quad \mathbf{BA} = \begin{pmatrix} 0 & 1 \\ 1 & 0 \end{pmatrix} \quad \mathbf{AC} = \begin{pmatrix} 0 & -2 \\ 2 & 0 \end{pmatrix} \quad \mathbf{CB} = \begin{pmatrix} -2 & 0 \\ 0 & 2 \end{pmatrix}$$

$$\mathbf{A}^2 = \begin{pmatrix} -1 & 0 \\ 0 & -1 \end{pmatrix} \quad \mathbf{B}^2 = \begin{pmatrix} 1 & 0 \\ 0 & 1 \end{pmatrix} \quad \mathbf{C}^2 = \begin{pmatrix} 4 & 0 \\ 0 & 4 \end{pmatrix} \quad \mathbf{D}^2 = \begin{pmatrix} 1 & 2 \\ 0 & 1 \end{pmatrix}$$

2   rotation of $-90°$ about the origin; rotation of $90°$ about the origin; reflection in $x + y = 0$; reflection in $x = 0$; rotation of $180°$ about the origin

3   a) $\begin{pmatrix} 3 & 0 \\ 0 & 3 \end{pmatrix}\begin{pmatrix} -1 & 0 \\ 0 & -1 \end{pmatrix}$ enlargement and half-turn

   b) $\begin{pmatrix} 0 & -3 \\ -3 & 0 \end{pmatrix}\begin{pmatrix} 0 & -1 \\ -1 & 0 \end{pmatrix}$ enlargement and reflection

   c) $2\begin{pmatrix} \cos\frac{1}{3}\pi & \sin\frac{1}{3}\pi \\ \sin\frac{1}{3}\pi & -\cos\frac{1}{3}\pi \end{pmatrix}$ enlargement and reflection

   d) $13\begin{pmatrix} \cos\alpha & -\sin\alpha \\ \sin\alpha & \cos\alpha \end{pmatrix}$ enlargement and rotation, $\tan\alpha = \frac{12}{5}$

   e) $\begin{pmatrix} 2 & 0 \\ 0 & 2 \end{pmatrix}\begin{pmatrix} 1 & -1\frac{1}{2} \\ 0 & 1 \end{pmatrix}$ shear and enlargement

   f) $\begin{pmatrix} 2 & 0 \\ 0 & 3 \end{pmatrix}\begin{pmatrix} 0 & -1 \\ 1 & 0 \end{pmatrix}$ x- and y-stretches and quarter-turn

   g) $\begin{pmatrix} 1 & 1 \\ -1 & 3 \end{pmatrix} = \begin{pmatrix} 1 & 0 \\ -1 & 1 \end{pmatrix}\begin{pmatrix} 1 & 0 \\ 0 & 4 \end{pmatrix}\begin{pmatrix} 1 & 1 \\ 0 & 1 \end{pmatrix}$ shears and y-stretch

4   $\begin{pmatrix} 1 & 0 \\ 0 & -1 \end{pmatrix}$, $\begin{pmatrix} \cos 2\theta & \sin 2\theta \\ \sin 2\theta & -\cos 2\theta \end{pmatrix}$;   rotation of 2θ about the origin

5   rotation of α, reflection in $y = x\tan(\frac{1}{2}\alpha)$

6   e.g. $\mathbf{A} = \begin{pmatrix} -\dfrac{1}{2} & -\dfrac{\sqrt{3}}{2} \\ \dfrac{\sqrt{3}}{2} & -\dfrac{1}{2} \end{pmatrix}$, $\mathbf{B} = \begin{pmatrix} 0 & -1 \\ 1 & 0 \end{pmatrix}$,

   $\mathbf{C} = \begin{pmatrix} \dfrac{1}{2} & -\dfrac{\sqrt{3}}{2} \\ \dfrac{\sqrt{3}}{2} & \dfrac{1}{2} \end{pmatrix}$, $\mathbf{D} = \begin{pmatrix} \dfrac{1}{\sqrt{2}} & -\dfrac{1}{\sqrt{2}} \\ \dfrac{1}{\sqrt{2}} & \dfrac{1}{\sqrt{2}} \end{pmatrix}$, $\mathbf{E} = \begin{pmatrix} \dfrac{\sqrt{3}}{2} & -\dfrac{1}{2} \\ \dfrac{1}{2} & \dfrac{\sqrt{3}}{2} \end{pmatrix}$

7   a) $\begin{pmatrix} 0 & 1 \\ -1 & 0 \end{pmatrix}$ b) $\begin{pmatrix} \frac{1}{2} & 0 \\ 0 & \frac{1}{2} \end{pmatrix}$ c) $\begin{pmatrix} 1 & -3 \\ 0 & 1 \end{pmatrix}$ d) $\begin{pmatrix} -1 & 0 \\ 0 & 1 \end{pmatrix}$

8 $p = b/d, q = a - bc/d, r = c, s = d$; shear parallel to $y = 0$

9 a) $\begin{pmatrix} 2 & 0 \\ 0 & 2 \end{pmatrix}\begin{pmatrix} x \\ y \end{pmatrix} + \begin{pmatrix} 1 \\ 0 \end{pmatrix}$ **b)** $\begin{pmatrix} 2 & -2 \\ 2 & 2 \end{pmatrix}\begin{pmatrix} x \\ y \end{pmatrix} + \begin{pmatrix} 0 \\ -2 \end{pmatrix}$

c) $\begin{pmatrix} 0 & -1 \\ 3 & 0 \end{pmatrix}\begin{pmatrix} x \\ y \end{pmatrix} + \begin{pmatrix} -2 \\ 1 \end{pmatrix}$

10 a) $\begin{pmatrix} 0 & -1 \\ -1 & 0 \end{pmatrix}\begin{pmatrix} x \\ y \end{pmatrix} + \begin{pmatrix} 0 \\ 0 \end{pmatrix}$ **b)** $\begin{pmatrix} 0 & 1 \\ 1 & 0 \end{pmatrix}\begin{pmatrix} x \\ y \end{pmatrix} + \begin{pmatrix} 2 \\ -2 \end{pmatrix}$

c) $\begin{pmatrix} 0 & -1 \\ 1 & 0 \end{pmatrix}\begin{pmatrix} x \\ y \end{pmatrix} + \begin{pmatrix} 1 \\ -3 \end{pmatrix}$ **d)** $\begin{pmatrix} 0 & -1 \\ 1 & 0 \end{pmatrix}\begin{pmatrix} x \\ y \end{pmatrix} + \begin{pmatrix} -1 \\ -1 \end{pmatrix}$

e) $\begin{pmatrix} 1 & 0 \\ 0 & 1 \end{pmatrix}\begin{pmatrix} x \\ y \end{pmatrix} + \begin{pmatrix} -2 \\ -2 \end{pmatrix}$; rotation of $90°$ about $(0, -1)$

Exercise 16.3

1 a) $\begin{pmatrix} 1 & 0 & 0 \\ 0 & -1 & 0 \\ 0 & 0 & 1 \end{pmatrix}$ **b)** $\begin{pmatrix} 0 & 1 & 0 \\ 1 & 0 & 0 \\ 0 & 0 & 1 \end{pmatrix}$

c) $\begin{pmatrix} \cos 2\alpha & \sin 2\alpha & 0 \\ \sin 2\alpha & -\cos 2\alpha & 0 \\ 0 & 0 & 1 \end{pmatrix}$ **d)** $\frac{1}{3}\begin{pmatrix} 1 & -2 & -2 \\ -2 & 1 & -2 \\ -2 & -2 & 1 \end{pmatrix}$ **e** $\begin{pmatrix} -1 & 0 & 0 \\ 0 & 1 & 0 \\ 0 & 0 & -1 \end{pmatrix}$

f) $\begin{pmatrix} 0 & 0 & 1 \\ 0 & -1 & 0 \\ 1 & 0 & 0 \end{pmatrix}$ **g)** $\frac{1}{3}\begin{pmatrix} -1 & 2 & 2 \\ 2 & -1 & 2 \\ 2 & 2 & -1 \end{pmatrix}$ **h)** $\begin{pmatrix} 1 & 0 & 0 \\ 0 & 0 & \mp1 \\ 0 & \pm1 & 0 \end{pmatrix}$

i) $\frac{1}{2}\begin{pmatrix} 0 & -\sqrt{2} & -\sqrt{2} \\ \sqrt{2} & 1 & -1 \\ \sqrt{2} & -1 & 1 \end{pmatrix}$ and its inverse   **j)** $\begin{pmatrix} \cos\theta & 0 & \sin\theta \\ 0 & 1 & 0 \\ -\sin\theta & 0 & \cos\theta \end{pmatrix}$

k) $\begin{pmatrix} 5 & 0 & 0 \\ 0 & 5 & 0 \\ 0 & 0 & 5 \end{pmatrix}$ **l)** $\begin{pmatrix} 1 & 0 & 0 \\ 0 & 0 & 0 \\ 0 & 0 & 1 \end{pmatrix}$

2 a) reflection in $y = 0$

  b) orthogonal projection on to $z = 0$

  c) $120°$ rotation about $x = y = z$

  d) quarter-turn about x-axis

  e) x and y stretches, z reflection

  f) enlargement $\times 3$ and central inversion

  g) rotation through $\tan^{-1}\frac{3}{4}$ about z-axis

  h) rotation through $\tan^{-1}\frac{4}{3}$ about x-axis

3 a) -1 **b)** 0 **c)** 1 **d)** 1 **e)** -6 **f)** -27 **g)** 1 **h)** 1

5 all points have images in the plane $x - 2y - z = 0$

6 $\begin{pmatrix} 0 & -1 & 0 \\ 1 & 0 & 0 \\ 0 & 0 & -1 \end{pmatrix} \begin{pmatrix} 1 & -1 & 1 & 1 & 1 & -1 & -1 & -1 \\ 1 & 1 & -1 & 1 & -1 & 1 & -1 & -1 \\ 1 & 1 & 1 & -1 & -1 & -1 & 1 & -1 \end{pmatrix}$

$= \begin{pmatrix} -1 & -1 & 1 & -1 & 1 & -1 & 1 & 1 \\ 1 & -1 & 1 & 1 & 1 & -1 & -1 & -1 \\ -1 & -1 & -1 & 1 & 1 & 1 & -1 & 1 \end{pmatrix}$

7 $\mathbf{p} = 2\begin{pmatrix} -1 \\ 2 \\ -1 \end{pmatrix} + 3\begin{pmatrix} 1 \\ -1 \\ 1 \end{pmatrix}$; $(0, 2, 3)$; $(0, 2, 3)$

8 $\begin{pmatrix} 3 \\ 1 \\ -\frac{8}{3} \end{pmatrix}, \begin{pmatrix} -2 \\ 3 \\ 4 \end{pmatrix}, \begin{pmatrix} -1 \\ 1 \\ \frac{7}{3} \end{pmatrix}; \begin{pmatrix} 3 & -2 & -1 \\ 1 & 3 & 1 \\ -\frac{8}{3} & 4 & \frac{7}{3} \end{pmatrix}$

9 rotation of $120°$ about the axis $x = y = z$; rotation of $180°$ about z-axis; reflection in $x = z$; enlargement, centre the origin, scale factor 2

$\mathbf{AB} = \begin{pmatrix} 0 & 0 & 1 \\ -1 & 0 & 0 \\ 0 & -1 & 0 \end{pmatrix}$ $\mathbf{BA} = \begin{pmatrix} 0 & 0 & -1 \\ -1 & 0 & 0 \\ 0 & 1 & 0 \end{pmatrix}$ $\mathbf{AC} = \begin{pmatrix} 1 & 0 & 0 \\ 0 & 0 & 1 \\ 0 & 1 & 0 \end{pmatrix}$

$\mathbf{CB} = \begin{pmatrix} 0 & 0 & 1 \\ 0 & -1 & 0 \\ -1 & 0 & 0 \end{pmatrix}$ $\mathbf{A}^2 = \begin{pmatrix} 0 & 1 & 0 \\ 0 & 0 & 1 \\ 1 & 0 & 0 \end{pmatrix}$ $\mathbf{B}^2 = \begin{pmatrix} 1 & 0 & 0 \\ 0 & 1 & 0 \\ 0 & 0 & 1 \end{pmatrix}$

$\mathbf{C}^2 = \begin{pmatrix} 1 & 0 & 0 \\ 0 & 1 & 0 \\ 0 & 0 & 1 \end{pmatrix}$ $\mathbf{D}^2 = \begin{pmatrix} 4 & 0 & 0 \\ 0 & 4 & 0 \\ 0 & 0 & 4 \end{pmatrix}$

10 a) $\begin{pmatrix} 1 & 0 & 0 \\ 0 & 1 & 0 \\ 0 & 0 & 3 \end{pmatrix}\begin{pmatrix} -1 & 0 & 0 \\ 0 & 0 & 1 \\ 0 & 1 & 0 \end{pmatrix}$ **b)** $\begin{pmatrix} 2 & 0 & 0 \\ 0 & 2 & 0 \\ 0 & 0 & 2 \end{pmatrix}\begin{pmatrix} 0 & 1 & 0 \\ 1 & 0 & 0 \\ 0 & 0 & 1 \end{pmatrix}$

c) $\begin{pmatrix} 1 & 0 & 0 \\ 0 & 1 & 0 \\ 0 & 0 & 0 \end{pmatrix}\begin{pmatrix} 0 & 1 & 0 \\ 1 & 0 & 0 \\ 0 & 0 & 1 \end{pmatrix}$

11 a) $\begin{pmatrix} \frac{1}{3} & 0 & 0 \\ 0 & \frac{1}{3} & 0 \\ 0 & 0 & \frac{1}{3} \end{pmatrix}$ **b)** $\begin{pmatrix} 0 & -1 & 0 \\ 1 & 0 & 0 \\ 0 & 0 & 1 \end{pmatrix}$ **c)** $\begin{pmatrix} 1 & 0 & 0 \\ 0 & 1 & 0 \\ 0 & 0 & -1 \end{pmatrix}$

12 e.g. $A = \begin{pmatrix} 0 & 1 & 0 \\ -1 & 0 & 0 \\ 0 & 0 & 1 \end{pmatrix}$, $B = \begin{pmatrix} \dfrac{1}{\sqrt{2}} & -\dfrac{1}{\sqrt{2}} & 0 \\ \dfrac{1}{\sqrt{2}} & \dfrac{1}{\sqrt{2}} & 0 \\ 0 & 0 & 1 \end{pmatrix}$

13 suspect $120°$ rotation about line $2x = y = 2z$

14 vectors $(0 \ \ 0 \ \ z)$

Exercise 16.4a

1 a) $\begin{pmatrix} \frac{1}{2} & 0 \\ 0 & 1 \end{pmatrix}$ **b)** $\begin{pmatrix} \frac{1}{3} & 0 \\ 0 & \frac{1}{3} \end{pmatrix}$ **c)** $\begin{pmatrix} -1 & 0 \\ 0 & 1 \end{pmatrix}$ **d)** $\begin{pmatrix} 0 & -1 \\ 1 & 0 \end{pmatrix}$

e) none **f)** $\begin{pmatrix} 0 & 1 \\ 1 & 0 \end{pmatrix}$ **g)** $\begin{pmatrix} \frac{5}{7} & -\frac{1}{7} \\ \frac{3}{7} & \frac{2}{7} \end{pmatrix}$ **h)** $\begin{pmatrix} \frac{1}{4} & 0 \\ \frac{1}{2} & 1 \end{pmatrix}$

2 a) I b) A

3 a) $(\frac{2}{5}, -\frac{1}{5})$ **b)** $(1, 1)$ **c)** $(1, -1)$ **d)** $(-1\frac{1}{2}, 1)$

4 $\begin{pmatrix} 1 & 1 \\ 1 & 1\frac{1}{2} \end{pmatrix}\begin{pmatrix} 1 & 1 & -1 & -1 \\ 1 & -1 & 1 & -1 \end{pmatrix} = \begin{pmatrix} 2 & 0 & 0 & -2 \\ 2\frac{1}{2} & -\frac{1}{2} & \frac{1}{2} & -2\frac{1}{2} \end{pmatrix}$

5 a) rotation of $-90°$ about the origin; $\begin{pmatrix} 0 & -1 \\ 1 & 0 \end{pmatrix}$, $\begin{pmatrix} 0 & 1 \\ -1 & 0 \end{pmatrix}$

b) reflection in $y = 0$; $\begin{pmatrix} 1 & 0 \\ 0 & -1 \end{pmatrix}$, $\begin{pmatrix} 1 & 0 \\ 0 & -1 \end{pmatrix}$

c) enlargement, centre $(0, 0)$, scale factor $\frac{1}{3}$; $\begin{pmatrix} 3 & 0 \\ 0 & 3 \end{pmatrix}$, $\begin{pmatrix} \frac{1}{3} & 0 \\ 0 & \frac{1}{3} \end{pmatrix}$

6 $\begin{pmatrix} \frac{3}{2} & -\frac{5}{2} \\ -1 & 2 \end{pmatrix}$ **a)** $-7, 6$ **b)** $2, -1$ **c)** $-6\frac{1}{2}, 5$ **d)** $1\frac{1}{2}, -2$

7 a) $(-34, 15)$ **b)** $(-\frac{2}{7}, \frac{15}{7})$ **c)** none **d)** $(\frac{7}{4}, -\frac{19}{4})$

8 a) $\begin{pmatrix} -7\frac{1}{2} & -4 \\ 3\frac{1}{2} & 2 \end{pmatrix}$ **b)** $\begin{pmatrix} -4 & 3\frac{1}{2} \\ 2 & -1\frac{1}{2} \end{pmatrix}$

9 $d = a, bc = 1 - a^2$; half-turn and reflections

10 $ad \neq bc$

11 a) $\begin{pmatrix} -8 & 19 \\ 3 & -7 \end{pmatrix}$ **b)** $\begin{pmatrix} -1 & -3 \\ 1 & 2\frac{1}{2} \end{pmatrix}$

12   reflection in $x = 0$; rotation of $120°$ about the origin; reflection in $y = -x/\sqrt{3}$; ± 1, $\begin{pmatrix} \sqrt{3} \\ -1 \end{pmatrix}$, $\begin{pmatrix} 1 \\ \sqrt{3} \end{pmatrix}$

Exercise 16.4b

1 $\frac{1}{6}\begin{pmatrix} 29 & 31 & 13 \\ 1 & -1 & -1 \\ 9 & 9 & 3 \end{pmatrix}$ **2** $\begin{pmatrix} 1 & -2 & 1 \\ 0 & 1 & -1 \\ 0 & 0 & 1 \end{pmatrix}$ **3** $\frac{1}{2}\begin{pmatrix} 2 & 3 & 1 \\ -6 & -8 & 2 \\ 14 & 19 & -1 \end{pmatrix}$

4 $\frac{1}{3}\begin{pmatrix} 1 & -2 & -1 \\ -1 & 5 & 3 \\ 3 & -12 & -6 \end{pmatrix}$ **5** $\begin{pmatrix} 1 & 1 & 1 \\ 1 & 2 & 2 \\ 1 & 2 & 3 \end{pmatrix}$ **6** $\begin{pmatrix} 1 & -a & ac-b \\ 0 & 1 & -c \\ 0 & 0 & 1 \end{pmatrix}$

7 $(1-a)^{-1}(1+2a)^{-1}\begin{pmatrix} 1+a & -a & -a \\ -a & 1+a & -a \\ -a & -a & 1+a \end{pmatrix}$

8 $(a^2 + b^2 + c^2 - 1 - 2abc)^{-1}\begin{pmatrix} c^2-1 & a-bc & b-ac \\ a-cb & b^2-1 & c-ab \\ b-ac & c-ab & a^2-1 \end{pmatrix}$

9 **a)** $\begin{pmatrix} -1 & 0 & 0 \\ 0 & 1 & 0 \\ 0 & 0 & 1 \end{pmatrix}$ **b)** $\begin{pmatrix} 1 & 0 & 0 \\ 0 & 0 & 1 \\ 0 & -1 & 0 \end{pmatrix}$ **c)** $\begin{pmatrix} 1 & 0 & 0 \\ 0 & \frac{1}{2} & 0 \\ 0 & 0 & 1 \end{pmatrix}$

10 $\begin{pmatrix} 2 & -4 & 2 \\ 2 & 2 & -1 \\ -4 & -4 & 5 \end{pmatrix} = \text{inverse} \times 6$; $\begin{pmatrix} -1 \\ -\frac{1}{2} \\ 1\frac{1}{2} \end{pmatrix}$

11 $\begin{pmatrix} 4 & -1 & -2 \\ -22 & 2 & 12 \\ -8 & 1 & 4 \end{pmatrix}$, $\begin{pmatrix} -2 & \frac{1}{2} & 1 \\ 11 & -1 & -6 \\ 4 & -\frac{1}{2} & -2 \end{pmatrix}$; $\begin{pmatrix} -5\frac{1}{2} \\ 30 \\ 10\frac{1}{2} \end{pmatrix}$

12 $\frac{1}{2}\begin{pmatrix} 1 & -1 & 1 \\ 1 & 1 & -1 \\ -1 & 1 & 1 \end{pmatrix}$; $\begin{pmatrix} a \\ b \\ c \end{pmatrix} = \begin{pmatrix} 2 \\ 1 \\ 3 \end{pmatrix}$

13 $k = 1$; $\begin{pmatrix} 2 & -4 & -2 \\ 2 & -4 & -3 \\ -2 & 5 & 3 \end{pmatrix}$

14 $\mathbf{M}^{-1} = \begin{pmatrix} -25 & 4 & 3 \\ 26 & -4 & -3 \\ -33 & 5 & 4 \end{pmatrix}$; $(\mathbf{M}^T)^{-1} = (\mathbf{M}^{-1})^T$

15 $\dfrac{1}{\Delta}\begin{pmatrix} A & C & B \\ B & A & C \\ C & B & A \end{pmatrix}$, where $A = a^2 - bc$, etc.; $\Delta = \Sigma a^3 - 3abc = 0$ when

$a + b + c = 0$, or $a = b = c$

18 $\begin{pmatrix} \cos\alpha\cos\beta & -\cos\alpha\sin\beta & \sin\alpha \\ \sin\alpha\cos\beta & \sin\alpha\sin\beta & -\cos\alpha \\ -\sin\beta & \cos\beta & 0 \end{pmatrix}$; matrix orthogonal; rotation

Exercise 16.5a

1 **a)** -5 **b)** 54 **c)** 0 **d)** 2670 **e)** 2232
f) 0 **g)** 0

2 $\begin{vmatrix} a & b \\ a^n & b^n \end{vmatrix}$ has factor $a - b$

3 **a)** $4ab$ **b)** 1 **c)** $\cos 2\theta$ **d)** 1 **e)** $-1 - t^4$
f) $(ps - rq)(x - y)z$

4 e.g. $\begin{vmatrix} pa + qb & ra + sb \\ pc + qd & rc + sd \end{vmatrix} = \Delta \times (ps - qr)$

5 $\begin{vmatrix} a & b \\ c & d \end{vmatrix} + \begin{vmatrix} a & y \\ c & q \end{vmatrix} + \begin{vmatrix} x & b \\ p & d \end{vmatrix} + \begin{vmatrix} x & y \\ p & q \end{vmatrix}$

7 $\lambda = -1, 7$ **8** $x = 0, -1$

9 $x = -2\frac{1}{2}, y = \frac{1}{2}$

Exercise 16.5b

1 **a)** -1 **b)** -3 **c)** -21 **d)** -19320
2 **a)** 16 **b)** 0
3 $a^3 + b^3 + c^3 - 3abc$, $a^3 + b^3 + c^3 - 3abc$, $3abc - a^3 - b^3 - c^3$;
$a^3 + b^3$, $1 - 2x^3 + x^6$
4 $-2, 8, 9$
5 any pair of rows or columns are linearly dependent
6 **a)** $2abc$ **b)** 0 **c)** 0

Exercise 16.5c

1 **a)** 276 **b)** $-ab(a - b)(1 - a)(1 - b)$
c) $(a + b + c)^3$
d) $-(a + b + c)(a + b\omega + c\omega^2)(a + b\omega^2 + c\omega)$

2 $(a + b + c)(a + b\omega + c\omega^2)(a + b\omega^2 + c\omega) = (a + b + c)(\sum a^2 - \sum bc)$

3 $(a - 1)(a^2 + a + 1), \quad (1 + a + a^2)^2(1 - a)^2; \quad 0$

4 **a)** $(b - a)(a - c)(c - b)$ **b)** $(a + b + c)(b - a)(a - c)(c - b)$
   c) $\sum bc - \sum a^2$ **d)** $(b - a)(a - c)(c - b)(bc + ca + ab)$
   e) 0

5 -1

6 $(2a + b)(a - b)^2, \quad -(2a + b)(a - b)^2, \quad 0$

7 $(a + x + y)(a - x)(a - y); \quad 0, \pi/2, 2\pi/3, \pi$

8 **b)** 49

9 e.g. $\begin{vmatrix} 1 & 0 \\ 0 & 1 \end{vmatrix} + \begin{vmatrix} 1 & 1 \\ 1 & 1 \end{vmatrix} = \begin{vmatrix} 2 & 1 \\ 1 & 2 \end{vmatrix}$

10 $\begin{vmatrix} a^2 + bc & ac & c^2 \\ ab & bc + ac & bc \\ ab & ab & ac + b^2 \end{vmatrix} = \begin{vmatrix} a^2 + c^2 & ab & ac \\ ab & b^2 + c^2 & bc \\ ac & bc & a^2 + b^2 \end{vmatrix}$

11 $\begin{vmatrix} a & p & 0 \\ b & q & 0 \\ c & r & 0 \end{vmatrix}^2 = 0$

12 $|\mathbf{a}+\mathbf{x} \quad \mathbf{b}+\mathbf{y} \quad \mathbf{c}+\mathbf{z}| = |\mathbf{a\ b\ c}| + |\mathbf{a\ b\ z}| + |\mathbf{x\ b\ c}| + |\mathbf{x\ b\ z}|$
   $+ |\mathbf{a\ y\ z}| + |\mathbf{x\ y\ c}| + |\mathbf{x\ y\ z}|$

13 **a)** $\begin{vmatrix} 0 & c & b \\ c & 0 & a \\ b & a & 0 \end{vmatrix}^2 = 4a^2b^2c^2$ **b)** $(b - a)^3$ **c)** -8
   d) $(a + b + c + d)(a - b + c - d)(a + b - c - d)(a - b - c + d)$
   e) 0

14 e.g. $\begin{vmatrix} 1 & 2 & 3 & 4 \\ 2 & 3 & 4 & 5 \\ 3 & 4 & 5 & 6 \\ 1 & 1 & 1 & 1 \end{vmatrix}$

Exercise 16.6

1 $4, 2, 1$ **2** $1, 3, -2$ **3** $3, -5, -1$

4 infinite number of solutions   **5** no solutions

6 $\begin{pmatrix} 1 & 0 & 0 \\ 0 & 1 & 0 \\ 0 & 0 & 1 \end{pmatrix}, \begin{pmatrix} 1 & 0 & 0 \\ 0 & 1 & 0 \\ 0 & 0 & 1 \end{pmatrix};$ **a)** $5, 1, 0$ **b)** $-5, -7, -16$

7 **a)** point $(1, 0, 0)$ **b)** line $x = 1, y = z$

8 **a)** point $(2, -1, 0)$ **b)** no solution, planes form a prism

9 **a)** the line of intersection of the planes   **b)** point $(1\frac{1}{2}, 2\frac{1}{2}, 7)$ **c)** no solution

10 **a)** $x = 1, y = 1, z = 0$

b) $\lambda = 2;$ $\begin{pmatrix} x \\ y \\ z \end{pmatrix} = \begin{pmatrix} 0 \\ \frac{4}{5} \\ -\frac{1}{5} \end{pmatrix} + t \begin{pmatrix} 5 \\ 11 \\ -9 \end{pmatrix}$

11 $\lambda = \frac{1}{3}$: no solution, planes form a prism;

$\lambda = 3$: infinite number of solutions, along the line

$\begin{pmatrix} x \\ y \\ z \end{pmatrix} = \begin{pmatrix} -6 \\ 3 \\ 0 \end{pmatrix} + t \begin{pmatrix} -7 \\ 2 \\ 1 \end{pmatrix}$

other values of λ: the point $\left(\dfrac{6}{1 - 3\lambda}, \dfrac{-3 - 3\lambda}{1 - 3\lambda}, \dfrac{6}{1 - 3\lambda} \right)$

12 $\lambda = 0$ and $\mu \neq -2;$ $\begin{pmatrix} x \\ y \\ z \end{pmatrix} = \begin{pmatrix} -1 \\ -2 \\ 0 \end{pmatrix} + t \begin{pmatrix} 2 \\ 5 \\ 1 \end{pmatrix}; x + 2y - 12z = -5$

13 $\beta = -1; \lambda, -3\lambda, \lambda + 3$

14 $k = 1$, no solutions; $k = -2$ $(\lambda, 5\lambda + 8, -\lambda - 2)$

15 $3x = a + b + c,$ $3y = a + b\omega^2 + c\omega,$ $3z = a + b\omega + c\omega^2$

16 $a + b + c = 0$ $\Rightarrow$ $x = y = z$; or $a = b = c$ $\Rightarrow$ $x + y + z = 0$

17 $\lambda = 3, -1, -1;$ $\lambda = 3, x{:}y{:}z = 1{:}0{:}1;$ $\lambda = -1, x{:}y{:}z = 3{:}4{:}-9$

19 $\alpha \neq 5;$ if $\alpha = 5,$ $\begin{pmatrix} x \\ y \\ z \end{pmatrix} = \begin{pmatrix} 2 \\ 1 \\ 0 \end{pmatrix} + t \begin{pmatrix} 9 \\ 11 \\ -5 \end{pmatrix};$

for other values of $\alpha,$ $x = \dfrac{\alpha - 23}{5(\alpha - 5)},$ $y = \dfrac{8 - 6\alpha}{5(\alpha - 5)},$ $z = \dfrac{\alpha - 3}{\alpha - 5}$

20 when $\lambda = 1$ the three equations all represent the same plane.

Exercise 16.7a

4 the third row of the matrix becomes $(0\,0\,0)$; if the third component of the vector is then 0 there are an infinite number of solutions, otherwise there is no solution

5 **a)** no solutions   **b)** $-4\frac{1}{2}, 19, 7\frac{1}{2}$

 c) $-1, 1, 1$ **d)** $\frac{1}{2}, 0, -\frac{1}{2}$

6 $-1, -1, 1$

7 **a)** $5t, -12t, t$ **b)** no solutions

 c) $5t + 5, -12t - 10, t$ **d)** no solutions

8 $x = 5.23, y = -6.91, z = -2.33$

9 $-4, -\frac{22}{7}, \frac{23}{7};$ $\begin{pmatrix} 1 \\ -2 \\ 3 \end{pmatrix}$

10 a) $-\frac{1}{2}, \frac{1}{2}, -\frac{1}{2}$ **b)** $\lambda, 0, \lambda$

Exercise 16.7b

1 a) $\begin{pmatrix} 1 & -1\frac{1}{2} \\ -1 & 2 \end{pmatrix}$ **b)** $\frac{1}{3}\begin{pmatrix} 4 & -5 \\ 5 & -7 \end{pmatrix}$ **c)** $\dfrac{1}{ad-bc}\begin{pmatrix} d & -b \\ -c & a \end{pmatrix}$

2 a) singular   **b)** $\begin{pmatrix} -2 & \frac{1}{2} & 1 \\ 11 & -1 & -6 \\ 4 & -\frac{1}{2} & -2 \end{pmatrix}$ **c)** $\frac{1}{12}\begin{pmatrix} 1 & 5 & 3 \\ 4 & -4 & 0 \\ 7 & -1 & -3 \end{pmatrix}$

d) $\frac{1}{8}\begin{pmatrix} 11 & 2 & -5 \\ -4 & 0 & 4 \\ -1 & 2 & -1 \end{pmatrix}$ **e)** $\frac{1}{780}\begin{pmatrix} 72 & -32 & 44 \\ 132 & -37 & -71 \\ -60 & 70 & 50 \end{pmatrix}$

3 $\frac{1}{4}\begin{pmatrix} 6 & 0 & -2 \\ 2 & -1 & 1 \\ 0 & -1 & 3 \end{pmatrix}$; $-2, 3\frac{1}{4}, 8\frac{1}{4}$

4 e.g. $\begin{pmatrix} 0 & 1 \\ 1 & 0 \end{pmatrix}, \begin{pmatrix} 1 & 0 \\ 0 & \frac{1}{6} \end{pmatrix}, \begin{pmatrix} 1 & -3 \\ 0 & 1 \end{pmatrix}$;   reflection in $y = x$;   one-way stretch factor $\frac{1}{6}$ in the y-direction; shear with x-axis invariant.

5 $\begin{pmatrix} 1 & 0 & 0 \\ 0 & 1 & 0 \\ 0 & -2 & 1 \end{pmatrix}, \begin{pmatrix} 1 & 0 & 0 \\ -4 & 1 & 0 \\ 0 & 0 & 1 \end{pmatrix}, \begin{pmatrix} 1 & 0 & 3 \\ 0 & 1 & 0 \\ 0 & 0 & 1 \end{pmatrix}$;

$\begin{pmatrix} 1 & -6 & 3 \\ -4 & 1 & 0 \\ 0 & -2 & 1 \end{pmatrix}$;   1

6 a) one-way stretch, factor 3 in the x-direction
b) reflection in $x = y$
c) shear in the x-direction, with $z = 0$ invariant

7 a) $\begin{pmatrix} 1 & 0 & 0 \\ 0 & 4 & 0 \\ 0 & 0 & 1 \end{pmatrix}$ **b)** $\begin{pmatrix} 1 & 0 & 0 \\ 3 & 1 & 0 \\ 0 & 0 & 1 \end{pmatrix}$ **c)** $\begin{pmatrix} 0 & 0 & 1 \\ 0 & 1 & 0 \\ 1 & 0 & 0 \end{pmatrix}$

8 $\begin{pmatrix} 1 & 0 & 1 & 0 \\ -\frac{7}{2} & -\frac{1}{2} & -\frac{7}{2} & \frac{3}{2} \\ \frac{1}{2} & \frac{1}{2} & \frac{1}{2} & -\frac{1}{2} \\ 1 & 1 & 2 & -1 \end{pmatrix}$

Miscellaneous problems

1 $(x, y, z) \xrightarrow{\mathbf{Q_1}} (x, -z, y); (x, y, z) \xrightarrow{\mathbf{Q_2}} (z - 1, y, -x + 1);$

   $(x, y, z) \xrightarrow{\mathbf{Q_2Q_1}} (y - 1, -z, -x + 1);$

   $x = y, x = z - 1, y = 0, z = 1;$
   reflections in three planes would have a fixed point
2 closure   3 $r^2 = s^2 = e$
4 $86y = 41x + 176$
6 $k = 2; x = 6 - 5z, y = 3z - 2; X = 2 + 2z, Y = 8 - 8z, Z = 8 - 9z$
7 Let $\mathbf{l} = (l\,m\,n)$, $\mathbf{x} = (x\,y\,z)$, $\mathbf{a} = (a\,b\,c)$. Equations state that $\mathbf{a} = \mathbf{l} \times \mathbf{x}$, so
   we must have $\mathbf{a}$, $\mathbf{l}$ perpendicular, i.e. $al + bm + cn = 0$; also $\mathbf{a}$, $\mathbf{x}$
   perpendicular, so $ax + by + cz = 0$. A vector parallel to $\mathbf{a} \times \mathbf{l}$ satisfies
   $lx + my + nz = 0$, i.e. $(bn - cm, cl - an, am - bl)(l^2 + m^2 + n^2)^{-1}$
8 least 6 (three zeros in same row or column); greatest 11 (no two zeros in
   same row or column)
9 $(abc)^{-1}(b - c)(c - a)(a - b) \sum a^{-r}b^{-s}c^{-t}(r + s + t = n - 1)$
10 no
11 -8
12 $a^3 - 2ab^2, a^4 - 3a^2b^2 + b^4$

Revision exercises

Chapter 11

1 $8:1$

2 $\mathbf{t} = \frac{1}{3}\mathbf{a} + \frac{2}{3}\mathbf{b}, \mathbf{m} = \frac{1}{2}\mathbf{a} + \mathbf{b}, \mathrm{OB}:\mathrm{BK} = 1:1$

3 $\frac{1}{3}\mathbf{a} + \frac{2}{3}\mathbf{b} - \mathbf{c}; r(\frac{1}{3}\mathbf{a} - \mathbf{b} + \frac{2}{3}\mathbf{c}), s(\frac{1}{3}\mathbf{a} + \frac{2}{3}\mathbf{b} - \mathbf{c}); \frac{3}{5}, \frac{3}{5}$

4 distance $\dfrac{3(a^2 - 2l^2)}{4(2a + 3l)}$ from common base

5 $5h/16$ 6 $28a/3\pi$ 7 $\frac{4}{5}, \frac{2}{5}$

8 $\mathbf{r} = \begin{pmatrix} 2 \\ 1 \\ 1 \end{pmatrix} + t\begin{pmatrix} -1 \\ 2 \\ 1 \end{pmatrix}; (4, -3, -1); x - y + 3z = 4$

9 $\frac{3}{2}\mathbf{b} + \frac{1}{2}\mathbf{c}$

11 $\frac{1}{7}\sqrt{629}$

12 components $(-a\omega \cos\theta, a\omega(2 - \sin\theta)); (-4a\omega^2, 0)$

Chapter 12

1 $3\sqrt{2}$

2 $\frac{1}{2}(1 + \sqrt{3}) + \frac{1}{2}(1 - \sqrt{3})i$

3 $1, -2.50$; radius 1, centre the origin

4 $0, 2^{1/6}\{\cos(\frac{7}{12}\pi + \frac{2}{3}k\pi) + i\sin(\frac{7}{12}\pi + \frac{2}{3}k\pi)\}, \quad k = 0, 1, 2$

5 $r_1^2, 2\theta_1; 1/r_1, -\theta_1; r_1, -\theta_1; r_1r_2, \theta_1 + \theta_2;$
$r_1/r_2, \theta_1 - \theta_2; r_1/r_2, \theta_1 + \theta_2$

7 **a)** line-segment from -1 to i
  b) the line $y = x + 1$, excluding the line segment in **a)**
  c) semi-circle through $-1, -1 + i, i$

8 $z^2 - 2z + 4 = 0; a = -2, b = 12$

10 **a)** circle, centre i, radius 1
  b) $y = -\frac{1}{2}$

11 **b)** $32\cos^6\theta - 48\cos^4\theta + 18\cos^2\theta - 1$

12 **c)** $|z + 4i| = 4, x^2 + (y + 4)^2 = 16$
  d) semi-circle through $4i, -2i, 2\sqrt{3} - 2i$

13 $z^2 - 2z + 4$; 2, $-\pi/3$; 4, $-2\pi/3$; 8, π; roots -1, $1 \pm \sqrt{3}i$; A' is
  $(\sqrt{3}/2 - 1) + \frac{5}{2}i$

14 $1, \cos\frac{2}{5}k\pi \pm i\sin\frac{2}{5}k\pi, \quad k = 1, 2$

15 **a)** $u^2 + v^2 = 1$
  b) $-1, \frac{1}{2} \pm \sqrt{3}i/2; z = \frac{1}{2}, \frac{1}{2} \pm \sqrt{3}i/2$

16 $0, \frac{1}{4}\pi, \frac{3}{4}\pi, \pi, \frac{5}{4}\pi, \frac{7}{4}\pi, 2\pi$

17 $\cos\theta + i\sin\theta, \qquad \cos 2\theta + i\sin 2\theta, \qquad \cos(\theta - \pi/2) + i\sin(\theta - \pi/2),$
  $2\sqrt{2}\{\cos(\theta + \pi/4) + i\sin(\theta + \pi/4)\}$
  a) anti-clockwise round circle $|z - 2 + 3i| = 1$, from $3 - 3i$ to $1 - 3i$
  b) clockwise round circle $|z| = 1$, from 1 to -1
  c) anti-clockwise right round the circle $|z| = 1$, starting from 1
  d) anti-clockwise round circle $|z| = 1$, from $-i$ to i
  e) anti-clockwise round circle $|z| = 2\sqrt{2}$, from $2 + 2i$ to $-2 - 2i$

Chapter 13

1 **a)** $\frac{2}{3}(y^3 - 1) = x - \sin x \cos x$
  b) $y = \frac{3}{4}(\cos 2x - \sin 2x) + \frac{1}{4}$

2 $x = Ae^t + Be^{-3t} + 3e^{-t}; \quad x = 3e^{-t} - 2e^{-3t}$

3 **a)** $x = t$
   b) $y = Ae^{-\tan^{-1}x} + \tan^{-1}x - 1$

4 **a)** $x = At^3 + \frac{1}{2}t^5$
   b) $y = (1 - 2x)e^{-2x}$

5 $y = Ax + \dfrac{B}{x} + \ln x; \quad y = x - \dfrac{1}{x} + \ln x$

6 **a)** $y = Ae^{-3x} + Be^{2x} - x - \frac{1}{6}$
   b) $(2a, a^2)$

8 $\sqrt{(P/k)}, \dfrac{m}{2k}\ln\dfrac{4}{3}$

9 $\dfrac{dP}{dt} = \frac{1}{50}P - \frac{4}{5}\sqrt{P}$ **b)** $947, 890$ $(P = 1600)$

10 $z = \tan(x + c); \quad y = \tan x - x$

11 $\dfrac{12000}{v} = 1000\dfrac{dv}{dt}; v^2 = 100 + 24t; 4\,\text{s}$

12 $y = 20e^{-(t^2 + 2t)/(1 + t)^2}; 20\,\text{g}$

Chapter 14

1 $2\pi/5\,\text{m s}^{-1}, 1\frac{2}{3}\,\text{s}, 8\pi^2/25\,\text{N}$
2 $600\,l^2\,\text{J}; 800(40 + l) = 600\,l^2; 48\,\text{m}$
3 $\sqrt{(2gx(5\,l - 2x)/l)}, \frac{5}{2}\sqrt{(gl)}$
4 $3V/2, 3mV^2/4, 3V/4, 3mV$
5 $1.58\,\text{m}$
6 $\frac{1}{8}(1 - e)(3 + e^2)u, \frac{1}{4}(1 + e)^2u, e > 3 - 2\sqrt{2}$
7 $20e^6\,\text{m}, 4e\,\text{s}; e = \frac{1}{3}, 65\,\text{m}$
8 $2x - 4y + a = 0$
9 At $116°$ to $Ox, +180\,\text{N m}; 80x + 39y = 180; a = 75, b = -32, F = 60$
10 $P\sqrt{29}, \frac{5}{2}; a$
11 $5\,mg/\sqrt{13}$
12 $\frac{1}{3}h\sqrt{(9 + 16\tan^2\alpha)}$
14 $\frac{3}{4}mg\tan\alpha, \frac{1}{4}mg$
15 $\frac{5}{2}W$
16 $-m\omega^2(x\mathbf{i} + y\mathbf{k}); m\omega^2 z(y\mathbf{i} - x\mathbf{j})$

17 $143\,ma^2, \sqrt{\left(\dfrac{124g}{143a}\right)}$

18 **a)** $0.2\,\text{m}, 2\,\text{s}$ **b)** $0.9\,\text{N m}$ **c)** $6.32\,\text{N}, 51.6°$

19 $\sqrt{\left(\dfrac{GM}{R + h}\right)}$

20 $3.77\,\text{m s}^{-2}; 5.05\,\text{km s}^{-1}, 3.28\,\text{km s}^{-1}$

Chapter 15

1. $\mu_X = 3.25, \sigma_X{}^2 = 0.854;$ $\mu_Y = 3.4, \sigma_Y{}^2 = 2.24;$ $\mu_Z = -0.15,$
    $\sigma_Z{}^2 = 3.094$
2. $0.5, 0.036; 0.365; 0.982$
3. $20, 200$ **a)** 0.594 **b)** 0.287
4. 1007.9 grams; 996.5 grams; 970.8 grams (or 4854 grams for the whole bumper pack); 2.6%
5. 3.9, 3.7; Poisson distribution; 74
6. $0.5, 0.29; 0, 0.41$
7. $\frac{1}{3}, \frac{2}{3}; \frac{1}{3}, 1; \frac{1}{6}; 2$
8. $0.023; 4; 0.855$
9. **a)** B, expected loss of 5p
    b) $n = 49$ (using a two-tailed test)
10. significant at the 1% level that the new lubricant has increased spinning time
11. 3600; the claim is just valid at the 5% significance level.
12. **a)** $89.82, 90.18$ **b)** $898.2, 901.8$
13. **a)** $152, 4.9$ **b)** 0.66
14. **b)** $y = 1.035x + 0.411$
15. 0.706

Chapter 16

1. **a)** $\begin{pmatrix} k^2 & 2k \\ 0 & k^2 \end{pmatrix}, \begin{pmatrix} k^3 & 3k^2 \\ 0 & k^3 \end{pmatrix}, \begin{pmatrix} k^n & nk^{n-1} \\ 0 & k^n \end{pmatrix}$

2. $\mathbf{u}$ and $\mathbf{v}$ perpendicular; 'squash' on to the line $y = x \tan \alpha$;

    $\begin{pmatrix} \sin^2 \alpha & -\cos \alpha \sin \alpha \\ -\cos \alpha \sin \alpha & \cos^2 \alpha \end{pmatrix}$

3. **a)** $x = 0$ **b)** $(3, -\frac{1}{2}, -1\frac{1}{2})$ **c)** $y = z$

4. $k = 0, \begin{pmatrix} 1 \\ 1 \\ -1 \end{pmatrix};$ $k = 1, \begin{pmatrix} 2 \\ 1 \\ -1 \end{pmatrix};$ $k = 2, \begin{pmatrix} -1 \\ 1 \\ 1 \end{pmatrix}$

    a) $\mathbf{r} = \lambda \begin{pmatrix} 1 \\ 1 \\ -1 \end{pmatrix}$ **b)** $\mathbf{r} = \lambda \begin{pmatrix} 2 \\ 1 \\ -1 \end{pmatrix}$ **c)** $2x - y + 3z = 0$

5. $\begin{pmatrix} \cos \beta & 0 & -\sin \beta \\ 0 & 1 & 0 \\ \sin \beta & 0 & \cos \beta \end{pmatrix};$ $\begin{pmatrix} \cos n\alpha & 0 & -\sin n\alpha \\ 0 & 1 & 0 \\ \sin n\alpha & 0 & \cos n\alpha \end{pmatrix}$

6 $\begin{pmatrix} 3 & 2 & 3 \\ 1 & 3 & 2 \\ 2 & 1 & 1 \end{pmatrix}$; $\begin{pmatrix} 8 \\ 6 \\ 4 \end{pmatrix}$, $\begin{pmatrix} 15 \\ 11 \\ 10 \end{pmatrix}$, $\begin{pmatrix} 23 \\ 17 \\ 14 \end{pmatrix}$

7 3; $\frac{1}{3}\begin{pmatrix} 2 & -7 & -10 \\ 1 & 1 & 1 \\ -1 & 5 & 8 \end{pmatrix}$

8 $(a - b)(b - c)(c - a)(1 - a)(1 - b)(1 - c)$

9 a) $-(a - b)(b - c)(c - a)$

b) $\begin{pmatrix} 0 & 0 & 0 \\ 0 & 0 & 0 \\ 1 & 0 & 0 \end{pmatrix}$, $\begin{pmatrix} 0 & 0 & 0 \\ 0 & 0 & 0 \\ 0 & 0 & 0 \end{pmatrix}$

10 non-zero determinant; $\frac{1}{17}\begin{pmatrix} 7 & 2 & -3 \\ -5 & 1 & 7 \\ -6 & 8 & 5 \end{pmatrix}$; $x = 1, y = -2, z = 4$

11 $(2, -1, 3)$; $(3\frac{1}{3}, \frac{1}{3}, 1\frac{2}{3})$

12 a) $a = 3$, $\mathbf{x} = \lambda\begin{pmatrix} -2 \\ 3 \\ 1 \end{pmatrix}$ **b)** $a = -4\frac{1}{2}$, $\mathbf{x} = \lambda\begin{pmatrix} 3 \\ 2 \\ 2 \end{pmatrix}$

13 $-1, -2, 3$; $1/\sqrt{2}$

14 $\begin{pmatrix} x \\ y \\ z \end{pmatrix} = \begin{pmatrix} 0 \\ -1 \\ 0 \end{pmatrix} + \lambda\begin{pmatrix} 1 \\ -2 \\ 1 \end{pmatrix}$

a) line **b)** plane **c)** planes have a common line

Contents of Book 1

Preface

Notation

1 Introduction
1.1 Logical symbols
1.2 Sets and Boolean algebra
1.3 Coordinates
1.4 Straight lines
1.5 Rational and irrational numbers
1.6 Surds
* 1.7 Linear transformations and matrices
* 1.8 Multiplication of matrices

2 Functions
2.1 Functions and their graphs
2.2 Quadratic functions
* 2.3 Polynomial functions and the remainder theorem
2.4 Indices and power functions
2.5 Logarithms and logarithmic functions
* 2.6 Continuity and limits
* 2.7 Rational functions

3 Rates of change: differentiation
3.1 Average velocity and instantaneous velocity
3.2 Differentiation
3.3 Standard methods of differentiation
3.4 Higher derivatives
3.5 Alternative notation: tangents and normals
3.6 Kinematics
3.7 Reverse processes
3.8 Stationary values: maxima and minima
* 3.9 Further differentiation
* 3.10 Implicit functions and parameters
* 3.11 Rates of change; approximations and errors
* 3.12 Further kinematics

4 Areas: integration
4.1   The area beneath a curve: the definite integral
4.2   The generation of area: indefinite integrals
4.3   Calculation of definite integrals
* 4.4   Numerical integration
* 4.5   Other summations: volumes of revolution
* 4.6   Mean values
* 4.7   Integration by substitution

5 Trigonometric functions
5.1   Sine, cosine and tangent
5.2   Simple equations
5.3   Cotangent, secant, cosecant and the use of Pythagoras' theorem
* 5.4   Sine and cosine rules
5.5   Compound angles: $a \cos \theta + b \sin \theta$
5.6   Multiple angles
5.7   Factor formulae
5.8   Radians, general solutions and small angles
5.9   Differentiation of trigonometric functions
5.10 Integration of trigonometric functions
* 5.11 Inverse trigonometric functions

6 Sequences and series
6.1   Sequences and series
6.2   Arithmetic progressions
6.3   Geometric progressions
* 6.4   Finite series: the method of differences
* 6.5   Mathematical induction
* 6.6   Iteration: recurrence relations
* 6.7   Solution of equations: iterative methods
* 6.8   The Newton–Raphson method
* 6.9   Interlude: Fibonacci numbers and prime numbers
6.10 Permutations and combinations
6.11 Pascal's triangle: the binomial theorem

7 Probability and statistics
7.1   Trials, events and probabilities
7.2   Compound events
* 7.3   Conditional probability
* 7.4   Probability distributions: three standard types
7.5   Statistics: location and spread
7.6   Frequency distributions
7.7   The Normal distribution

8 Introduction to vectors
8.1 Vectors and vector addition
8.2 Position vectors and components
8.3 Scalar products
* 8.4 Lines, planes and angles
8.5 Differentiation of vectors
8.6 Motion under gravity: projectiles
8.7 Motion in a circle
* 8.8 Relative motion

9 Introduction to mechanics
9.1 Historical introduction: Newton's first law
9.2 Force, mass and weight: Newton's second law
9.3 Reactions: Newton's third law
9.4 Friction
9.5 Impulse and momentum
9.6 Work, energy and power

10 Further calculus
10.1 The logarithmic function, $\ln x$
10.2 The exponential function, e^x
10.3 Interlude: infinite series
10.4 Maclaurin series
* 10.5 Integration by parts: reduction formulae
* 10.6 Partial fractions
* 10.7 General integration

Appendix: introduction to coordinate geometry
A1 Straight lines
A2 Circles
* A3 Parabolas
* A4 Ellipses and hyperbolas
* A5 Polar coordinates

Revision exercises

Answers to exercises

Index

Index

acceleration
  in Cartesian coordinates, 22
  in polar coordinates, 23
algebra, fundamental theorem of, 54
Apollonius, circle of, 71
Argand diagram, 35
argument of complex number, 43

binomial distribution, 208
  limit of, 226
bollard, friction at, 126

Cantor, G., 388
cardioid, 100
central limit theorem, 232
centre of mass, 6
centroid, 5, 27
Ceva's theorem, 27
circle of Apollonius, 29
complementary function, 134
complex numbers, 29
  argument of, 43
  conjugate of, 52
  modulus of, 40
  roots of, 59, 63
confidence intervals, 269
conjugate of complex number, 52
continuum, non-enumerability of, 390
correlation, 275
  coefficient, 280
  Kendall's rank coefficient, 297
  significance, 289
  Spearman's rank coefficient, 294
couple, 173
  moment of, 173
covariance, 280
Cramer's rule, 369
cycloid, 26

de Moivre's theorem, 57

dependence, linear, 331
determinants, 339, 344, 351
  cofactors of, 354
  minors of, 354
  multiplication of, 359
differential equations, 105
  complementary function, 134
  general solution, 106, 134
  linear, 128
  order, 105
  particular integral, 134
  step-by-step solution, 108
dimensions, 195
distribution
  binomial, 208, 226
  continuous probability, 212
  function, 213
  gamma, 302
  Normal, 223
  Poisson, 235
dynamics
  of a system, 182
  of rotating bodies, 190

e, $e^{i\theta}$, 76
echelon form of matrix, 375
eigenvalue, 317
eigenvector, 317
elastic energy, 147
elasticity, 146, 163
  modulus of, 146
equations, systems of linear, 364
estimates, unbiased, 259
Euler, L., 78
expectation, 203
  algebra, 248
expected value, 203

forces
  conservative, 148

forces (*contd.*)
  equilibrium of, 175
  equivalent systems of, 170
  external, 175
  internal, 175
  moments of, 166, 169
  parallel, 171
  systems of, 170
functions
  cumulative probability, 213
  distribution, 213
  hyperbolic, 80
  inverse hyperbolic, 84
  of complex variable, 88
  probability density, 215
  reciprocal, 95
fundamental theorem of algebra, 54

gamma distribution, 302
gravity, centre of, 172
gyration, radius of, 186

Hooke's law, 146
hyperbolic functions, 80
  inverse, 84

i, 30
impact, Newton's law of, 163
integrating factor, 130
inverse hyberbolic functions, 84

Kendall's rank correlation coefficient,
  297
Kepler, J., 24, 27

linear dependence, 331
linear equations, systems of, 364
  Cramer's rule, 369
  systematic reduction, 375
linear transformations, 313
  determinants of, 314
  eigenvectors of, 317
  in three dimensions, 327
  in two dimensions, 313
loci in Argand diagram, 68

matrices, 205
  addition of, 306
  adjoint, 343

matrices (*contd.*)
  conformable, 307
  determinants of, 344
  echelon form, 375
  elementary, 376
  identity, 309
  inverse, 338
  multiplication of, 306
  orthogonal, 350
  pivotal condensation, 375
  singular, 331
  transpose, 310
  triangular form, 375
  unit, 309
Menelaus' theorem, 27
modulus
  of complex number, 40
  of elasticity, 146
moment
  of couple, 173
  of force, 166, 169
moment of inertia, 183
  parallel axis theorem, 187
  perpendicular axis theorem, 188

Napoleon's theorem, 104
Newton, I., 25
  law of impact, 163
non-enumerability of continuum, 390
Normal distribution, 223
  derivation of, 225
numbers
  complex, 29
  countable, 387
  rational, 388
  real, 389
  transcendental, 391

orthocentre, 27
orthogonal families, 121
Osborn's rule, 82
oscillation, 150
  amplitude of, 150
  frequency of, 150
  period of, 150
  simple harmonic, 155

parallelepiped, 20

parameters of distribution, 219
particular integral, 134
pendulum, simple, 154
pivotal condensation, 375
points
    of balance, 1
    of subdivision, 3
Poisson distribution, 235
    mean and variance, 239
probability
    density function, 215
    function, cumulative, 213
    generating function, 205
    generator, 205
    models, 202
probability distributions, 202
    binomial, 208
    continuous, 212
    joint, 249
    Normal, 223
    Poisson, 235

radius of gyration, 186
random variables, 248
rank correlation
    Kendall's coefficient, 297
    Spearman's coefficient, 295
rational numbers, 388
real numbers, 389
rectangular hyperbola, 82
regression, 276

restitution, coefficient of, 163
rotation, 183

samples, 247
    from Normal distribution, 257
    large and small, 257
scalar triple product, 19
significance, statistical, 262
similarity, spiral, 48
simple harmonic motion, 155
Spearman's coefficient, 294
spiral similarity, 48
standard deviation, 203

transcendental numbers, 391
transformations
    combined, 322, 332
    enlargement, 315, 328
    linear, 313
    reflection, 315, 328
    rotation, 315, 328
    shear, 315
    translation, 324
translations, 324
trigonometric series, 77

vector product, 16
vector triple product, 19
velocity
    in Cartesian coordinates, 22
    in polar coordinates, 23